Tensor and Vector ANALYSIS

With Applications to Differential Geometry

C. E. Springer

Dover Publications, Inc.
Mineola, New York

Bibliographical Note

This Dover edition, first published in 2012, is an unabridged republication of the work originally published by The Ronald Press, New York, in 1962.

Library of Congress Cataloging-in-Publication Data

Springer, C. E. (Charles Eugene), 1903–1999.
   Tensor and vector analysis : with applications to differential geometry / C.E. Springer. — Dover ed.
     p. cm.
   Originally published: New York : Ronald Press, 1962.
   Summary: "Assuming only a knowledge of basic calculus, this text presents an elementary and gradual development of tensor theory. From this treatment, the traditional material of courses on vector analysis is deduced as a particular case. In addition, the book forms an introduction to metric differential geometry. 1962 edition"— Provided by publisher.
   Includes index.
   ISBN-13: 978-0-486-49801-0 (pbk.)
   ISBN-10: 0-486-49801-8 (pbk.)
    1. Calculus of tensors. 2. Vector analysis. 3. Geometry, Differential. I. Title.

QA433.S68 2012
515'.63—dc23

2012019030

Manufactured in the United States by Courier Corporation
49801802   2013
www.doverpublications.com

To Lucile

Preface

This volume aims to provide an elementary introduction to tensor theory and to derive from it as a particular case the usual content of courses in vector analysis. The approach presented here and successfully used in the classroom has persuaded the author that there is considerable merit in treating vectors as tensors of the first order.

The book is intended for students who have had a course in elementary calculus. A particular effort has been made here to approach each new concept in a gradual manner. The tensor notation itself is introduced only after the student is prepared for it through use of a familiar notation. Moreover, the book begins with the concept of groups of coordinate transformations as exemplified by translations and rotations to enable the student to grasp the essential nature of tensors (and, in particular, of vectors) as sets of objects which are attached to each of a group of coordinate systems and which transform in a characteristic manner under a change of coordinates.

The important concept of invariance is often given little attention in conventional treatments of vector analysis. Here this concept is made more meaningful and more easily assimilated because the fact of invariance is apparent by virtue of the notation of tensor analysis. Students in vector analysis are normally introduced to only two types of curvilinear coordinates—cylindrical and spherical—and undue difficulty is experienced in finding components of vectors in these systems. With the tensor background offered here, the determination of vector components, even in an arbitrary system of non-orthogonal curvilinear coordinates, is almost automatic. The facility to be gained from the tensor viewpoint is often of significant advantage in applied mathematics.

In addition to its use in a course in vector analysis, the book is designed to furnish an introduction to metric differential geometry. Geometric concepts are interwoven throughout with the development of tensor theory. The concluding chapter uses tensor theory to develop the differential equations of geodesics on a surface in several different ways and to exhibit further differential geometry which is important for the study of the motion of a particle on a surface. This chapter also contains the author's results on union curves.

The reader is advised that equations are numbered serially in each chapter, and that a particular equation is referred to by chapter number and equation number. Thus, for instance, the reference (8–4) indicates Equation 4 in Chapter 8. On the other hand, a reference such as Section 8–4 means Section 4 in Chapter 8.

The author is indebted to Professor Bernard Epstein of Yeshiva University for his valuable suggestions during the writing of the manuscript.

C. E. SPRINGER

Norman, Oklahoma
February, 1962

Contents

CHAPTER	PAGE
1 Coordinate Transformations and Mappings	3

Two Aspects, 3 · A Change of Notation, 6 · Rotations in Three Dimensions, 7 · The Kronecker Delta, 10

2 Loci in Three-Space 12

One-Dimensional Extent, 12 · Two-Dimensional Extent, 13 · Some Differential Geometry of Space Curves, 15 · Some Differential Geometry of Surfaces, 21

3 Transformation of Coordinates in Space; Differentiation . 26

Linear Transformation, 26 · Transformation to Curvilinear Coordinates, 26 · Partial Differentiation, 28 · Derivative of a Determinant, 30 · Cramer's Rule, 32 · Product of Determinants, 33

4 Tensor Algebra 36

Cogredience and Contragredience, 36 · First View of a Tensor, 38 · Operations of Tensor Algebra, 40 · Transitivity, Symmetry, Skew-Symmetry, 42

5 Tensor Analysis 45

The Fundamental Quadratic Form, 45 · Covariant and Contravariant Tensors of the First Order, 48 · A Quadratic Form from a Tensor Product, 51 · Definition of a General Tensor, 52 · Inner Product of Two Vectors, 52 · Associate Tensors, 54

6 Vector Analysis 57

Length of a Vector, 57 · Angle Between Two Vectors; Orthogonal Vectors, 58 · Some Applications, 61 · Geometric Meaning of Contravariant and Covariant Components of a Vector, 65 · Alternative (Reciprocal) Geometrical Interpretation of Contravariant and Covariant Components of a Vector, 68

CHAPTER

7 Vector Algebra … 72

Base Vectors, 72 · Products of Vectors, 74 · Linear Dependence, 79 · Vector Equation of a Line, 80 · Applications in Mechanics, 82 · Vector Methods in Geometry, 84

8 Differentiation of Vectors … 88

Vector Functions of a Scalar Variable, 88 · Frenet Formulas for Space Curves, 90 · Application in Mechanics, 93 · Motion in a Plane, 93 · Law of Transformation for Velocity Components, 98 · Vector Functions of Two Scalar Parameters, 100 · Riemannian Metric, 101 · Extrinsic and Intrinsic Geometry, 103 · Surface Normal and Tangent Plane, 105 · Local and Global Geometry, 105

9 Differentiation of Tensors … 109

Equivalence of Forms; Christoffel Symbols, 109 · The Riemann-Christoffel Tensor, 112 · Covariant Differentiation; Parallelism of Vectors, 116 · Covariant Derivative of Covariant Tensors, 118 · Covariant Derivative of a General Tensor, 120 · Tensors Which Behave as Constants, 121

10 Scalar and Vector Fields … 124

Fields, 124 · Divergence of a Vector Field; the Laplacian, 125 · The Curl of a Vector Field, 128 · Physical Components, 130 · Some Vector Identities Involving Divergence and Curl, 131 · Frenet Formulas in General Coordinates, 132 · The Acceleration Vector, 134 · Equations of Motion, 135 · The Lagrange Form of the Equations of Motion, 137

11 Integration of Vectors … 147

Line Integrals, 147 · Vector Form of Line Integrals, 153 · Surface and Volume Integrals, 156 · Green's Theorem in the Plane, 162 · Simply and Multiply Connected Regions, 167 · Independence of the Path of Integration, 174 · Test for Independence of Path, 176 · Green's Theorem in Three-Space (The Divergence Theorem), 181 · An Application of the Divergence Theorem, 186 · The Theorem of Stokes (The Curl Theorem), 189 · Applications of the Curl Theorem, 194

12 Geodesic and Union Curves … 200

Two-Dimensional Curved Space, 200 · Geodesics as Curves of Shortest Distance, 201 · The Second Fundamental Form of a Surface, 207 · Normal Curvature of a Surface, 212 · Curvature Formulas, 214 · Geodesic Curvature, 218 · Geodesics as Auto-

| CHAPTER | PAGE |

Parallel Curves, 221 · A Generalization of the Theorem of Meusnier, 227 · Union Curves on a Surface, 230 · Union Curves and Dynamical Trajectories, 233

Index **239**

Tensor AND Vector
ANALYSIS

1

Coordinate Transformations and Mappings

1-1. Two Aspects. The reader will recall two transformations from analytical geometry in the plane. The first is the translation which can be represented by the equations

1) $$\bar{x} = x + h,$$
$$\bar{y} = y + k,$$

and the second is the rotation described by the equations

2) $$\bar{x} = x \cos \theta - y \sin \theta,$$
$$\bar{y} = x \sin \theta + y \cos \theta.$$

Equations (1) may be interpreted in two ways. In the first way a point P has cartesian coordinates (x,y) referred to the x- and y-axes through the origin O, and the same point P also has coordinates $(\bar{x},\bar{y})$ referred to a set of $\bar{x}$- and $\bar{y}$-axes parallel to the x- and y-axes and passing through the point $\bar{O}$. Notice that the origin O has coordinates (h,k) referred to the "barred" axes, while the point $\bar{O}$ has coordinates $(-h,-k)$ in the xy system. In this first aspect of equations (1), any point P remains fixed but its identification or "name" changes when the frame of reference is changed, and equations (1) serve as a sort of dictionary which reveals the "name" $(\bar{x},\bar{y})$ if the "name" (x,y) is known. One speaks, therefore, of a transformation of coordinates effected by equations (1), which is the result of a translation of the coordinate axes.

A second interpretation of equations (1) is the following. There is only one set of coordinate axes through a fixed origin O. The point $P(x,y)$ is mapped or transported into a different position $\bar{P}$ with coordinates $(\bar{x},\bar{y})$ referred to the single set of xy axes through O. One speaks of equations (1) as representing a mapping of the points of the plane by a translation determined by the parameters h and k. Notice that if both h and k are not zero, then no point in the finite plane can remain fixed under the mapping defined by equations (1). In projective geometry, the mapping of (1) is styled an *affine* mapping because finite points map into finite points and so-called points at infinity map into points at infinity. Observe that h and k are parameters, each of which could be assigned an arbitrary numerical value. Hence, one could say that equations (1) represent a doubly infinite set of mappings.

The two aspects discussed here have been referred to as the "alias" and "alibi" aspects. Under the "alias" connotation points remain unchanged but their coordinates (names) change. Under the "alibi" aspect, the points are moved to different positions.

Whether equations (1) are regarded as a transformation of axes or a mapping of points, these equations exemplify a concept, that of a *group*, which will now be explained.

Let the translation designated by T_1 be taken as

$$T_1: \quad \bar{x} = x + h_1,$$
$$\bar{y} = y + k_1,$$

and let the translation T_2 be given by

$$T_2: \quad \bar{\bar{x}} = \bar{x} + h_2,$$
$$\bar{\bar{y}} = \bar{y} + k_2.$$

The mapping of point $P(x,y)$ into $\bar{P}(\bar{x},\bar{y})$ by T_1 is indicated by $\bar{P} = T_1(P)$. Then, by T_2 the point $\bar{P}(\bar{x},\bar{y})$ goes into $\bar{\bar{P}}(\bar{\bar{x}},\bar{\bar{y}})$, so one may write $\bar{\bar{P}} = T_2(\bar{P})$, or equally well $\bar{\bar{P}} = T_2[T_1(P)]$, or $\bar{\bar{P}} = T_2T_1(P)$. One then speaks of mapping point $P(x,y)$ into $\bar{\bar{P}}(\bar{\bar{x}},\bar{\bar{y}})$ directly by the "product" T_2T_1 of the mappings T_1 and T_2. If the coordinates $\bar{x}$ and $\bar{y}$ of T_1 are substituted for $\bar{x}$ and $\bar{y}$ of T_2, there follows a third mapping which may be designated by T_3 represented by

$$T_3 = T_2T_1: \quad \bar{\bar{x}} = x + (h_1 + h_2),$$
$$\bar{\bar{y}} = y + (k_1 + k_2).$$

Observe that $h_1 + h_2$ and $k_1 + k_2$ are among the infinitude of choices for the numbers h and k in the set of mappings (1) which may be denoted simply by T. It should be clear now that the set T has the property that if any mapping of the set (say T_1) is followed by any mapping of the set (say T_2), then the resulting product mapping (T_2T_1) is a member of the set T. Thus, the first group property is satisfied by the set T.

Next, it is seen that the inverse of T, denoted by T^{-1}, is obtained by solving equations (1) to obtain

$$T^{-1}: \quad x = \bar{x} + (-h),$$
$$y = \bar{y} + (-k),$$

so that for any mapping of the set T with parameters h and k, the inverse T^{-1} is given by the parameters $-h$ and $-k$. Since h and k in the set T range over all real numbers, $-h$ and $-k$ determine a mapping of the set. Hence, a second group property is enjoyed by the set T, that is, for any mapping of the set the inverse of it is present. A third group property holds for the set T. If h and k are both chosen to be zero, the resulting mapping takes every point into itself, and this is called the identity mapping which may be denoted by I. It is apparent then that the identity mapping is present in the set T. Because the set T of mappings (1) enjoys the three properties just enumerated, T constitutes an instance of a group. Since the effect of any mapping of T is a translation, one may speak of the group operation as translation. [For an abstract group the operations must be associative, but this property is always satisfied by the mappings considered here, i.e., $T_3(T_2T_1) = (T_3T_2)T_1$].

The concept of a group of mappings will now be made more precise in the following

Definition: A set of mappings, finite or infinite in number, is called a group of mappings, if it has the following properties.

(a) *Closure.* The product of every two mappings of the set is a mapping of the set.
(b) *Inverse.* For every mapping of the set there exists an inverse mapping which is in the set.

It should be noted that since for any mapping M of the set, the inverse (denoted by M^{-1}) is in the set, so that the product $MM^{-1} = M^{-1}M = I$, which is the identity mapping, is in the set. The identity mapping I leaves all points fixed.

As was seen in the group of translations in the plane, the "product" M_2M_1 of two mappings M_1 and M_2 means that mapping M_1 is effected and then M_2 is carried out. In the set of translations the order of effecting two translations is immaterial, so $T_1T_2 = T_2T_1$. For some mappings the order *is* material. If the mapping M_1M_2 is the same as M_2M_1, for every two mappings of the group, the group is called commutative, or abelian. If the commutative property is not satisfied, the group is called non-commutative.

Several concepts have now been introduced with reference to equations (1) as an instance. The reader should test his understanding at this point by considering the following

EXERCISES

1) Give two interpretations (i.e., the two aspects) of the set described by equations (2).

2) What is the parameter in equations (2)? What range of values should be assigned to the parameter?

3) Designate the set of transformations (or mappings) in equations (2) by R. Let R_1 be the particular rotation through angle θ_1, and R_2 the rotation through angle θ_2. Show that the product R_2R_1 is a member (say R_3) of the set R.

4) What value of the parameter in R gives the identity?

5) Find the equations which represent the inverse R^{-1}.

6) Show that the set R in equations (2) constitute a one-parameter group with rotation as the group operation. Is this an affine mapping?

7) Show that the set of mappings $\bar{x} = ax$, $\bar{y} = by$, with the two parameters a and b ranging over the real numbers (except zero) form a group. Is this an affine mapping?

8) Show that the mappings given by $\bar{x} = x \cos\theta + y \sin\theta$, $\bar{y} = x \sin\theta - y \cos\theta$ do *not* form a group. Which group properties do not hold?

9) If S is a mapping such that $SS(P) = P$, that is $S^2 = I$, where I designates the identity mapping, show that $S = S^{-1}$. Give an example of this type of mapping.

10) Show that the inverse of the product of two mappings S and R is the product of the inverses in reverse order, that is, show that $(RS)^{-1} = S^{-1}R^{-1}$. Deduce that $(RST)^{-1} = T^{-1}S^{-1}R^{-1}$.

11) If T and R are described by equations (1) and (2), respectively, write the equations which describe the mapping TR. Does $TR = RT$?

1–2. A Change of Notation. In order to prepare for the notation of tensor analysis, the two-parameter group in equations (1) is written in the form

$$3) \quad T: \quad \begin{aligned} \bar{x}_1 &= x_1 + a_1, \\ \bar{x}_2 &= x_2 + a_2, \end{aligned}$$

or in the equivalent form

$$4) \quad T: \quad \bar{x}_i = x_i + a_i \quad (i = 1,2).$$

If i is allowed to take the range 1, 2, 3, equations (4) represent the three-parameter group of translations in three-dimensional space. If i takes the range 1, 2, $\cdots$, n, one speaks of the n-parameter group of translations in n-dimensional space. The superiority of the new notation is apparent.

Equations (4), even with i running from 1 to n, furnish a very special group of mappings. A more general situation is described by

5) $$\bar{x}_i = f_i(x_1, \cdots, x_n; a_1, \cdots, a_r) \qquad (i = 1, \cdots, n)$$

where the functions $f_1, \cdots, f_n$ meet certain requirements which allow equations (5) to represent an r-parameter group in n-space. Note that in equations (3), $i = 2$, $f_1 \equiv x_1 + a_1$, $f_2 \equiv x_2 + a_2$, and $r = 2$.

1–3. Rotations in Three Dimensions. Instead of considering the general situation posed by equations (5), it is more meaningful here to consider the form which equations (5) assume for the extension to three dimensions of the rotation given by equations (2). Consider points of space referred to a fixed system of orthogonal cartesian axes, which form a right-handed system (Fig. 1). This means that if the head of a right-

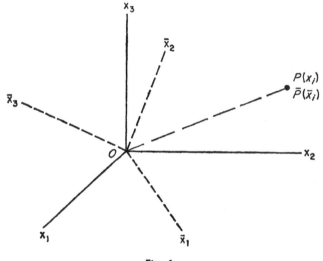

Fig. 1

handed screw were placed in the x_1x_2-plane pointing in the direction of the x_3-axis, a rotation from x_1 to x_2 would cause the screw to advance in the positive x_3-direction. Similarly, a motion from x_2 to x_3 would advance a right-handed screw in the positive x_1-direction.

Now consider a rotation of axes to a new right-handed system $\bar{x}_1, \bar{x}_2, \bar{x}_3$. A point P has coordinates x_i ($i = 1, 2, 3$) referred to the original axes, and coordinates $\bar{x}_i$ referred to the barred axes. It is necessary to find the equations of transformation which relate x_i and $\bar{x}_i$. Let $\bar{x}_1$ have direction cosines λ_i ($i = 1, 2, 3$) relative to the $x_1x_2x_3$-axes, and let $\bar{x}_2$ and $\bar{x}_3$ have direction cosines μ_i and ν_i, respectively, with respect to the original ($x_1x_2x_3$) axes. Now the distance from the origin O to P is invariant under the rotation. The projection of OP on the x_1-axis is x_1. The projection of OP on the $\bar{x}_i$ axis is $\bar{x}_i$. The sum of the projections $\bar{x}_i$ on the x_1 axis is equal to the projection x_1 of OP on the x_1 axis. Hence,

6)
$$x_1 = \lambda_1\bar{x}_1 + \mu_1\bar{x}_2 + \nu_1\bar{x}_3,$$
$$x_2 = \lambda_2\bar{x}_1 + \mu_2\bar{x}_2 + \nu_2\bar{x}_3,$$
$$x_3 = \lambda_3\bar{x}_1 + \mu_3\bar{x}_2 + \nu_3\bar{x}_3.$$

The last two equations were obtained by projecting OP and also the barred coordinates of P onto the x_2- and x_3-axes. Equations (6) represent the desired transformation relating the two systems of coordinates of any point P in space. The manner of development shows that OP is invariant under the rotation of axes about the fixed point O, but this can readily be verified by showing that

$$\overline{OP}^2 = (x_1)^2 + (x_2)^2 + (x_3)^2 = (\bar{x}_1)^2 + (\bar{x}_2)^2 + (\bar{x}_3)^2.$$

In the calculation it must be observed that $\lambda_1^2 + \lambda_2^2 + \lambda_3^2 = 1$, $\mu_1^2 + \mu_2^2 + \mu_3^2 = 1$, $\nu_1^2 + \nu_2^2 + \nu_3^2 = 1$, and that $\mu_1\nu_1 + \mu_2\nu_2 + \mu_3\nu_3 = 0$, $\nu_1\lambda_1 + \nu_2\lambda_2 + \nu_3\lambda_3 = 0$, $\lambda_1\mu_1 + \lambda_2\mu_2 + \lambda_3\mu_3 = 0$, which relations are known to obtain because of the orthogonality of each set of axes.

The inverse of equations (6) is readily obtained by multiplying the first by λ_1, the second by λ_2, the third by λ_3, and then adding. By use of the orthogonality relations $\lambda_1^2 + \lambda_2^2 + \lambda_3^2 = 1$, $\lambda_1\mu_1 + \lambda_2\mu_2 + \lambda_3\mu_3 = 0$, and those obtained from these by cyclic permutation of λ,μ,ν, one finds

7)
$$\bar{x}_1 = \lambda_1 x_1 + \lambda_2 x_2 + \lambda_3 x_3,$$
$$\bar{x}_2 = \mu_1 x_1 + \mu_2 x_2 + \mu_3 x_3,$$
$$\bar{x}_3 = \nu_1 x_1 + \nu_2 x_2 + \nu_3 x_3,$$

where the second and third equations result by using μ_i and ν_i in turn in place of λ_i.

A change of notation is now in order for equations (7). If the matrix of the direction cosines in equations (7), namely,

8)
$$\begin{pmatrix} \lambda_1 & \lambda_2 & \lambda_3 \\ \mu_1 & \mu_2 & \mu_3 \\ \nu_1 & \nu_2 & \nu_3 \end{pmatrix}$$

is written in the form

9) $$\begin{pmatrix} \lambda_{11} & \lambda_{12} & \lambda_{13} \\ \lambda_{21} & \lambda_{22} & \lambda_{23} \\ \lambda_{31} & \lambda_{32} & \lambda_{33} \end{pmatrix},$$

then equations (7) take the form

10) $$\begin{aligned} \bar{x}_1 &= \lambda_{11}x_1 + \lambda_{12}x_2 + \lambda_{13}x_3, \\ \bar{x}_2 &= \lambda_{21}x_1 + \lambda_{22}x_2 + \lambda_{23}x_3, \\ \bar{x}_3 &= \lambda_{31}x_1 + \lambda_{32}x_2 + \lambda_{33}x_3. \end{aligned}$$

Observe now that the first of the last three equations can be written as

$$\bar{x}_1 = \sum_{j=1}^{3} \lambda_{1j}x_j,$$

and that all three equations (10) are expressed by

$$\bar{x}_i = \sum_{j=1}^{3} \lambda_{ij}x_j \qquad (i = 1, 2, 3).$$

As a further simplification in notation, the summation symbol $\sum$ will be dropped, and it will be understood that a repeated subscript will require a summation unless specific mention is made to the contrary. Thus, for instance, $a_i x_i$ will mean $a_1 x_1 + a_2 x_2 + a_3 x_3$ (if the range of i is 1, 2, 3). With this convention (which will be found to be extremely useful in the tensor analysis), equations (10) take the form

11) $$\bar{x}_i = \lambda_{ij} x_j,$$

where it is understood that i ranges from 1 to 3, and j is summed from 1 to 3 in each of the three equations.

If a rotation R, given by equations (11), is followed by a translation T, given by equations (4) for $i = 1, 2, 3$, the transformation equations for the displacement of a rigid body in three-space are obtained, namely

12) $$\bar{x}_i = \lambda_{ij} x_j + a_i.$$

One may think of the x_i axes as fixed in space and the $\bar{x}_i$ axes as fixed in a rigid body. The effect of $x'_i = \lambda_{ij} x_j$ is a rotation with fixed point O, and the subsequent addition of a_i to obtain $\bar{x}_i = x'_i + a_i$ effects a translation of the body. The parameters λ_{ij} and a_i may be functions of a parameter, say t, where t may represent time.

Apparently there are twelve parameters in equations (12), but they are not independent because of the six orthogonality relations involving the λ_{ij}. The six-parameter group G_6 represented by equations (12) is the group of motions of a rigid body in three-space. The six parameters may be counted otherwise as follows. Three parameters are used in locating the center of gravity of the body, two independent parameters are needed to specify an axis through the center of gravity, and one parameter is necessary to measure rotation about the axis. The metric geometry of euclidean three-space involves the study of invariants of configurations under the group G_6 of mappings given by equations (12).

1–4. The Kronecker Delta. A useful notation to be used in the sequel is now introduced by considering the orthogonality relations for the coefficients λ_{ij} in equations (11). Observe from (8) and (9) that because $\lambda_1^2 + \lambda_2^2 + \lambda_3^2 = 1$, it follows that $\lambda_{11}^2 + \lambda_{12}^2 + \lambda_{13}^2 = 1$, which can be shortened to $\sum_{i=1}^{3} \lambda_{1i}\lambda_{1i} = 1$, or further to $\lambda_{1i}\lambda_{1i} = 1$. Also, it is true that

13) $$\lambda_{ki}\lambda_{ki} = 1,$$

where $k = 1, 2,$ or 3, and i is summed. From (8) and (9), since, for instance, $\lambda_1\lambda_2 + \mu_1\mu_2 + \nu_1\nu_2 = 0$, it follows that $\sum_{l=1}^{3} \lambda_{l1}\lambda_{l2} = 0$, or $\lambda_{i1}\lambda_{i2} = 0$. (Notice that the "dummy index" which indicates summation can be any letter, so long as it is repeated.) Now instead of using 1 and 2 in $\lambda_{i1}\lambda_{i2} = 0$, one may use 2 and 3, or 3 and 1, so it is true that

14) $\quad \lambda_{ij}\lambda_{ik} = 0 \quad (j \neq k), \quad \lambda_{ik}\lambda_{ik} = 1 \quad (j = k),$

where k is not summed. The second set of equations in (14) come from (13) with i and k interchanged. It is possible to write equations (14) in the form

15) $$\lambda_{ij}\lambda_{ik} = \delta_{jk},$$

when δ_{jk}, called a Kronecker delta, is defined as follows:

$$\delta_{jk} = 1 \text{ if } j = k \text{ (not summed), and } \delta_{jk} = 0 \text{ if } j \neq k.$$

Of course, the index i sums in equations (15). A similar Kronecker delta will be found very convenient in the tensor analysis.

EXERCISES

1) If R_λ: $\bar{x}_i = \lambda_{ij}x_j$ is a rotation of orthogonal axes, and R_μ: $\bar{\bar{x}}_k = \mu_{ki}\bar{x}_i$ is a second such rotation, prove that the product $R_\mu R_\lambda$ describes a third rotation of the same type. (Hint: Let $\nu_{kj} = \mu_{ki}\lambda_{ij}$, and show that $\nu_{kj}\nu_{kl} = \delta_{jl}$. Note that $\mu_{ki}\lambda_{ij}\mu_{ks}\lambda_{sl} = \mu_{ki}\mu_{ks}\lambda_{ij}\lambda_{sl} = \delta_{is}\lambda_{ij}\lambda_{sl} = \lambda_{ij}\lambda_{il} = \delta_{jl}$.)

2) Carry out the details to verify that the distance OP is invariant under the mapping $\bar{x}_i = \lambda_{ij}x_j$, if $\lambda_{ij}\lambda_{ik} = \delta_{jk}$. (This should be done as follows: $\overline{OP}^2 =$

$\sum \bar{x}_i{}^2 = \bar{x}_i\bar{x}_i = \delta_{ij}\bar{x}_i\bar{x}_j = \delta_{ij}(\lambda_{ik}x_k)(\lambda_{jl}x_l) = \delta_{ij}\lambda_{ik}\lambda_{jl}x_kx_l = \lambda_{ik}\lambda_{il}x_kx_l = \delta_{kl}x_kx_l = x_kx_k = \overline{OP}^2.$)

3) What is the value of δ_{ii}, where i is summed?

4) If $x_i = a_{ij}\bar{x}_j$, a general homogeneous linear transformation for which the determinant $|a_{ij}| \neq 0$, has the inverse $\bar{x}_j = b_{jk}x_k$, show that $a_{ij}b_{jk} = \delta_{ik}$.

5) Verify that the mappings R_1 and R_2 given by

$$R_1: \begin{array}{l} 3\bar{x}_1 = x_1 - 2x_2 + 2x_3 \\ 3\bar{x}_2 = -2x_1 + x_2 + 2x_3 \\ 3\bar{x}_3 = -2x_1 - 2x_2 - x_3 \end{array} \qquad R_2: \begin{array}{l} 7\bar{x}_1 = 3\bar{x}_1 - 6\bar{x}_2 + 2\bar{x}_3 \\ 7\bar{x}_2 = 6\bar{x}_1 + 2\bar{x}_2 - 3\bar{x}_3 \\ 7\bar{x}_3 = 2\bar{x}_1 + 3\bar{x}_2 + 6\bar{x}_3 \end{array}$$

are rotations by showing that the orthogonality relations are satisfied.

6) Carry out the substitutions to show that $R_2R_1 \neq R_1R_2$ in Exercise 5. This means that the group of rotations in space is not commutative. The reader may wish to verify that both R_2R_1 and R_1R_2 are members of the set of rotations about the fixed point O.

2

Loci in Three-Space

2-1. One-Dimensional Extent. The reader will recall that the symmetric equations of a straight line through the point (a_1, a_2, a_3) with direction numbers $\lambda_1, \lambda_2, \lambda_3$ are

$$\frac{x_1 - a_1}{\lambda_1} = \frac{x_2 - a_2}{\lambda_2} = \frac{x_3 - a_3}{\lambda_3}.$$

If each of these variable ratios is equated to the parameter t, the equations of the line can be written in the form

1) $\qquad x_i = a_i + \lambda_i t \qquad (i = 1, 2, 3).$

It is evident then from equations (1) that if the cartesian coordinates x_i are linear functions of a parameter, the set of points obtained as the parameter ranges over all real numbers is a straight line. It is assumed well known that the parameter t in (1) is proportional to the distance between the fixed point a_i and the variable point x_i, and that t is equal to this distance in case the λ_i are direction cosines, the condition for which is $\lambda_i \lambda_i = 1$. (Remember that the index i is summed from 1 to 3.) One may style the point set determined by equations (1) as a linear one-dimensional extent immersed in a euclidean space of three dimensions. Equations (1) may also be called the imbedding equations for a line in three-space.

A one-dimensional extent with different structure is realized if the x_i are specified by non-linear functions of a single parameter. In this case the point set which the x_i range over as the parameter varies is called a curve. The range of the parameter must be prescribed in order to have the point set described completely. In general, the functions may be arbitrary. However, if one investigates the differential geometry of the curve, he may require that the parameter range over some continuous subset of the real numbers, and that not all three of the derivatives dx_i/dt vanish at any point on the curve (see Section 2-3).

A curve may be represented by the parametric equations

2) $$x_i = f_i(t).$$

It is often simpler to write $x_i(t)$ instead of $f_i(t)$ to have the equations of the curve in the form $x_i = x_i(t)$.

Example. Consider the curve represented by $x_1 = at$, $x_2 = bt^2$, $x_3 = ct^3$, where $abc \neq 0$. This is called a cubic curve because it intersects any plane in three points (see Exercise 1 in Section 2-2).

A parametric representation of a curve is not unique. Indeed, there are infinitely many representations, as is shown by the fact that t may be replaced by an arbitrary single-valued function of a new parameter r, say $t = \phi(r)$, so that equations (2) become

3) $$x_i = f_i[\phi(r)] = g_i(r).$$

If it is required that not all dx_i/dr vanish, that is

$$\frac{dx_i}{dr} = \frac{df_i}{dt}\frac{dt}{dr} = \frac{df_i}{dt}\frac{d\phi(r)}{dr}$$

do not all vanish, then $d\phi(r)/dr$ cannot vanish. This means that the equation $t = \phi(r)$ must have a unique inverse.

2-2. Two-Dimensional Extent. By analogy with the straight line point set, a plane may be defined by taking the space coordinates x_i as linear functions of two independent parameters, say u_1 and u_2, where these parameters range independently over all real numbers. The equations may take the form

4) $$x_i = c_{i1}u_1 + c_{i2}u_2 + c_{i3} \qquad (i = 1, 2, 3),$$

where the c_{ik} are constants. Equations (4) are called the imbedding equations of a two-dimensional linear extent in three-space. It is assumed that at least one pair of the three equations in (4) can be solved for u_1 and u_2. If this is done and the expressions for u_1 and u_2 are substituted into the remaining equation, there results a linear equation in x_i which is of the type

5) $$a_i x_i + h = 0.$$

Observe that if no restriction is placed upon the coordinates x_i of a point, it has three degrees of freedom (in three-space) and may take any position in space. However, if the x_i satisfy a condition such as (5), the number of degrees of freedom is reduced to two, for equation (5) can be solved for one of the three variables x_i in terms of the other two, and then only these two variables may take on arbitrary values. Thus, equation (5) represents a two-dimensional extent which is linear.

If x_i must satisfy two equations like (5), say

6) $$a_i x_i + h = 0, \qquad b_i x_i + k = 0,$$

the number of degrees of freedom for the point x_i is further reduced, and x_i must lie in the intersection of the point sets determined by equations (6). Of course, if the planes in (6) are parallel, they have no intersection.

If the x_i coordinates are specified as non-linear functions of two independent parameters u_1 and u_2, a two-dimensional extent of more complex structure is obtained. The point set so realized is called a surface. Its equations may be taken as

7) $$x_i = x_i(u_1, u_2) \qquad (i = 1, 2, 3).$$

Note here that if $x_i(u_1, u_2)$ satisfy a linear relation $a_i x_i + h = 0$ identically in u_1 and u_2, then equations (7) represent a two-dimensional linear space (a plane) instead of a two-dimensional curved space (a *curved surface*). Equations (7) cover those in (4) as a particular case, so that a plane is also a surface.

It is assumed that at least one pair of the three equations (7) can be solved for u_1, u_2. Let the solutions for u_1, u_2 be substituted into the remaining equation. The result is, in general, a non-linear equation in x_i, which may be written as

8) $$f(x_1, x_2, x_3) = 0.$$

The points x_i of space which satisfy (8) have two degrees of freedom. If the x_i must satisfy two independent equations of type (8), say

9) $$f(x_1, x_2, x_3) = 0, \qquad g(x_1, x_2, x_3) = 0,$$

then the points x_i lie on the intersection of the two surfaces in (9), which in the most common cases is a curve. Suppose equations (9) can be solved for x_2 and x_3 to obtain $x_2 = \phi(x_1)$, $x_3 = \psi(x_1)$. Then a possible parametric form for the curve (9) would be

$$x_1 = x_1, \qquad x_2 = \phi(x_1), \qquad x_3 = \psi(x_1),$$

where x_1 is the parameter on the one-dimensional extent.

EXERCISES

1) Show that the curve $x_1 = at$, $x_2 = bt^2$, $x_3 = ct^3$ ($abc \neq 0$) intersects any plane $h_i x_i + k = 0$ in three points.

2) Show that the curve $x_1 = a \cos \theta$, $x_2 = a \sin \theta$, $x_3 = b\theta$ lies on a circular cylinder. This curve is called a helix. Find another surface upon which the curve lies.

3) If $x_1 = a \sin u_1 \cos u_2$, $x_2 = a \sin u_1 \sin u_2$, $x_3 = a \cos u_1$, find the surface equation in the form $f(x_1,x_2,x_3) = 0$.

4) Show that if $x_i = x_i(u_1,u_2)$ and $u_2 = \phi(u_1)$, the point set which x_i generates is a curve (which includes a straight line as a particular instance).

5) Do the equations $x_1 = u_1 + u_2 + 3$, $x_2 = \sin(u_1 + u_2) + 4$, $x_3 = (u_1 + u_2)^2 - 2(u_1 + u_2)$ represent a surface? Why?

6) Show that $x_1 = u_1^2 + 2u_2^2 - 1$, $x_2 = 2u_1^2 - u_2^2 + 2$, $4x_3 = 1 - 8u_1^2 + 7u_2^2$ represent a plane. See the remark following equations (7).

2-3. Some Differential Geometry of Space Curves. It will be of use in later developments to know some fundamental facts about space curves.

LENGTH OF ARC. Let a curve C be represented by $x_i = x_i(t)$, where t is a general parameter. Instead of t, the arc length s, measured on C from a fixed point to a variable point P, may be used as parameter. If x_i and $x_i + dx_i$ are neighboring points in euclidean space, the distance ds between them is found from

10) $$(ds)^2 = (dx_1)^2 + (dx_2)^2 + (dx_3)^2 = dx_i\, dx_i.$$

From $x_i = x_i(t)$, $dx_i = (dx_i/dt)\, dt$, and equation (10) gives

11) $$(ds)^2 = \frac{dx_i}{dt} \frac{dx_i}{dt} (dt)^2,$$

from which

12) $$\frac{ds}{dt} = \left(\frac{dx_i}{dt}\frac{dx_i}{dt}\right)^{1/2},$$

where the positive sign is chosen for ds/dt if s increases with t. From (12), the relation between the parameters s and t is

13) $$s = \int_{t_0}^{t} \left(\frac{dx_i}{dt}\frac{dx_i}{dt}\right)^{1/2} dt = \phi(t),$$

where $t = t_0$ at the point where $s = 0$. Note that t can be identified as arc length if and only if $(dx_i/dt)(dx_i/dt) = 1$.

TANGENT LINE. For what follows let the curve C be defined by $x_i = x_i(s)$. A point $\bar{P}(\bar{x}_i)$ distant ϵ along C from $P(x_i)$ has parameter $s + \epsilon$. Taylor's series may be employed to express the $\bar{x}_i$ coordinates in terms of coordinates x_i and their derivatives, evaluated at $P(x_i)$, in the form

14) $$\bar{x}_i = x_i + x'_i \epsilon + \frac{1}{2!} x''_i \epsilon^2 + \frac{1}{3!} x'''_i \epsilon^3 + \cdots \qquad (i = 1, 2, 3),$$

where the prime is used to indicate differentiation with respect to s. The equations of any line through $P(x_i)$ are [see equations (1)]

15) $$\xi_i = x_i + \lambda_i \sigma,$$

where ξ_i are current coordinates on the line and λ_i are direction cosines so that σ is the distance from $P(x_i)$ to the point ξ_i. If the line (15) passes through $\bar{P}(\bar{x}_i)$,

$$\bar{x}_i - x_i = \lambda_i \sigma,$$

and by (14) it follows that

$$\lambda_i \sigma = x'_i \epsilon + \tfrac{1}{2} x''_i \epsilon^2 + \cdots$$

or

16) $$\lambda_i \frac{\sigma}{\epsilon} = x'_i + \frac{1}{2} x''_i \epsilon + \cdots .$$

It is known that $\lim_{\epsilon \to 0} \sigma/\epsilon = 1$. Hence, the limit in (16) as $\epsilon \to 0$ yields $\lambda_i = x'_i$. Thus, the tangent line (15) to the curve C at $P(x_i)$ is represented by

17) $$\xi_i = x_i(s) + x'_i(s) t,$$

where t is a general parameter independent of s, but note that t is the distance from $P(x_i)$ to the variable point ξ_i on the tangent line. One should observe that if both s and t vary independently, equations (17) represent the surface generated by the tangent lines to the curve C. It is understood that not all three of $x'_i(s)$ are zero at any point of the curve. With all three $x'_i(s)$ zero at a point the tangent direction would not be defined there.

TANGENT PLANE, OSCULATING PLANE, ORDER OF CONTACT. The equation of any plane in space may take the form

18) $$A_i X_i + D = 0,$$

where A_i are the direction cosines of a normal line to the plane. The relation $A_i A_i = 1$ therefore holds. If the plane passes through the point $P(x_i)$ on C, then $D = -A_i x_i$. Use of this expression for D in (18) yields

19) $$A_i(X_i - x_i) = 0.$$

The plane (19) has, in general, only one point in common with the curve C. This situation is described by saying that the plane has zero order of contact with C at P. The next step is to force the plane to have first order contact with C.

It is known from analytical geometry that the distance δ from a point $\bar{x}_i$ to a plane $A_i(X_i - x_i) = 0$ is obtained merely by substituting the coordinates $\bar{x}_i$ for X_i in the equation. (Remember that $A_i A_i = 1$ which means that the equation of the plane is in normal form.) The result of the substitution of $\bar{x}_i$ for X_i is

$$\delta = A_i(\bar{x}_i - x_i)$$

which, by use of the Taylor series (14), takes the form

20) $\quad \delta = A_i(x'_i\epsilon + \tfrac{1}{2}x''_i\epsilon^2 + \tfrac{1}{6}x'''_i\epsilon^3 + \tfrac{1}{24}x^{iv}_i\epsilon^4 + \cdots).$

This last result shows that if the sum $A_i x'_i \neq 0$, the distance δ is of the first order relative to the arc length ϵ from P to $\bar{P}$ on C.

Consider next the geometrical significance of the vanishing of $A_i x'_i$. In analytical geometry it is shown that the cosine of the angle between two space lines with direction cosines α_i and β_i is given by $\cos\theta = \alpha_i\beta_i$. If the lines are orthogonal, $\alpha_i\beta_i = 0$. Therefore, the condition $A_i x'_i = 0$ means that the tangent line to C at P is perpendicular to the normal A_i to the plane through P. Hence, the plane $A_i(X_i - x_i) = 0$ contains the tangent line to C at P and is therefore called a tangent plane to C. Notice that there are still infinitely many planes which contain the tangent to C at P. A double infinitude of planes pass through P. By demanding that the plane through P contain also the tangent line to C at P, the number of degrees of freedom of the plane is reduced from two to one.

So far one can say that if $A_i x'_i = 0$ but $A_i x''_i \neq 0$, the distance δ from $\bar{P}$ to the plane is of the second order in ϵ, or that the plane has two-point contact (or *first-order* contact) with the curve C at P.

The next step is to suppose that $A_i x'_i = 0$, $A_i x''_i = 0$, but $A_i x'''_i \neq 0$. The distance δ is now of the third order relative to ϵ. But what is the geometrical significance of the condition $A_i x''_i = 0$? This means that the normal A_i to the plane is perpendicular to a line through P with direction x''_i. This condition reduces the number of degrees of freedom of the plane to zero so only one plane is determined which passes through P and contains the lines through P with directions x'_i and x''_i. The line with direction x''_i is called the *principal normal* to C at P, and the tangent plane which contains the principal normal is called the *osculating plane* to the curve C at P. In general, $A_i x'''_i \neq 0$ for points on C, but if points on C exist where $A_i x^{(n)}_i = 0$ for $n > 2$, it is said that the plane *superosculates* the curve at such points, or has contact of higher than the second order at these points. It should be observed here that in case C is a straight line $x_i(s) = a_i + b_i s$, where a_i and b_i are constants, with $b_i b_i = 1$, and $x'_i = b_i$, $x^{(n)}_i = 0$, for $n > 1$. Hence, any plane through the line superosculates the line. In this case, there is a singly infinite set of tangent planes, but the osculating plane is indeterminate.

The line through P which is perpendicular to the osculating plane is called the *binormal* to the curve at P. Let α_i, β_i, γ_i denote the direction cosines of the tangent, principal normal, and binormal, respectively, at

P on C. Then the normal plane to C at P has the equation

21) $$(X_i - x_i)\alpha_i = 0,$$

the plane perpendicular to the principal normal, called the *rectifying* plane, is given by

22) $$(X_i - x_i)\beta_i = 0,$$

and the osculating plane has the equation

23) $$(X_i - x_i)\gamma_i = 0.$$

It has been established that $\alpha_i = x'_i(s)$, so the normal plane may have the form $(X_i - x_i)x'_i = 0$. It was shown that β_i are proportional to $x''_i(s)$, so the rectifying plane may be written as $(X_i - x_i)x''_i = 0$. In order to express γ_i in terms of $x_i(s)$, the equation of the osculating plane will be obtained in the following way. Begin with the general equation of a plane through P, that is

24) $$A_i(X_i - x_i) = 0.$$

For the osculating plane it was found that

24') $$A_i x'_i = 0, \qquad A_i x''_i = 0.$$

Solve these last two equations for the ratios $A_1:A_2:A_3$ and introduce the result into (24). The outcome is an equation independent of the A_i. It can be written in the determinant form

25) $$\begin{vmatrix} X_1 - x_1 & X_2 - x_2 & X_3 - x_3 \\ x'_1 & x'_2 & x'_3 \\ x''_1 & x''_2 & x''_3 \end{vmatrix} = 0.$$

The left-hand member of (25) is called the *eliminant* of equations (24) and (24'). Equation (25) is a necessary and sufficient condition that equations (24), (24') have a solution for A_i in addition to the obvious trivial solution $0, 0, 0$. The proof of the necessity of the condition was effected in arriving at (25), but the proof of the sufficiency of condition (25) is omitted.

From equation (25) of the osculating plane to C at x_i, together with (23), it is seen that γ_i are proportional to the three second-order determinants

26) $$\begin{vmatrix} x'_2 & x'_3 \\ x''_2 & x''_3 \end{vmatrix}, \qquad \begin{vmatrix} x'_3 & x'_1 \\ x''_3 & x''_1 \end{vmatrix}, \qquad \begin{vmatrix} x'_1 & x'_2 \\ x''_1 & x''_2 \end{vmatrix}.$$

Recall that to change direction numbers p_i into direction cosines one divides each of the p_i by the square root of the sum of the squares of p_i. That is, if p_i are direction numbers, $p_i/(p_k p_k)^{1/2}$ are direction cosines. Since β_i are proportional to x''_i, the values of β_i are given by

$$27) \qquad \beta_i = \frac{x''_i}{\pm\sqrt{x''_k x''_k}} \qquad (i = 1, 2, 3;\ k \text{ summed}).$$

In order to render the numbers in (26) into direction cosines, the Lagrange identity will be helpful. For any two sets of triples, say (a_i) and (b_i), the identity is

$$27') \qquad \begin{vmatrix} a_2 & a_3 \\ b_2 & b_3 \end{vmatrix}^2 + \begin{vmatrix} a_3 & a_1 \\ b_3 & b_1 \end{vmatrix}^2 + \begin{vmatrix} a_1 & a_2 \\ b_1 & b_2 \end{vmatrix}^2$$
$$\equiv (a_1^2 + a_2^2 + a_3^2)(b_1^2 + b_2^2 + b_3^2) - (a_1 b_1 + a_2 b_2 + a_3 b_3)^2.$$

On placing the numbers in (26) in the left-hand side of (27') (x'_i for a_i and x''_i for b_i), the right-hand side gives

$$28) \qquad (x'_i x'_i)(x''_k x''_k) - (x'_i x''_i)^2.$$

Now $x'_i x'_i \equiv \alpha_i \alpha_i = 1$, and $x'_i x''_i = 0$ because $\alpha_i \beta_i = 0$, so that the sum of the squares of the determinants in (26) is simply $x''_k x''_k$. Therefore,

$$29) \qquad \gamma_i = \frac{1}{\pm\sqrt{x''_k x''_k}} (x'_2 x''_3 - x''_2 x'_3,$$
$$x'_3 x''_1 - x''_3 x'_1,\ x'_1 x''_2 - x''_1 x'_2).$$

For any given curve $C: x_i = x_i(s)$, the $\alpha_i,\ \beta_i,\ \gamma_i$ can be calculated, and from (21), (22), (23) the equations of the normal, rectifying, and osculating planes can be determined. For the last two planes it is immaterial which sign in (27) and (29) is chosen. However, it will be found convenient later to choose three positive directions along the legs of the trihedron constituted by the tangent, principal normal, and binormal at each point of the curve, and to make the choice of directions so that this trihedron is oriented in a right-handed manner as are the axes of reference.

A NEW NOTATION. A more concise form can be written for the osculating plane, and for other equations to appear later, by use of the quantities e^{ijk} defined as follows:

$e^{ijk} =$ 0 when any two of i, j, k take on the same value,
$\quad = $ 1 when i, j, k have the values 1, 2, 3 or any even permutation of them, i.e., 2, 3, 1 or 3, 1, 2,
$\quad = -1$ when i, j, k have values which are an odd permutation of 1, 2, 3, i.e., 1, 3, 2 or 2, 1, 3 or 3, 2, 1.

The reader should write all non-zero terms of the sum in the equation

30) $$e^{ijk}(X_i - x_i)x'_j x''_k = 0$$

to verify that it is the equation (25) of the osculating plane. Observe also that the three determinantal coefficients in (26) may be written in one expression as

$$e^{ijk} x'_j x''_k \qquad (i = 1, 2, 3).$$

For instance, when $i = 1$, $e^{123} x'_2 x''_3 + e^{132} x'_3 x''_2 = x'_2 x''_3 - x''_2 x'_3$.

EXERCISES

1) Find the direction cosines of the tangent to the curve $x_1 = a \cos \theta$, $x_2 = a \sin \theta$, $x_3 = b\theta$ at any point. Show that the angle between the tangent and the generator of the cylinder upon which the curve lies is the same at all points of the curve.

2) Find the parametric equations of the helix in Exercise 1 with the arc length s as parameter.

3) Find the equations of the tangent line to the helix in Exercise 1 at the point where $\theta = \pi/4$.

4) Show that the equation of the osculating plane to the helix in Exercise 1 at any point θ is $b(x_1 \sin \theta - x_2 \cos \theta - a\theta) + ax_3 = 0$.

5) Show that equation (25) is the osculating plane to a curve with general parameter, where the primes no longer indicate differentiation with respect to arc length.

6) Show that the parallels through the origin to the binormals of the helix in Exercise 1 are generators of the cone $a^2(x_1^2 + x_2^2) = b^2 x_3^2$.

7) Show that the principal normal to a helix is perpendicular to the cylinder on which it lies.

8) For the cubic $x_1 = t$, $x_2 = t^2$, $x_3 = t^3$, find the equations of the normal, rectifying, and osculating planes at any point t.

9) Make use of the definition of e^{ijk} to show that

$$e^{ijk} a_{i1} a_{j2} a_{k3} \equiv \begin{vmatrix} a_{11} & a_{12} & a_{13} \\ a_{21} & a_{22} & a_{23} \\ a_{31} & a_{32} & a_{33} \end{vmatrix}.$$

10) (a) Find the value of p for which the three planes $3x_1 + x_2 - x_3 = 0$, $x_1 - x_2 + 2x_3 = 0$, $2x_1 + 3x_2 + px_3 = 0$ have a point in common other than the origin, and show that for this value of p the three planes have a line in common. (b) Three planes $a_i x_i = 0$, $b_i x_i = 0$, $c_i x_i = 0$ are *linearly dependent* if there exist constants k_1, k_2, k_3, not all zero, such that $k_1(a_i x_i) + k_2(b_i x_i) + k_3(c_i x_i) = 0$ identically in x_1, x_2, x_3. Find a set of multiples k_1, k_2, k_3 for the planes $3x_1 + x_2 - x_3 = 0$, $x_1 - x_2 + 2x_3 = 0$, $8x_1 + 12x_2 - 19x_3 = 0$, and thus show that the three planes are linearly dependent.

2–4. Some Differential Geometry of Surfaces.

Although a surface may be represented by $f(x_1,x_2,x_3) = 0$, or by the Monge form $x_3 = \phi(x_1,x_2)$, it is more convenient for most purposes to describe the surface by equations in the Gauss form

31) $$x_i = x_i(u_1,u_2), \quad (i = 1, 2, 3)$$

where the parameters u_1, u_2 are independent. If u_2 is set equal to a function of u_1, say $u_2 = g(u_1)$, then x_i are functions of a single variable and hence represent a curve on the surface. It will be assumed that the functions $x_i(u_1, u_2)$ possess partial derivatives to any order necessary in a given discussion. From (31) the following matrix of first partial derivatives can be calculated:

32) $$\begin{pmatrix} \dfrac{\partial x_1}{\partial u_1} & \dfrac{\partial x_2}{\partial u_1} & \dfrac{\partial x_3}{\partial u_1} \\[1em] \dfrac{\partial x_1}{\partial u_2} & \dfrac{\partial x_2}{\partial u_2} & \dfrac{\partial x_3}{\partial u_2} \end{pmatrix}.$$

The three second-order determinants which can be formed from this matrix may be expressed by

33) $$e^{ijk} \frac{\partial x_j}{\partial u_1} \frac{\partial x_k}{\partial u_2} \quad (i = 1, 2, 3).$$

For instance, for $i = 1$,

34) $$e^{1jk} \frac{\partial x_j}{\partial u_1} \frac{\partial x_k}{\partial u_2} = e^{123} \frac{\partial x_2}{\partial u_1} \frac{\partial x_3}{\partial u_2} + e^{132} \frac{\partial x_3}{\partial u_1} \frac{\partial x_2}{\partial u_2} = \begin{vmatrix} \dfrac{\partial x_2}{\partial u_1} & \dfrac{\partial x_3}{\partial u_1} \\[1em] \dfrac{\partial x_2}{\partial u_2} & \dfrac{\partial x_3}{\partial u_2} \end{vmatrix}.$$

If x_1 and x_2 in equations (31) are both functions of x_3, the surface degenerates to a curve. Consider first x_2 and x_3. If $x_2(u_1,u_2)$ is a function of $x_3(u_1,u_2)$, then x_2 and x_3 are described as *functionally dependent*. It is important to develop a test to determine dependence or independence of two functions. In order to do this, suppose that x_2 and x_3 are functionally dependent, so that a function $\phi(x_2,x_3)$ exists for which

$$\phi(x_2,x_3) \equiv 0$$

where x_2 and x_3 are both functions of u_1 and u_2. Take the partial derivatives of $\phi \equiv 0$ with respect to u_1 and u_2 in turn to have

$$\frac{\partial \phi}{\partial x_2} \frac{\partial x_2}{\partial u_1} + \frac{\partial \phi}{\partial x_3} \frac{\partial x_3}{\partial u_1} = 0,$$

$$\frac{\partial \phi}{\partial x_2} \frac{\partial x_2}{\partial u_2} + \frac{\partial \phi}{\partial x_3} \frac{\partial x_3}{\partial u_2} = 0,$$

which are two linear homogeneous equations in $\partial \phi/\partial x_2$ and $\partial \phi/\partial x_3$. These equations have non-trivial solutions only in case the eliminant vanishes, that is,

$$\begin{vmatrix} \dfrac{\partial x_2}{\partial u_1} & \dfrac{\partial x_3}{\partial u_1} \\ \dfrac{\partial x_2}{\partial u_2} & \dfrac{\partial x_3}{\partial u_2} \end{vmatrix} = 0.$$

Hence, a *necessary* condition that x_2 and x_3 be functionally related is that the determinant in (34) vanish. This determinant is called the *jacobian* of x_2 and x_3 with respect to u_1 and u_2, and it may be denoted by J_{23} or by

$$\frac{\partial (x_2,x_3)}{\partial (u_1,u_2)}.$$

It should be emphasized that J_{23} may be identically zero in u_1 and u_2, but, because x_2 and x_3 are functions of u_1 and u_2, J_{23} may be a function of these parameters. In this case $J_{23} = 0$ determines u_2 as a function of u_1 which gives a curve on the surface. Only at points of this curve are x_2 and x_3 dependent. Similarly, the vanishing of J_{13}, the jacobian of x_1 and x_3 with respect to u_1 and u_2, is a necessary condition for the functional dependence of x_1 and x_3. It can be shown (see Exercise 7) that if two of the three jacobians in (33) are zero, the third one is also zero. It can also be demonstrated that the vanishing of J_{23}, for instance, is a *sufficient* condition for x_2 and x_3 to be functionally dependent, but the reader is referred to any textbook on advanced calculus for this.

To sum up, it can be stated that a necessary and sufficient condition that equations (31) determine a surface and not a curve is that at least one of the three jacobian determinants in (33) be not identically zero. Values of u_1 and u_2 for which all three jacobians vanish correspond to *singular points* on the surface or to singular points of the particular parametric representation employed in (31). Such points, which may be isolated or along singular curves, require special consideration. It will be assumed that the region of the surface under study will not contain singular points. See Exercise 10 for an additional remark on this point.

A change of parameter was effected (2–3)[1] on a space curve. Likewise, a change of parameters from (u_1,u_2) to $(\bar{u}_1,\bar{u}_2)$ can be made on a surface. If $u_\alpha = u_\alpha(\bar{u}_1,\bar{u}_2)$ $(\alpha = 1, 2)$ are substituted into (31), the x_i functions of u_1, u_2 become $\bar{x}_i$ functions of $\bar{u}_1, \bar{u}_2$. Of course, it is assumed that u_1 and u_2 are independent functions of $\bar{u}_1$ and $\bar{u}_2$, so that, in view of

---

[1] As stated in the Preface, this will be understood to mean Chapter 2, equation 3.

the foregoing discussion on functional independence, the jacobian

$$35) \quad \frac{\partial(u_1, u_2)}{\partial(\bar{u}_1, \bar{u}_2)} \equiv \begin{vmatrix} \dfrac{\partial u_1}{\partial \bar{u}_1} & \dfrac{\partial u_2}{\partial \bar{u}_1} \\ \dfrac{\partial u_1}{\partial \bar{u}_2} & \dfrac{\partial u_2}{\partial \bar{u}_2} \end{vmatrix} \neq 0.$$

Under condition (35) the equations $u_\alpha = u_\alpha(\bar{u}_1, \bar{u}_2)$ can be solved locally for the inverse $\bar{u}_\alpha = \bar{u}_\alpha(u_1, u_2)$. It should be observed that the determinant in (35) may vanish along some curve on the surface. This determinant may be a function of $\bar{u}_1$, $\bar{u}_2$, say $J(\bar{u}_1, \bar{u}_2)$, and the vanishing of J gives a singular curve which will be avoided.

The u_1 coordinate curves on the surface are the curves on which u_1 alone varies. That is, u_2 = constant represents a one-parameter family of curves on the surface. Similarly, u_1 = constant represents a curve (for each value of the constant) on which only u_2 varies. Thus, u_1 = constant is the equation of the u_2 curves. One speaks of a point on the surface with coordinates u_1, u_2 as the point $u_\alpha(\alpha = 1, 2)$. Through each point u_α there is one and only one curve of each of the families u_1 = constant, and u_2 = constant.

TANGENT PLANE. The tangent plane at the point x_i on the surface S (that is, at the point u_α) may be defined as the plane determined by the tangents to the coordinate curves u_1 = constant, u_2 = constant through the point. Take the tangent plane in the form

$$36) \quad A_i(X_i - x_i) = 0,$$

where A_i are to be determined. Because the A_i direction is perpendicular to the tangent to the u_1-curve with direction numbers $\partial x_i/\partial u_1$, it follows that

$$37) \quad A_i \frac{\partial x_i}{\partial u_1} = 0.$$

[Remember that the derivatives of x_i with respect to arc length are the direction cosines of the tangent to the u_1-curve, and note that $\partial x_i/\partial u_1$ are proportional to dx_i/ds which are $(\partial x_i/\partial u_1)(du_1/ds)$.] Similarly, the A_i direction is perpendicular to the tangent to the u_2 curve with direction numbers $\partial x_i/\partial u_2$, so

$$38) \quad A_i \frac{\partial x_i}{\partial u_2} = 0.$$

Elimination of the A_i from equations (36), (37), and (38) leads to the equation of the tangent plane to S in the form

$$39) \qquad e^{ijk}(X_i - x_i)\frac{\partial x_j}{\partial u_1}\frac{\partial x_k}{\partial u_2} = 0.$$

Observe from (39) that the direction of the normal line to the surface $S: x_i = x_i(u_1, u_2)$ is given by the three quantities

$$40) \qquad e^{ijk}\frac{\partial x_j}{\partial u_1}\frac{\partial x_k}{\partial u_2}.$$

Elimination of the parameters u_1, u_2 from equations (31) yields the equation of the surface in the form

$$41) \qquad f(x_1, x_2, x_3) = 0.$$

If x_i from (31) are substituted into (41) an identity in u_1, u_2 results. Partial differentiation of equation (41) then gives

$$42) \qquad \frac{\partial f}{\partial x_1}\frac{\partial x_1}{\partial u_1} + \frac{\partial f}{\partial x_2}\frac{\partial x_2}{\partial u_1} + \frac{\partial f}{\partial x_3}\frac{\partial x_3}{\partial u_1} = 0,$$

and

$$43) \qquad \frac{\partial f}{\partial x_1}\frac{\partial x_1}{\partial u_2} + \frac{\partial f}{\partial x_2}\frac{\partial x_2}{\partial u_2} + \frac{\partial f}{\partial x_3}\frac{\partial x_3}{\partial u_2} = 0.$$

Both (42) and (43) can be expressed by

$$44) \qquad \frac{\partial f}{\partial x_i}\frac{\partial x_i}{\partial u_\alpha} = 0 \qquad (\alpha = 1, 2).$$

Notice that the Greek letter is used for the range 1, 2 while the Latin letter i is summed over 1, 2, 3. If the two homogeneous equations (44) are solved for the ratios $(\partial f/\partial x_1):(\partial f/\partial x_2):(\partial f/\partial x_3)$, one finds that the $\partial f/\partial x_i$ are proportional to the three jacobians (40), so that

$$\frac{\partial f}{\partial x_i} = Ke^{ijk}\frac{\partial x_j}{\partial u_1}\frac{\partial x_k}{\partial u_2},$$

where K is a proportionality factor. Hence, by (40) the direction of the normal to the surface at x_i is given by

$$45) \qquad \frac{\partial f}{\partial x_i} \qquad (i = 1, 2, 3),$$

evaluated at x_i.

EXERCISES

1) Find the equation of the tangent plane to the helicoidal surface $x_1 = u_1 \cos u_2$, $x_2 = u_1 \sin u_2$, $x_3 = cu_2$ at any point u_α.

2) Find the equation of the tangent plane to the sphere $x_1 = a \sin u_1 \cos u_2$, $x_2 = a \sin u_1 \sin u_2$, $x_3 = a \cos u_1$ at any point by use of (39), and show that this plane is perpendicular to the radius of the sphere to the point of tangency.

3) Show that the tangent plane to $f(x_1,x_2,x_3) = 0$ at x_i is $(X_i - x_i) \partial f/\partial x_i = 0$.

4) Find the tangent plane to $x_3 = \phi(x_1,x_2)$ at x_i. Use the result in Exercise 3, or consider the surface in Gauss form as $x_1 = x_1$, $x_2 = x_2$, $x_3 = \phi(x_1,x_2)$ with x_1, x_2 as the parameters.

5) Write the equation of the tangent plane at $\bar{x}_i$ to the cone $x_1^2 + x_2^2 - 4x_3^2 = 0$, and show that the plane does not vary as the point of tangency moves along a generator of the cone.

6) Show that a surface with Gauss equations

$$x_1 = u_1 \cos u_2, \qquad x_2 = u_1 \sin u_2, \qquad x_3 = f(u_1)$$

is obtained by revolving the plane curve given by

$$x_3 = f(u_1), \qquad x_2 = 0$$

about the x_3 axis. Any surface of revolution can be represented by the Gauss equations shown here, provided the function $f(u_1)$ is properly specified. The curves $u_2 = $ constant are called *meridian* curves on the surface.

7) Prove that the vanishing of any two of the jacobians in (33) implies the vanishing of the third.

8) Show that the normal line to the surface $f(x_1,x_2,x_3) = 0$ at $\bar{x}_i$ is given by $\xi_i = \bar{x}_i + t(\partial f/\partial \bar{x}_i)$, where ξ_i are current coordinates and t is a parameter.

9) Find the equations of the normal line to the elliptic paraboloid $x_1^2 + 4x_2^2 = 8x_3$ at the point $\bar{x}_i$, and show that there are five normals to this surface through a given point.

10) Consider the parametric representation for a sphere in Exercise 2. Show that the points $(0,0, \pm a)$ on the sphere are singular points for that particular representation, but that the same surface points $(0,0, \pm a)$ are not singular points for the parametric representation

$$x_1 = u_1, \qquad x_2 = u_2, \qquad x_3 = \pm(a^2 - u_1^2 - u_2^2)^{1/2}.$$

3

Transformation of Coordinates in Space; Differentiation

3–1. Linear Transformation. A rotation of axes followed by a translation, as in (1–12) is a linear transformation of a particular type in that the coefficients are subject to the orthogonality conditions. A general linear transformation in three-space can be represented by

1) $$\bar{x}_i = \lambda_{ij} x_j + a_i,$$

where the only restriction on the λ_{ij} is that the determinant $|\lambda_{ij}| \neq 0$, which condition allows the inverse of equations (1) to be found. It should be emphasized that transformation of coordinates and not mapping of points is the object of study here. The points of space remain fixed but their identification tags change as the frame of reference is changed.

In order to study transformations of coordinates one begins with some system of coordinates and transforms to another. Although not necessary, it may be more convenient and meaningful to begin with a fixed orthogonal cartesian system of axes. In order to anticipate a convention of tensor analysis relative to summing upper and lower indices, it is well to change the notation. Instead of x_i for the three coordinates, x^i will be used henceforth, so that a point P with coordinates x^1, x^2, x^3 will be referred to as the point x^i. Do not confuse x^2, for instance, with the square of x, which would be written as $(x)^2$, if the occasion should arise. Because the square of a quantity such as $(x^1)^2$ will not be encountered often, little confusion should arise incident to the new convention.

3–2. Transformation to Curvilinear Coordinates. The equations

2) $$x^i = x^i(\bar{x}^1, \bar{x}^2, \bar{x}^3) \qquad (i = 1, 2, 3)$$

define a general transformation from the orthogonal cartesian coordinates x^i to an arbitrary set of coordinates $\bar{x}^i$. It is understood that the three functions on the right-hand side of equations (2), together with their first partial derivatives, are single-valued and continuous in the domain D of space under consideration. It is further assumed that the jacobian of the transformation (2) does not vanish at any point of the domain D, so that equations (2) can be solved locally for the inverse in the form

3) $$\bar{x}^i = \bar{x}^i(x^1, x^2, x^3).$$

As in the discussion of the jacobian (2-35) relative to transformation of coordinates on a surface, it should be understood here that the third-order jacobian determinant

$$\left|\frac{\partial x^i}{\partial \bar{x}^j}\right| \qquad (i,j = 1, 2, 3)$$

is, in general, a function of $\bar{x}^i$, the vanishing of which represents a set of singular points where a unique inverse of equations (2) may not exist. The domain D will not contain such singular points. For any point $\bar{x}^i$ at which the jacobian is not zero, there exists a domain about the point in which the jacobian is non-zero, because of the stipulated continuity of the first partial derivatives. In such a domain the inverse of equations (2) can be effected. This explains the use of the term "locally" relative to the inverse of (2).

Observe that the three equations

4) $$\bar{x}^1 = c^1, \qquad \bar{x}^2 = c^2, \qquad \bar{x}^3 = c^3,$$

where the c^i are constants, represent three surfaces through the point (c^1, c^2, c^3) referred to the x^i system. These are, in general, curved surfaces, and their equations in cartesian coordinates x^i are given by

5) $$\bar{x}^1(x^1,x^2,x^3) = c^1, \qquad \bar{x}^2(x^1,x^2,x^3) = c^2, \qquad \bar{x}^3(x^1,x^2,x^3) = c^3,$$

which is just an alternative notation for the surfaces

$$f(x,y,z) = a, \qquad g(x,y,z) = b, \qquad h(x,y,z) = c.$$

Now if the c^i in (4) and (5) are regarded as parameters instead of constants, then equations (5) represent three one-parameter families of surfaces, and any point of space has one surface of each family passing through it. If, for instance, c^3 is held constant, one surface is selected from the family, but the c^1 and c^2 vary over this surface. From equations (2) it can be seen that if $\bar{x}^3 = $ constant, the resulting equations

6) $$x^i = x^i(\bar{x}^1, \bar{x}^2, \text{constant})$$

are the imbedding equations of a curved two-dimensional extent, i.e., a surface. (The more unconventional form of description used here for a surface becomes useful in considering subspaces of a hyperspace.) The surface $\bar{x}^2$ = constant is the surface on which $\bar{x}^1$ and $\bar{x}^3$ vary, and $\bar{x}^1$ = constant has the surface coordinates $\bar{x}^2$ and $\bar{x}^3$.

Next observe that if, say, $\bar{x}^2$ and $\bar{x}^3$ are both assigned constant values, then only $\bar{x}^1$ varies and the equations (6) become

7) $$x^i = x^i(\bar{x}^1, \text{constant}, \text{constant}),$$

which define a one-dimensional curved extent embedded in three-dimensional euclidean space. This curve may be styled an $\bar{x}^1$-curve. Similarly, there are the $\bar{x}^2$- and $\bar{x}^3$-curves. Through each point of space there is an $\bar{x}^1$-curve, an $\bar{x}^2$-curve, and an $\bar{x}^3$-curve.

The equations x^1 = constant, x^2 = constant, x^3 = constant give three one-parameter families of surfaces (planes) parallel to the original coordinate planes, and there are three such planes through each point of space. The equation of the plane x^3 = constant, for instance, in the $\bar{x}^i$ system is seen from equations (3) to be

8) $$\bar{x}^i = \bar{x}^i(x^1, x^2, \text{constant}).$$

Example. Consider a transformation from orthogonal cartesian coordinates to spherical coordinates as an illustration of the foregoing discussion.

Solution: A usual notation for transformation from rectangular to spherical coordinates is

$$x^1 = \rho \sin\phi \cos\theta, \qquad x^2 = \rho \sin\phi \sin\theta, \qquad x^3 = \rho \cos\phi,$$

but to conform to the notation introduced above, this is written in the form

9) $$x^1 = \bar{x}^1 \sin \bar{x}^2 \cos \bar{x}^3, \qquad x^2 = \bar{x}^1 \sin \bar{x}^2 \sin \bar{x}^3, \qquad x^3 = \bar{x}^1 \cos \bar{x}^2.$$

Observe that $\bar{x}^1 = c^1$ (with c^1 a constant) is a sphere of radius c^1 on which $\bar{x}^2$ and $\bar{x}^3$ represent co-latitude and longitude, that the surface $\bar{x}^2 = c^2$ is a cone with semivertical angle c^2, and that $\bar{x}^3 = c^3$ is a plane passing through the x^3-axis of the cartesian system.

The jacobian of the transformation (9) is found to be $(\bar{x}^1)^2 \sin \bar{x}^2$. This is zero for points on the x^3-axis, so something peculiar is to be expected at such points. Actually, one does not find a unique surface of each o the three families through a point of the x^3-axis. Indeed, all of the planes x^3 = constant (as the constant changes) pass through the x^3-axis, so this axis is a line of singular points. Notice that these are not singular points of space, but singular points of the particular analytical representation by spherical coordinates. The situation here is analogous to that of polar coordinates in the plane in which the origin is a singular point, for at the origin the angle coordinate is indeterminate.

3-3. Partial Differentiation. The reader will recall the formula for the total differential df of a function $f(x^1, x^2, x^3)$ given by

$$df = \frac{\partial f}{\partial x^1} dx^1 + \frac{\partial f}{\partial x^2} dx^2 + \frac{\partial f}{\partial x^3} dx^3,$$

which may be written in more concise form as

10) $$df = \frac{\partial f}{\partial x^i} dx^i,$$

where the i sums from 1 to 3. Now the total differentials for the functions in equations (2) are

11) $$dx^i = \frac{\partial x^i}{\partial \bar{x}^j} d\bar{x}^j.$$

Next, recall from the elementary calculus that if f is a function of y, and y, in turn, is a function of z, then f is a function of z, and by the well-known chain rule

$$\frac{df}{dz} = \frac{df}{dy} \frac{dy}{dz}.$$

Further, if f is a function of x and y, and x and y, in turn, are functions of r and s, then f is a function of r and s, and the partial derivative, for instance, of f with respect to r is given by

$$\frac{\partial f}{\partial r} = \frac{\partial f}{\partial x} \frac{\partial x}{\partial r} + \frac{\partial f}{\partial y} \frac{\partial y}{\partial r}.$$

Consider the partial derivative of the functions x^i in equations (2) with respect to say, $\bar{x}^1$. The expression $\partial x^i/\partial \bar{x}^1$ for this is calculated by holding $\bar{x}^2$ and $\bar{x}^3$ constant. One could write $\partial x^i/\partial \bar{x}^j$, where i and j could take any one of the values 1, 2, 3. Nine functions are therefore represented by $\partial x^i/\partial \bar{x}^j$. From equations (3), nine functions are likewise represented by $\partial \bar{x}^i/\partial x^j$ as i and j range over 1, 2, 3. The i and j here are referred to as *free* indices. What would $\partial x^i/\partial \bar{x}^i$ mean? This gives only one function because the index i is not free but summed. One has therefore

$$\frac{\partial x^i}{\partial \bar{x}^i} = \frac{\partial x^1}{\partial \bar{x}^1} + \frac{\partial x^2}{\partial \bar{x}^2} + \frac{\partial x^3}{\partial \bar{x}^3}.$$

Notice that if x^i are functions of $\bar{x}^i$ and, in turn, the $\bar{x}^i$ are functions of $\tilde{x}^i$, the partial derivative of x^i with respect to $\tilde{x}^j$ is given by

12) $$\frac{\partial x^i}{\partial \tilde{x}^j} = \frac{\partial x^i}{\partial \bar{x}^k} \frac{\partial \bar{x}^k}{\partial \tilde{x}^j}.$$

In (12) the i and j are free indices while k is a dummy index designating that a sum must take place.

Now, if the expressions for $\bar{x}^i$ from equations (3) are substituted into equations (2) the results must be identities, so that the expressions on the left sides of the equations

13) $$x^i - x^i(\bar{x}^1, \bar{x}^2, \bar{x}^3) \equiv 0$$

are independent of the x^i. Hence, the derivative of (13) with respect to, say, x^j, must be zero. Hence, on making use of the fact expressed in (12), one has

14) $$\frac{\partial x^i}{\partial x^j} - \frac{\partial x^i}{\partial \bar{x}^k}\frac{\partial \bar{x}^k}{\partial x^j} = 0.$$

Because x^1, x^2, x^3 are independent, one sees that $\partial x^i/\partial x^j$ is 1 when $i = j$ and zero otherwise. A Kronecker delta is useful here. Define $\delta_j{}^i = 0$, if $i \neq j$, and $\delta_j{}^i = 1$, if $i = j$. Equations (14) then take the form

15) $$\frac{\partial x^i}{\partial \bar{x}^k}\frac{\partial \bar{x}^k}{\partial x^j} = \delta_j{}^i.$$

3–4. Derivative of a Determinant. Consider first a determinant of the second order

16) $$a \equiv \begin{vmatrix} a_1{}^1 & a_2{}^1 \\ a_1{}^2 & a_2{}^2 \end{vmatrix},$$

wherein the elements $a_j{}^i$ are functions of $x^i (i = 1, 2, 3)$. (It is hoped that allowing the letter a to represent both the determinant and its value will cause no confusion.) Because the well-known expansion of a is $a_1{}^1 a_2{}^2 - a_2{}^1 a_1{}^2$, it follows that

$$\frac{\partial a}{\partial x^i} = a_1{}^1 \frac{\partial a_2{}^2}{\partial x^i} + a_2{}^2 \frac{\partial a_1{}^1}{\partial x^i} - a_2{}^1 \frac{\partial a_1{}^2}{\partial x^i} - a_1{}^2 \frac{\partial a_2{}^1}{\partial x^i},$$

which can be recast in the form of second-order determinants as

17) $$\frac{\partial a}{\partial x^i} = \begin{vmatrix} \dfrac{\partial a_1{}^1}{\partial x^i} & a_2{}^1 \\ \dfrac{\partial a_1{}^2}{\partial x^i} & a_2{}^2 \end{vmatrix} + \begin{vmatrix} a_1{}^1 & \dfrac{\partial a_2{}^1}{\partial x^i} \\ a_1{}^2 & \dfrac{\partial a_2{}^2}{\partial x^i} \end{vmatrix}.$$

It appears that the derivative of the determinant can be obtained by writing the sum of two determinants with a different column of derivatives in each.

In order to reduce the labor of writing, denote the partial derivative $\partial a_j{}^i / \partial x^k$ by $a^i{}_{j,k}$. With this notation, equation (17) becomes

18) $$a_{,k} = \begin{vmatrix} a^1{}_{1,k} & a_2{}^1 \\ a^2{}_{1,k} & a_2{}^2 \end{vmatrix} + \begin{vmatrix} a_1{}^1 & a^1{}_{2,k} \\ a_1{}^2 & a^2{}_{2,k} \end{vmatrix}.$$

It is readily verified that the partial derivative with respect to x^k of

19) $$a \equiv \begin{vmatrix} a_1{}^1 & a_2{}^1 & a_3{}^1 \\ a_1{}^2 & a_2{}^2 & a_3{}^2 \\ a_1{}^3 & a_2{}^3 & a_3{}^3 \end{vmatrix}$$

is given by

20) $$a_{,k} = \begin{vmatrix} a^1{}_{1,k} & a_2{}^1 & a_3{}^1 \\ a^2{}_{1,k} & a_2{}^2 & a_3{}^2 \\ a^3{}_{1,k} & a_2{}^3 & a_3{}^3 \end{vmatrix} + \begin{vmatrix} a_1{}^1 & a^1{}_{2,k} & a_3{}^1 \\ a_1{}^2 & a^2{}_{2,k} & a_3{}^2 \\ a_1{}^3 & a^3{}_{2,k} & a_3{}^3 \end{vmatrix} + \begin{vmatrix} a_1{}^1 & a_2{}^1 & a^1{}_{3,k} \\ a_1{}^2 & a_2{}^2 & a^2{}_{3,k} \\ a_1{}^3 & a_2{}^3 & a^3{}_{3,k} \end{vmatrix}.$$

In order to reduce the writing further, recall that the expansion of a determinant is equal to the sum of the products of the elements of a row (or column) by their corresponding cofactors. If the cofactor of any element $a_j{}^i$ is denoted by $A_i{}^j$, the value of a is $a^i{}_{(j)} A_i{}^{(j)}$, where the parentheses on j are used to indicate that j is held fixed while a sum takes place on i from 1 to 3. Thus, $a = a_1{}^i A_i{}^1 = a_2{}^i A_i{}^2 = a_3{}^i A_i{}^3$. Notice that the expansion of the determinant could also be written as $a = a_j{}^{(i)} A^j{}_{(i)}$ for any i. Recall further that if the elements of a row (or column) of a determinant are multiplied by the cofactors of the corresponding elements of some other row (or column) and the products added, the result is zero. That is, for instance,

21) $$a_j{}^i A_k{}^j = 0 \qquad (i \neq k).$$

Hence, one realizes the possibility of writing

22) $$a_j{}^i A_k{}^j = \delta_k{}^i a.$$

For a given value of i, the sum in (22) is zero if $k \neq i$, and equal to a if $k = i$. With the improved notation, the derivative $a_{,k}$ in (20) can be expressed by

23) $$a_{,k} = a^1{}_{1,k} A_1{}^1 + a^2{}_{1,k} A_2{}^1 + a^3{}_{1,k} A_3{}^1 + a^1{}_{2,k} A_1{}^2 + a^2{}_{2,k} A_2{}^2 \\ + a^3{}_{2,k} A_3{}^2 + a^1{}_{3,k} A_1{}^3 + a^2{}_{3,k} A_2{}^3 + a^3{}_{3,k} A_3{}^3,$$

or by

24) $$a_{,k} = a^i{}_{1,k} A_i{}^1 + a^i{}_{2,k} A_i{}^2 + a^i{}_{3,k} A_i{}^3$$

or, finally, by

25) $$a_{,k} = A_i{}^j a^i{}_{j,k},$$

where both i and j are summed from 1 to 3. By an argument similar to the development of the compact result in (25), it can readily be shown that if $a \equiv |a_j{}^i|$ is a determinant of order n with the elements $a_j{}^i$ functions of $x^1, \cdots, x^n$, then the derivative of a with respect to x^k is

26) $$a_{,k} = A_i{}^j a^i{}_{j,k},$$

where the i and j are dummy or umbral indices which sum from 1 to n, and k is a free index which ranges over $1, \cdots, n$. In this general case, notice that there is a sum of n determinants in (20), each determinant being of order n and having one column differentiated. The entire sum is then given by (26).

It may be of interest to see the general case proved as follows. Write the determinant as

$$a = |a_j{}^i| = e^{i_1 i_2 \cdots i_n} a_{i_1}{}^1 a_{i_2}{}^2 \cdots a_{i_n}{}^n,$$

where the e symbol is $+1$ if the n superscripts assume an even permutation of the numbers from 1 to n, -1 if an odd permutation, and zero otherwise. Differentiation with respect to x^k gives

$$a_{,k} = e^{i_1 i_2 \cdots i_n} \left(\frac{\partial a_{i_1}{}^1}{\partial x^k} a_{i_2}{}^2 \cdots a_{i_n}{}^n + a_{i_1}{}^1 \frac{\partial a_{i_2}{}^2}{\partial x^k} \cdots a_{i_n}{}^n + \cdots \right)$$

$$= \frac{\partial a_{i_1}{}^1}{\partial x^k} A_1{}^{i_1} + \frac{\partial a_{i_2}{}^2}{\partial x^k} A_2{}^{i_2} + \cdots + \frac{\partial a_{i_n}{}^n}{\partial x^k} A_n{}^{i_n}$$

$$= \frac{\partial a_i{}^j}{\partial x^k} A_j{}^i = A_j{}^i a^j{}_{i,k}.$$

3–5. Cramer's Rule. The notation exhibited by formula (22) is useful in solving a system of linear equations. Let it be required to solve the system

27) $$a_j{}^i x^j = b^i \qquad (i = 1, \cdots, n)$$

for the x's, where j sums from 1 to n in each of the n equations. First, notice that i is a free index, so multiplication of both members of (27) by $A_i{}^k$ effects an automatic sum on i to give

28) $$A_i{}^k a_j{}^i x^j = b^i A_i{}^k.$$

By (22), $A_i{}^k a_j{}^i = \delta_j{}^k a$, so (28) becomes

29) $$a \delta_j{}^k x^j = b^i A_i{}^k.$$

Because of the definition of $\delta_j{}^k$, the only surviving non-zero term in the sum $\delta_j{}^k x^j$ is x^k. Hence, after division by a, (29) yields the solution

30) $$x^k = b^i \frac{A_i{}^k}{a}.$$

The foregoing is merely Cramer's rule for solving a set of linear equations with the determinant of the coefficients non-vanishing. A further simplification of equation (30), introduced here for later use, is the following. Let a'^k_i denote the result of dividing the cofactor of $a_k{}^i$ by the value a of the determinant, that is, let

31) $$a'^k_i = \frac{A^k_i}{a}.$$

Then, equation (30) has the simpler form

32) $$x^k = a'^k_i b^i,$$

and equations (22) take the form

33) $$a_j{}^i a'^j_k = \delta^i_k.$$

Now let the b^i in (27) be replaced by $\bar{x}^i$. Equations (27) become

34) $$\bar{x}^i = a_j{}^i x^j,$$

which are of the form of equations (1-11) with λ_{ij} replaced by $a_j{}^i$. It is assumed that $a \equiv |a_j{}^i| \neq 0$. From the result in (32) the inverse of (34) is

35) $$x^k = a'^k_i \bar{x}^i.$$

Observe that the coefficients a'^k_i in (35) may be defined by equations (33), for if (33) is multiplied by $A_i{}^l$, there results

$$A_i{}^l a_j{}^i a'^j_k = A_i{}^l \delta^i_k = A_k{}^l,$$

or, since $A_i{}^l a_j{}^i = \delta^l_j a$,

$$a \delta^l_j a'^j_k = A_k{}^l$$

with the result that

$$a'^l_k = \frac{A_k{}^l}{a}$$

in accord with (31).

3-6. Product of Determinants. By direct calculation it can be verified that the product of two determinants $|a_{ij}|$ and $|b_{ij}|$ of the second order is

36) $$|a_{ij}||b_{kl}| \equiv \begin{vmatrix} a_{11} & a_{12} \\ a_{21} & a_{22} \end{vmatrix} \begin{vmatrix} b_{11} & b_{12} \\ b_{21} & b_{22} \end{vmatrix} = \begin{vmatrix} a_{11}b_{11} + a_{12}b_{21} & a_{11}b_{12} + a_{12}b_{22} \\ a_{21}b_{11} + a_{22}b_{21} & a_{21}b_{12} + a_{22}b_{22} \end{vmatrix}$$
$$= \begin{vmatrix} a_{1i}b_{i1} & a_{1i}b_{i2} \\ a_{2j}b_{j1} & a_{2j}b_{j2} \end{vmatrix} \equiv \begin{vmatrix} c_{11} & c_{12} \\ c_{21} & c_{22} \end{vmatrix} = |c_{rs}|.$$

(The notation is changed from $a_i{}^j$ to a_{ij} for no good reason.) Notice that the element c_{11} in the product determinant is the inner product of the elements a_{11}, a_{12} (in the first row of the $|a_{ij}|$ determinant) and the elements b_{11}, b_{21} (in the first column of the $|b_{ij}|$ determinant). In general, the element c_{ij} in the resulting determinant is the inner product $a_{il}b_{lj}$ of the ith row by the jth column of the respective determinants. This holds for determinants of any order. Therefore, the product of two nth order determinants can be indicated by

$$|a_{ij}|\,|b_{kl}| = |a_{rp}b_{ps}|.$$

Consider the transformation $T: \bar{x}_i = a_{ij}x_j$, followed by $S: \bar{\bar{x}}_k = b_{ki}\bar{x}_i$. The product ST is then given by

$$\bar{\bar{x}}_k = b_{ki}(a_{ij}x_j) = b_{ki}a_{ij}x_j.$$

The jacobians of transformations T and S are $|a_{ij}|$ and $|b_{ki}|$, respectively. The jacobian of the product transformation ST is the determinant $|b_{ki}a_{ij}|$, which is the product of the jacobians of S and T.

EXERCISES

1) Calculate the jacobian of the transformation (9) relating spherical coordinates and rectangular cartesian coordinates.

2) The transformation

$$x^1 = \bar{x}^1 \cos \bar{x}^2, \qquad x^2 = \bar{x}^1 \sin \bar{x}^2, \qquad x^3 = \bar{x}^3$$

relates cylindrical coordinates $\bar{x}^i$ to rectangular cartesian coordinates x^i. Calculate the jacobian for this transformation.

3) Find the jacobian of the transformation

$$x^1 = \bar{x}^1 + \bar{x}^2,$$
$$x^2 = \bar{x}^1 - \bar{x}^2,$$
$$x^3 = (\bar{x}^1)^2 + (\bar{x}^2)^2 + (\bar{x}^3)^2.$$

What are the singular points of the $\bar{x}$ system of coordinates?

4) What is the jacobian of the transformation $x^i = a_j{}^i\bar{x}^j$? What is the jacobian of the inverse $\bar{x}^i = a'_k{}^i x^k$? Show that the product of these two jacobians is 1.

5) Find the value of the product

$$\begin{vmatrix} 1 & 2 & 3 \\ 0 & 1 & -1 \\ 2 & 1 & 0 \end{vmatrix} \begin{vmatrix} 0 & 2 & -1 \\ 1 & 1 & 0 \\ 2 & 3 & 1 \end{vmatrix}$$

in two ways.

6) If f, g, h in the determinant

$$D \equiv \begin{vmatrix} f & g & h \\ f' & g' & h' \\ f'' & g'' & h'' \end{vmatrix}$$

are functions of t and the primes indicate differentiation with respect to t, show that

$$D' = \begin{vmatrix} f & g & h \\ f' & g' & h' \\ f''' & g''' & h''' \end{vmatrix}.$$

7) Show that

$$\frac{\partial \bar{x}^i}{\partial x^j} \frac{\partial x^j}{\partial \bar{x}^k} = \delta_k{}^i.$$

8) Evaluate $\delta_j{}^i A_p{}^{jk}$.

9) Evaluate $\delta_j{}^i \delta_k{}^j$.

10) Show that $\bar{\lambda}^i = \lambda^j(\partial \bar{x}^i/\partial x^j)$ implies $\lambda^p = \bar{\lambda}^q(\partial x^p/\partial \bar{x}^q)$. [*Hint:* Multiply the equation $\bar{\lambda}^i = \lambda^j(\partial \bar{x}^i/\partial x^j)$ by $\partial x^r/\partial \bar{x}^i$ and use (15).]

11) Make the transformation $x^i = h_j{}^i \bar{x}^j$ on the homogeneous polynomial $f \equiv a_{ij} x^i x^j$ to transform it to $\bar{f} \equiv \bar{a}_{ij} \bar{x}^i \bar{x}^j$. It is to be seen that $\bar{a}_{kl} = a_{ij} h_k{}^i h_l{}^j$. Take the determinant of both sides of the last equation to show that $|\bar{a}_{kl}| = |a_{ij}|\Delta^2$, where Δ is the jacobian $|h_j{}^i|$ of the transformation. The determinant $|a_{ij}|$ is called the *discriminant* of the form f.

12) Let J and J^* denote the jacobians of transformations (2) and (3), respectively. Prove that $JJ^* = 1$.

13) If J_1 is the jacobian of a transformation from x^i to y^i, and J_2 is the jacobian of a transformation from y^i to z^i, show that the jacobian J_3 of the product transformation from x^i to z^i is given by $J_3 = J_1 J_2$.

4

Tensor Algebra

4–1. Cogredience and Contragredience. Before considering the analysis of the general transformation $x^i = x^i(\bar{x}^1, \cdots, \bar{x}^n)$, where the x^i are arbitrary functions of $\bar{x}^i$, it will be helpful to consider further the linear transformation

1) $$x^i = a_j{}^i \bar{x}^j, \qquad (i,j = 1, 2, \cdots, n),$$

with its inverse

2) $$\bar{x}^i = a'_j{}^i x^j,$$

and the consequence

3) $$a_j{}^i a'_k{}^j = \delta_k{}^i.$$

Consider a second set of variables u_i, which may be styled *dual* coordinates, and let them transform into $\bar{u}_i$ by the equations

4) $$\bar{u}_j = a_j{}^i u_i,$$

with the inverse

5) $$u_j = a'_j{}^i \bar{u}_i,$$

and the same consequence (3). It is assumed that the jacobian (sometimes called the *modulus*) of transformations (1) and (4) does not vanish.

A *form* is defined as a homogeneous polynomial. In Exercise 11, Chapter 3, a *quadratic* form $a_{ij} x^i x^j$ was introduced. A *linear* form

6) $$\lambda_i x^i \qquad (i = 1, \cdots, n)$$

is the present object of study. (The reader may consider n to be 3 if he chooses.) Let the transformation (1) be performed on the x^i in (6) to obtain

7) $$\lambda_i a_j{}^i \bar{x}^j.$$

If it is required that the form be invariant, $\lambda_i x^i$ must go into

8) $$\bar{\lambda}_j \bar{x}^j.$$

Hence, it is evident from (7) that

9) $$\bar{\lambda}_j = \lambda_i a_j{}^i,$$

which has the inverse $\lambda_i = a'_i{}^j \bar{\lambda}_j$ of the type of (5). Observe that the transformation (1) on the x^i variables induces the transformation (9) on the coefficients λ_i in the form $\lambda_i x^i$. The sets of variables λ_i and x^i are said to be *contragredient* since the form $\lambda_i x^i$ goes into $\bar{\lambda}_i \bar{x}^i$ under the transformations (1) and the inverse of (9). The respective matrices of the λ_i and x^i transformations are inverses of each other.

Consider next a form of the type $\lambda^i u_i$ involving the dual variables u_i, and perform the transformation (5), i.e., $u_i = a'_i{}^j \bar{u}_j$ on the u_i to obtain

10) $$\lambda^i a'_i{}^j \bar{u}_j,$$

which may be written as

11) $$\bar{\lambda}^j \bar{u}_j,$$

where, as a consequence of the invariance of $\lambda^i u_i$,

12) $$\bar{\lambda}^j = \lambda^i a'_i{}^j.$$

Thus, from (12) it is seen that a transformation of the type (2) is induced on the λ^i if the form $\lambda^i u_i$ is to go into $\bar{\lambda}^i \bar{u}_i$ under the transformation (5) on the dual variables u_i.

Again, because of the invariance of $\lambda^i u_i$, the variables λ^i and u_i are said to be contragredient. Notice that the inverse of (12) may be written by observing the form of (1) and (2), or it may be obtained from (12) by solving directly for λ^i as follows. Multiply equation (12) by $a_j{}^k$, sum on j, and use $a_i{}^j a'_j{}^k = \delta_i{}^k$ to obtain

$$a_j{}^k \bar{\lambda}^j = \lambda^i a'_i{}^j a_j{}^k = \lambda^i \delta_i{}^k = \lambda^k,$$

or, on changing indices,

$$\lambda^i = a_j{}^i \bar{\lambda}^j,$$

which is precisely the form of equation (1). Thus, the λ^i and x^i transform in the same manner, type (1), and the λ_i and u_i transform in the same manner, type (5). The variables λ^i and x^i are called *cogredient* variables because they transform in the same manner. The variables λ_i and u_i are therefore also cogredient.

An interesting geometrical interpretation can be given the dual variables u_i relative to the point coordinates x^i, but here interest is centered upon the dual algebra. It is important to have in mind the following array of information which results from demanding that the forms $\lambda_i x^i$ and $\lambda^j u_j$ be invariant, which means that they go respectively into $\bar{\lambda}_i \bar{x}^i$ and $\bar{\lambda}^j \bar{u}_j$ under the given linear transformations (1) and (5).

Point variables	Transformation	Inverse
x^i	$x^i = a_j{}^i \bar{x}^j$	$\bar{x}^i = a'_j{}^i x^j$
Dual variables		
u_i	$u_i = a'_i{}^j \bar{u}_j$	$\bar{u}_i = a_i{}^j u_j$
Form coefficients		
λ_i	$\lambda_i = \bar{\lambda}_j a'_i{}^j$	$\bar{\lambda}_i = \lambda_j a_i{}^j$
λ^i	$\lambda^i = \bar{\lambda}^j a_j{}^i$	$\bar{\lambda}^i = \lambda^j a'_j{}^i$

Observe how the structure of the transformations and inverses is determined by the $a_j{}^i$ and $a'_j{}^i$. Although the general transformation $x^i = x^i(\bar{x}^1, \cdots, \bar{x}^n)$ is not linear, it will be found that it determines an associated set of linear transformations for which the present development is useful (see Section 5–2).

The coefficient variables λ_i and λ^i are called, respectively, *covariant* and *contravariant* variables. Notice particularly how they transform. When the $a_j{}^i$ are known in the equations $\lambda^i = \bar{\lambda}^j a_j{}^i$, and the $\bar{\lambda}^j$ are given, the λ^i can be calculated. Also, the $a'_i{}^j$ can be calculated from the $a_j{}^i$ so if $\bar{\lambda}_j$ are given in $\lambda_i = \bar{\lambda}_j a'_i{}^j$, the λ_i can be calculated directly.

Notice that $\lambda_i \lambda^i$ is itself an invariant because

$$\lambda_i \lambda^i = \bar{\lambda}_j a'_i{}^j \bar{\lambda}^k a_k{}^i = \bar{\lambda}_j \bar{\lambda}^k a_k{}^i a'_i{}^j = \bar{\lambda}_j \bar{\lambda}^k \delta_k{}^j = \bar{\lambda}_k \bar{\lambda}^k = \bar{\lambda}_i \bar{\lambda}^i.$$

It has been seen that $\lambda_i x^i$ and $\lambda^i u_i$ are invariants.

4–2. First View of a Tensor. Choose some system of coordinates x^i and consider the particular type of coordinate transformation given by equations (1). The notation is the same for n-space as for two- or three-space, so one may consider x^i as meaning $(x^1, \cdots, x^n)$. Such an ordered n-tuple of numbers is called a *point* in real n-dimensional space.

The form $a_{ij} x^i x^j$ which appeared earlier is quadratic, but the form $a_{ij} x^i y^j$ is linear in each of x^i and y^i. It is called a bilinear form. Another sort of form is $a_{ij}{}^k \lambda^i \mu^j \nu_k$. It is a cubic form, but it is linear in each of λ_i, μ^i, ν_i. It is desired now to build a form which will be simple enough to handle but which will be useful in explaining a first notion of a tensor. Take the form

13) $$f \equiv b_{ij}{}^k{}_l \xi^i \eta^j \zeta_k \tau^l,$$

where the coefficients $b_{ij}{}^k{}_l$ are functions of the point coordinates x^i, where ξ^i, η^i, τ^i are three sets of contravariant variables which transform like the x^i in (1), and where the set ζ_k transform like the u_i in (5). Under the transformation (1) the $b_{ij}{}^k{}_l$ functions of x^i transform into $\bar{b}_{pq}{}^r{}_s$ functions of

$\bar{x}^i$, and the form f in (13) will be required to be invariant, that is

14) $$f \equiv b_{ij}{}^k{}_l \xi^i \eta^j \zeta_k \tau^l = \bar{b}_{pq}{}^r{}_s \bar{\xi}^p \bar{\eta}^q \bar{\zeta}_r \bar{\tau}^s \equiv \bar{f}.$$

This requirement, together with the given laws of transformation of the $\xi^i, \eta^i, \zeta_i, \tau^i$, will induce a law of transformation for the functions $b_{ij}{}^k{}_l$. On using

15) $$\xi^i = \bar{\xi}^p a_p{}^i, \qquad \eta^j = \bar{\eta}^q a_q{}^j, \qquad \zeta_k = \bar{\zeta}_r a'_k{}^r, \qquad \tau^l = \bar{\tau}^s a_s{}^l$$

from the array of transformations in Section 4–1, one obtains

16) $$b_{ij}{}^k{}_l \xi^i \eta^j \zeta_k \tau^l = b_{ij}{}^k{}_l \bar{\xi}^p a_p{}^i \bar{\eta}^q a_q{}^j \bar{\zeta}_r a'_k{}^r \bar{\tau}^s a_s{}^l$$
$$= b_{ij}{}^k{}_l a_p{}^i a_q{}^j a'_k{}^r a_s{}^l \bar{\xi}^p \bar{\eta}^q \bar{\zeta}_r \bar{\tau}^s = \bar{b}_{pq}{}^r{}_s \bar{\xi}^p \bar{\eta}^q \bar{\zeta}_r \bar{\tau}^s,$$

so that the law of transformation induced on the $b_{ij}{}^k{}_l$ is

17) $$\bar{b}_{pq}{}^r{}_s = b_{ij}{}^k{}_l a_p{}^i a_q{}^j a'_k{}^r a_s{}^l.$$

Observe that the p, q, r, s are free indices which appear singly on both sides of (17), but that the i, j, k, l sum out on the right-hand side. The invariant form (13) is called a tensor of the fourth order, and the functions $b_{ij}{}^k{}_l$ are its components in the particular coordinate system chosen, and for a *linear* transformation of coordinates. (The more general transformation will be treated in the next chapter.) Because the components determine the tensor, one speaks elliptically of the components as the tensor. Note that the order of a tensor is the total number of free indices in the components.

The tensor with components $b_{ij}{}^k{}_l$, which was used as an example, is called a *mixed* tensor with three indices of covariance (lower indices) and one index of contravariance (upper index). Tensors of both higher and lower orders will appear later. As further instances, b_{ijk} could be the components of a third-order covariant tensor, and b^{ijkl} could be a fourth-order contravariant tensor, provided these functions transform in the proper tensor manner. The functions b_{ij} would be components of a covariant tensor of second order if

18) $$b_{ij} = \bar{b}_{pq} a'_i{}^p a'_j{}^q,$$

with the consequence that $b_{ij} \xi^i \eta^j$ would be an invariant form. In particular, the λ^i in Section 4–1 are components of a contravariant tensor of first order because $\lambda^i = \bar{\lambda}^p a_p{}^i$. Also, λ_i are components of a covariant tensor of first order since $\lambda_i = \bar{\lambda}_p a'_i{}^p$. A tensor of first order is called a *vector*. Hence, the object with components λ^i is a contravariant vector, and the object with components λ_i is a covariant vector.

A scalar function $b(x^i, \cdots, x^n)$ is a tensor of zero order, because the value of the function at a point is the same number regardless of the coordinate system. Under the point coordinate transformation (1), it is true that

19) $$b(x^1, \cdots, x^n) = \bar{b}(\bar{x}^1, \cdots, \bar{x}^n).$$

The function $b(x^1, \cdots, x^n)$, which is assumed to be single-valued, is said to determine a *scalar field*, because if the coordinates x^i of any point in the space are substituted into $b(x^1, \cdots, x^n)$, the value of the function at the point is found. Thus, the function attaches a number (scalar) to each point of space. Of course, the function varies, in general, from point to point in space, and an object of study is the manner of variation of the function as the point moves, say, on a curve $x^i = x^i(t)$, or perhaps on a surface $x^i = x^i(u^1, u^2)$ imbedded in the space. From (19) it is evident that if the value of the scalar function b is zero in the x^i coordinates, it is zero in any other coordinates system $\bar{x}^i$. This invariant property holds for a tensor of any order. Note in equations (17), for instance, that if the components $b_{ij}{}^k{}_l$ are all zero in the x^i coordinate system, then the components $\bar{b}_{pq}{}^r{}_s$ in the $\bar{x}^i$ system are likewise all zero. The vanishing of a tensor is therefore an invariant property.

Just as a tensor of zero order determines a scalar field, a tensor of first order determines a *vector field* in the space, and the study of such fields comprises vector analysis. It has appeared that the two tensors with components b_i and b^i are two separate entities, i.e., two different vectors, but it will be seen later that in a so-called "Riemannian" space a vector may be expressed in terms of either covariant or contravariant components. A vector field is determined by n functions at each point of the space (three functions at each point of three-space). These functions may be specified as either covariant or contravariant components of a vector.

On passing to tensors of higher than the first order, one speaks of *tensor fields*, and the study of tensor fields comprises tensor analysis. It may be noted here that the partial derivatives $\partial x^i / \partial \bar{x}^j$ of the x^i functions in the general transformation $x^i = x^i(\bar{x}^1, \bar{x}^2, \bar{x}^3)$ are also, in general, functions of $\bar{x}^i$, while for the simpler transformation $x^i = a_j{}^i \bar{x}^j$ of the present chapter $\partial x^i / \partial \bar{x}^j = a_j{}^i$ which are constants.

4–3. Operations of Tensor Algebra. The fundamental operations of tensor algebra are addition, multiplication, and contraction.

ADDITION. The sum of k tensors of the same order and kind is again a tensor of the same order and kind. This will be illustrated for the case $k = 2$. Consider the two tensors

$$A \equiv a_{ij}{}^{kl} \xi^i \eta^j \zeta_k \tau_l, \qquad B \equiv b_{ij}{}^{kl} \xi^i \eta^j \zeta_k \tau_l.$$

The sum is

$$C \equiv (a_{ij}{}^{kl} + b_{ij}{}^{kl}) \xi^i \eta^j \zeta_k \tau_l = c_{ij}{}^{kl} \xi^i \eta^j \zeta_k \tau_l,$$

where $c_{ij}{}^{kl} \equiv a_{ij}{}^{kl} + b_{ij}{}^{kl}$. Thus, tensor $C = A + B$. Subtraction of like tensors is immediate.

MULTIPLICATION. In multiplication the tensors do not have to be of the same order or kind. To illustrate, let

$$A \equiv a_{ij}\xi^i\eta^j, \qquad B \equiv b_k{}^l\zeta^k\tau_l.$$

The product

$$C = AB = a_{ij}b_k{}^l\xi^i\eta^j\zeta^k\tau_l = c_{ijk}{}^l\xi^i\eta^j\zeta^k\tau_l,$$

where $c_{ijk}{}^l \equiv a_{ij}b_k{}^l$. Thus, the components of a product of two tensors is the product of the components of the separate tensors. As a further example, the product of the two first order tensors λ^i and μ_j, given by the forms $\lambda^i\xi_i$ and $\mu_j\eta^j$, is given by the invariant $\lambda^i\mu_j\xi_i\eta^j$, so that the product tensor has components $b^i{}_j \equiv \lambda^i\mu_j$.

CONTRACTION. Consider a mixed tensor of the third order with components $b_{ij}{}^k$. If k is set equal to j, the result is $b_{ij}{}^j$, where the j is summed. One could write $b_{ij}{}^j \equiv c_i$. Because only one index is left free the resulting components are those of a covariant vector. Consider the transformation for the $b_{ij}{}^k$ components, namely,

$$b_{ij}{}^k = \bar{b}_{pq}{}^s a'{}_i^p a'{}_j^q a_s{}^k.$$

Now set $k = j$ to obtain

$$b_{ij}{}^j = \bar{b}_{pq}{}^s a'{}_i^p a'{}_j^q a_s{}^j = \bar{b}_{pq}{}^s a'{}_i^p \delta_s{}^q = \bar{b}_{pq}{}^q a'{}_i^p.$$

Thus, the $b_{ij}{}^j$ components transform as a covariant vector. It is said that the vector is obtained from the third-order tensor by *contracting* on j and k.

If one begins with the mixed tensor $b_i{}^j$, and contracts, the result is $b_i{}^i$ which is an invariant, which means that $b_i{}^i = \bar{b}_p{}^p$. Note that $b_i{}^j = \bar{b}_p{}^q a'{}_i^p a_q{}^j$. Contraction gives $b_i{}^i = \bar{b}_p{}^q a'{}_i^p a_q{}^i = \bar{b}_p{}^q \delta_q{}^p = \bar{b}_p{}^p$, which verifies the invariance.

It is evident from the examples that if a mixed tensor of order $h + k$, covariant of order h, and contravariant of order k, is contracted with respect to an upper and a lower index, the resulting tensor is of order $h + k - 2$. As a further example, the tensor with components $a_{ij}{}^{kl}{}_m$ can be contracted to give $a_{ij}{}^{kj}{}_m$, a mixed tensor of the third order. Further contraction of $a_{ij}{}^{kj}{}_m \equiv b_i{}^k{}_m$ gives $b_i{}^i{}_m \equiv c_m$, so the result is a tensor of first order, or a vector of components c_m.

Suppose the equations

$$b_{ij}{}^k = \bar{b}_{pq}{}^r a'{}_i^p a'{}_j^q a_r{}^k$$

are given and it is desired to find the inverse equations. To accomplish this, multiply both sides by $a_s{}^i a_t{}^j a'_k{}^m$ to obtain

$$a_s{}^i a_t{}^j a'_k{}^m b_{ij}{}^k = \bar{b}_{pq}{}^r (a'_i{}^p a_s{}^i)(a'_j{}^q a_t{}^j)(a_r{}^k a'_k{}^m) = \bar{b}_{pq}{}^r \delta_s{}^p \delta_t{}^q \delta_r{}^m$$
$$= \bar{b}_{sq}{}^r \delta_t{}^q \delta_r{}^m = \bar{b}_{st}{}^r \delta_r{}^m = \bar{b}_{st}{}^m.$$

Hence,

$$\bar{b}_{st}{}^m = b_{ij}{}^k a_s{}^i a_t{}^j a'_k{}^m.$$

On changing the free indices s, t, m to p, q, r the last equations may be written as

$$\bar{b}_{pq}{}^r = b_{ij}{}^k a_p{}^i a_q{}^j a'_k{}^r.$$

Caution: The reader should not misinterpret the meaning of the term "inverse." No process of division has taken place here. It is only shown that the transformation of the components of a tensor from "new" to "old" coordinates is of the same form as from "old" to "new." Although division of tensors has no meaning, there is a concept referred to as the *quotient law* of tensors, which is illustrated by the following example. Suppose the form $b_{ij}\xi^i\eta^j$ is known to be a tensor (i.e., invariant) for arbitrary choice of the vectors with components indicated by ξ^i and η^j. It will be shown that b_{ij} are therefore components of a covariant tensor of the second order. One has, by the hypothesis of invariance,

$$b_{ij}\xi^i\eta^j = \bar{b}_{rs}\bar{\xi}^r\bar{\eta}^s = \bar{b}_{rs}a'_i{}^r\xi^i a'_j{}^s\eta^j = \bar{b}_{rs}a'_i{}^r a'_j{}^s \xi^i\eta^j$$

which must hold for arbitrary ξ^i and η^j. Hence

$$b_{ij} = \bar{b}_{rs} a'_i{}^r a'_j{}^s,$$

which shows that b_{ij} are in fact components of a tensor.

4–4. Transitivity, Symmetry, Skew-Symmetry. It will be shown next for a particular tensor $H_{ij}{}^k$, that the law of transformation of tensor components is *transitive*. This means the following. Let $H_{ij}{}^k$ in x^i coordinates be transformed to $\bar{H}_{st}{}^m$ in $\bar{x}^i$ coordinates where the coordinate transformation [of type (1)] is $x^i = a_j{}^i \bar{x}^j$. Then let $\bar{H}_{st}{}^m$ be transformed to $\bar{\bar{H}}_{pq}{}^r$ by another transformation of coordinates of the type of (1), say, $\bar{x}^i = b_j{}^i \bar{\bar{x}}^j$. (Note that the product of the coordinate transformations is $x^i = a_j{}^i b_k{}^j \bar{\bar{x}}^k = c_k{}^i \bar{\bar{x}}^k$, with $c_k{}^i \equiv a_j{}^i b_k{}^j$.) It is necessary to know how $\bar{\bar{H}}_{pq}{}^r$ are related to $H_{ij}{}^k$. One has

$$\bar{H}_{st}{}^m = H_{ij}{}^k a_s{}^i a_t{}^j a'_k{}^m$$

and

$$\bar{\bar{H}}_{pq}{}^r = \bar{H}_{st}{}^m b_p{}^s b_q{}^t b'_m{}^r = H_{ij}{}^k a_s{}^i a_t{}^j a'_k{}^m b_p{}^s b_q{}^t b'_m{}^r$$
$$= H_{ij}{}^k (a_s{}^i b_p{}^s)(a_t{}^j b_q{}^t)(a'_k{}^m b'_m{}^r)$$
$$= H_{ij}{}^k c_p{}^i c_q{}^j c'_k{}^r.$$

As was to be expected, the components of $\bar{H}_{pq}{}^r$ are related to $H_{ij}{}^k$ (by using the product transformation $x^i = c_k{}^i \bar{x}^k$) in the same manner that $H_{st}{}^m$ are related to $H_{ij}{}^k$. A tensor of any order can be treated similarly to prove transitivity.

The concept of a *symmetric* tensor will now be set forth. If two covariant (or contravariant) indices in the components of a tensor of any order can be interchanged without changing the values of the components, the tensor is said to be *symmetric* with respect to those two indices. For instance, if $H_{ij}{}^{kl} = H_{ji}{}^{kl}$, the tensor is symmetric with respect to i and j. If symmetry holds for all pairs of indices, the tensor is called symmetric. For instance, if a_{ij} is a covariant tensor, and if $a_{ij} = a_{ji}$ then a_{ij} is a symmetric tensor. Is symmetry of tensorial character? That is, if a tensor, say a_{ijkl}, is symmetric with respect to i and j in one coordinate system x^i, is it symmetric in any other system $\bar{x}^i$? To show that symmetry is indeed invariant under transformation of coordinates, suppose that in the x^i system

$$a_{ijkl} = a_{jikl}.$$

Because the difference of two tensors is a tensor, it follows from $a_{ijkl} - a_{jikl} = 0$ that the difference tensor has all components zero in the x^i system. It is known that if a tensor has zero components in one coordinate system, it has zero components in all systems. Hence, $a_{ijkl} = a_{jikl}$ for arbitrary coordinate system.

If the interchange of any two indices of a tensor changes the sign of the components, the tensor is called *skew-symmetric* with respect to those indices. For instance, if $a_{ijk} = -a_{ikj}$, the tensor with components a_{ijk} is skew-symmetric relative to j and k. Proof of the tensorial character of skew-symmetry is similar to that given above for symmetry.

The reader should observe carefully the tensor character of an equation involving tensor components. If, say, three indices are free on one side of the equation, the same indices must appear as free on the other side, and all other subscripts and superscripts sum out. Note, too, that if a given index is repeated in a single term, the same index cannot be used to denote a second summation. Thus, for instance, $b_{ij}\xi^i \eta^j$ would be allowable, while $b_{ij}\xi^j \eta^j$ might indicate an error.

EXERCISES

[Assume that the transformation of coordinates is given by equations (1).]

1) Write the law of transformation for the tensor with components $b_i{}^{jk}$.

2) Show that if a tensor of the second order has all of its components zero in one coordinate system, the components are all zero in any other coordinate system.

3) Given $b_{ij} = \bar{b}_{pq} a'^p_i a'^q_j$, solve for $\bar{b}_{pq}$.

4) By contracting first on i and k and then on j and l show that the tensor with components $b_{ij}{}^{kl}$ becomes an invariant.

5) The symbols $\delta_{ij} = 1$, $i = j$, $= 0$, $i \neq j$, have the appearance of components of a covariant tensor of the second order. Show that δ_{ij} are *not* components of a tensor.

6) Show that $\delta_i{}^j$ *are* components of a mixed tensor.

7) Devise a suitable notation for the transformation law of components of a mixed tensor of order h in the covariant indices and of order k in the contravariant indices.

8) How many components does a fourth-order tensor have in a space of four dimensions?

9) Write the sums indicated by H^{ij}_{ijk} if the indices range over 1, 2.

10) Show that any covariant tensor of the second order can be expressed as the sum of two covariant tensors of the second order, one of them symmetric and the other skew-symmetric. [*Suggestion:* Consider the identity $H_{ij} \equiv \frac{1}{2}(H_{ij} + H_{ji}) + \frac{1}{2}(H_{ij} - H_{ji})$.]

5
Tensor Analysis

5-1. The Fundamental Quadratic Form. Consider the general transformation of coordinates

1) $$x^i = x^i(\bar{x}^1, \bar{x}^2, \bar{x}^3) \qquad (i = 1, 2, 3)$$

which was introduced in (3–2). The x^i coordinates may be considered for the present as orthogonal cartesian (although they may be used as general coordinates later) and the $\bar{x}^i$ are general curvilinear coordinates. The functions x^i in (1) are single-valued and continuous together with their first partial derivatives. The jacobian of the x^i functions with respect to the $\bar{x}^i$ variables is not zero, at least in the domain under consideration. Therefore, the inverse of (1) exists and may be written as

2) $$\bar{x}^i = \bar{x}^i(x^1, x^2, x^3).$$

It has been seen (3–11) that from (1)

3) $$dx^i = \frac{\partial x^i}{\partial \bar{x}^k} d\bar{x}^k,$$

4) $$\frac{\partial x^i}{\partial \bar{x}^j} \frac{\partial \bar{x}^j}{\partial x^k} = \delta_k{}^i,$$

and from (2)

5) $$\frac{\partial \bar{x}^i}{\partial x^j} \frac{\partial x^j}{\partial \bar{x}^k} = \delta_k{}^i.$$

It is known that, if x^i are orthogonal cartesian coordinates, the distance ds from the point $P(x^i)$ to the point $Q(x^i + dx^i)$ is obtained from

6) $$(ds)^2 = (dx^1)^2 + (dx^2)^2 + (dx^3)^2 = dx^i \, dx^i = \delta_{ij} \, dx^i \, dx^j.$$

The distance between P and Q possesses a meaning independent of the coordinate system used to express it. In the $\bar{x}$ coordinates the quantity $(ds)^2$ will, in general, not assume the "rectangular" form given by (6). A formula for $(ds)^2$ is desired which, in contrast to (6), is valid for all coordinate systems. The expression for $(ds)^2$ will be found next in the $\bar{x}^i$ system. By use of (3)

7) $$(ds)^2 = \delta_{ij}\, dx^i\, dx^j = \delta_{ij} \frac{\partial x^i}{\partial \bar{x}^k} \frac{\partial x^j}{\partial \bar{x}^l}\, d\bar{x}^k\, d\bar{x}^l = \frac{\partial x^i}{\partial \bar{x}^k} \frac{\partial x^i}{\partial \bar{x}^l}\, d\bar{x}^k\, d\bar{x}^l.$$

Denote the sum on i in the last member of (7) by $\bar{a}_{kl}$, that is, let

8) $$\bar{a}_{kl} \equiv \frac{\partial x^i}{\partial \bar{x}^k} \frac{\partial x^i}{\partial \bar{x}^l}.$$

It is evident from (8) that the $\bar{a}_{kl}$ are symmetric, i.e., $\bar{a}_{kl} = \bar{a}_{lk}$. Thus, for a general coordinate system in space, the square of the distance element $(ds)^2$ is represented by a quadratic differential form, that is,

9) $$ds^2 = \bar{a}_{kl}\, d\bar{x}^k\, d\bar{x}^l,$$

where, as is seen from (8), the $\bar{a}_{kl}$ are functions of position $(\bar{x}^i)$. If x^i are now considered as general coordinates, and not necessarily as cartesian, the bars may be dropped in (9), and one has the general arc length formula

10) $$ds^2 = a_{ij}\, dx^i\, dx^j.$$

The matrix of the coefficients a_{ij} in (10) is

11) $$(a_{ij}) \equiv \begin{pmatrix} a_{11} & a_{12} & a_{13} \\ a_{21} & a_{22} & a_{23} \\ a_{31} & a_{32} & a_{33} \end{pmatrix}.$$

If it happens that the x^i are orthogonal cartesian coordinates, the matrix (11) becomes simply

12) $$(a_{ij}) \equiv (\delta_{ij}) \equiv \begin{pmatrix} 1 & 0 & 0 \\ 0 & 1 & 0 \\ 0 & 0 & 1 \end{pmatrix}$$

and

13) $$ds^2 = \delta_{ij}\, dx^i\, dx^j = \sum_{i=1}^{3} (dx^i)^2.$$

Example. Find the expression for ds^2 and the elements in the matrix (a_{ij}) for spherical coordinates.

Solution: For the change from cartesian to spherical coordinates given by

$$x^1 = \bar{x}^1 \sin \bar{x}^2 \cos \bar{x}^3, \qquad x^2 = \bar{x}^1 \sin \bar{x}^2 \sin \bar{x}^3, \qquad x^3 = \bar{x}^1 \cos \bar{x}^2,$$

the expression for ds^2 is

14) $\quad (dx^1)^2 + (dx^2)^2 + (dx^3)^2 = 1(d\bar{x}^1)^2 + (\bar{x}^1)^2(d\bar{x}^2)^2 + (\bar{x}^1 \sin \bar{x}^2)^2(d\bar{x}^3)^2.$

If the $\bar{a}_{ij}$ are identified with spherical coordinates, the matrix

15) $\quad (\bar{a}_{ij}) \equiv \begin{pmatrix} 1 & 0 & 0 \\ 0 & (\bar{x}^1)^2 & 0 \\ 0 & 0 & (\bar{x}^1 \sin \bar{x}^2)^2 \end{pmatrix}.$

Now start with any general coordinate system in which $ds^2 = a_{ij} dx^i dx^j$ and effect the transformation (1) to new $(\bar{x}^i)$ coordinates in which $ds^2 = \bar{a}_{ij} d\bar{x}^i d\bar{x}^j$. If ds^2 is to remain invariant, it follows that

16) $\quad ds^2 = a_{ij} dx^i dx^j = \bar{a}_{kl} d\bar{x}^k d\bar{x}^l.$

Making use of (3) in (16) yields

17) $\quad ds^2 = a_{ij} dx^i dx^j = a_{ij} \dfrac{\partial x^i}{\partial \bar{x}^k} d\bar{x}^k \dfrac{\partial x^j}{\partial \bar{x}^l} d\bar{x}^l = \bar{a}_{kl} d\bar{x}^k d\bar{x}^l,$

from which follows

18) $\quad \left(a_{ij} \dfrac{\partial x^i}{\partial \bar{x}^k} \dfrac{\partial x^j}{\partial \bar{x}^l} - \bar{a}_{kl} \right) d\bar{x}^k d\bar{x}^l \equiv 0.$

Equation (18) must be an identity, that is, the left-hand member must be zero for all directions $d\bar{x}^k$. It follows that the coefficients in the quadratic form (18) must vanish, which gives

19) $\quad \bar{a}_{kl} = a_{ij} \dfrac{\partial x^i}{\partial \bar{x}^k} \dfrac{\partial x^j}{\partial \bar{x}^l}.$

This may be verified by using several directions $d\bar{x}^k$ in (18). For instance, one may use the $d\bar{x}^k$ as proportional to $(1, 0, 0)$ to obtain from (18)

20) $\quad a_{ij} \dfrac{\partial x^i}{\partial \bar{x}^1} \dfrac{\partial x^j}{\partial \bar{x}^1} - \bar{a}_{11} = 0.$

Another choice for $d\bar{x}^k$, say $(0, 1, 0)$, leads to the fact that (19) is true for $k = l = 2$. Continuing in this manner with further choices for $d\bar{x}^k$ will verify that (19) is true for all values of the free indices k and l. (Notice that a_{ij} are symmetric, i.e., $a_{ij} = a_{ji}$.)

Hence, equations (19) furnish the law of transformation for the coefficients in the fundamental quadratic differential form for ds^2 when an arbitrary change of coordinates (1) is made. Remember that the a_{ij} are, in general, functions of position x^i, so that (19) describes the manner in which a set of nine functions $a_{ij}(x^1, x^2, x^3)$ transform under a change of coordinates.

Definition: If any set of functions $b_{ij}(x^1,x^2,x^3)$ transform like the a_{ij} in (19), that is, if

21) $$\bar{b}_{kl}(\bar{x}^1,\bar{x}^2,\bar{x}^3) = b_{ij}(x^1,x^2,x^3)\frac{\partial x^i}{\partial \bar{x}^k}\frac{\partial x^j}{\partial \bar{x}^l},$$

the b_{ij} (and likewise the $\bar{b}_{kl}$) are called components of a covariant tensor of the second order. Notice that subscripts are used to indicate covariance. It will be seen that superscripts are always used to indicate contravariance of components, as in the following

Definition: If any set of functions $c^{ij}(x^1,x^2,x^3)$ transform by (1) into $\bar{c}^{kl}(\bar{x}^1,\bar{x}^2,\bar{x}^3)$, and if c^{ij} and $\bar{c}^{kl}$ are related by

$$\bar{c}^{kl} = c^{ij}\frac{\partial \bar{x}^k}{\partial x^i}\frac{\partial \bar{x}^l}{\partial x^j},$$

then c^{ij} and $\bar{c}^{kl}$ are components in their respective coordinate systems of a contravariant tensor of the second order.

The form for ds^2 exhibits the a_{ij} components for any coordinate system. The a_{ij} are called, therefore, the components of the covariant metric tensor of the space, and they determine the fundamental quadratic differential form for the space. Any space with such a metric is called a Riemannian space, and the geometry of the space is called a Riemannian geometry.

5–2. Covariant and Contravariant Tensors of the First Order. Before studying further the invariant $a_{ij}\,dx^i\,dx^j$ it will be well to consider the transformation law for tensors of the first order. A covariant tensor of the first order is introduced by the

Definition: If a set of three functions $\lambda_i(x^1,x^2,x^3)$ in any coordinate system x^i transform to $\bar{\lambda}_i(\bar{x}^1,\bar{x}^2,\bar{x}^3)$ in any other system $\bar{x}^i$ by equations (1) with the result that

22) $$\lambda_i(x^1,x^2,x^3) = \bar{\lambda}_j(\bar{x}^1,\bar{x}^2,\bar{x}^3)\frac{\partial \bar{x}^j}{\partial x^i},$$

or, equivalently,

23) $$\bar{\lambda}_i(\bar{x}^1,\bar{x}^2,\bar{x}^3) = \lambda_j(x^1,x^2,x^3)\frac{\partial x^j}{\partial \bar{x}^i},$$

the functions λ_i (and likewise $\bar{\lambda}_i$) are called components of a covariant tensor of the first order. A tensor of the first order is also called a vector, so (22) or (23) define a covariant vector. Actually, because a vector is determined at each point of space by the three components $\lambda_i(x^1,x^2,x^3)$ these functions are said to determine a vector field.

Definition: As in Chapter 4 a scalar point function $f(x^1,x^2,x^3)$ is called a tensor of zero order. The value of the function at any point is the same in any coordinate system, which fact is expressed by

24) $$f(x^1,x^2,x^3) = \bar{f}(\bar{x}^1,\bar{x}^2,\bar{x}^3).$$

A scalar field is determined by the function $f(x^1,x^2,x^3)$, which exhibits the single component of the zero-order tensor at any point x^i. An instance of a covariant tensor of the first order can be obtained by taking the three partial derivatives of f in (24). Thus,

25) $$\frac{\partial f}{\partial x^i} = \frac{\partial \bar{f}}{\partial \bar{x}^j} \frac{\partial \bar{x}^j}{\partial x^i},$$

which is the law (22), where $\lambda_i \equiv \partial f/\partial x^i$, and $\bar{\lambda}_j \equiv \partial \bar{f}/\partial \bar{x}^j$.

The particular type of vector field determined by the covariant vector with components $\partial f/\partial x^i$ in (25) is called the *gradient* vector field of the scalar field f, and is denoted by grad f or merely by ∇f in vector analysis.

Although taking the partial derivatives of the tensor of zero order in (24) does yield a tensor of the first order (25), one should not be deceived into concluding that the first or higher partial derivatives of a tensor constitute the components of a tensor of higher order. For instance, the set of nine second partial derivatives $\partial^2 f/\partial x^i \, \partial x^j$ of the scalar function $f(x^1,x^2,x^3)$ are not components of a second-order tensor for a general transformation of coordinates.

In order to make this fact clear, consider the partial derivative with respect to x^j of (25) which is first rewritten as

26) $$\frac{\partial f}{\partial x^i} = \frac{\partial \bar{f}}{\partial \bar{x}^k} \frac{\partial \bar{x}^k}{\partial x^i}.$$

One finds

27) $$\frac{\partial^2 f}{\partial x^j \, \partial x^i} = \frac{\partial^2 \bar{f}}{\partial \bar{x}^l \, \partial \bar{x}^k} \frac{\partial \bar{x}^k}{\partial x^i} \frac{\partial \bar{x}^l}{\partial x^j} + \frac{\partial \bar{f}}{\partial \bar{x}^k} \frac{\partial^2 \bar{x}^k}{\partial x^j \, \partial x^i}.$$

This must be compared with equation (21), or better its inverse, which is

28) $$b_{ij} = \bar{b}_{kl} \frac{\partial \bar{x}^k}{\partial x^i} \frac{\partial \bar{x}^l}{\partial x^j}.$$

If the last term on the right-hand side of (27) were zero, the quantities $\partial^2 f/\partial x^j \, \partial x^i$ would be components of a covariant tensor of the second order. But $\partial \bar{f}/\partial \bar{x}^k$ are not zero, in general, so the vanishing of

29) $$\frac{\partial^2 \bar{x}^k}{\partial x^j \, \partial x^i}$$

would mean that the transformation (1) is linear, a rather special type of transformation. The second derivatives in (29) would, of course, be zero for a transformation of orthogonal cartesian coordinates into orthogonal cartesian coordinates, which is a special case of a linear transformation. Transformations to curvilinear coordinates are not linear, and the transformations in Riemannian (curved) spaces, to be met later, are not linear. Therefore, in general, the derivative of a tensor is not a tensor.

Remark: A new type of differentiation, called *covariant* differentiation will be introduced in the sequel. It will be seen that the covariant derivative of a tensor is always a tensor of one higher order.

A contravariant tensor of the first order is now introduced.

Definition: If a set of functions $\mu^i(x^1,x^2,x^3)$ transform under the change of variables $x^i \to \bar{x}^i$, given by equations (1), in the following manner,

30) $$\mu^i(x^1,x^2,x^3) = \bar{\mu}^j(\bar{x}^1,\bar{x}^2,\bar{x}^3)\frac{\partial x^i}{\partial \bar{x}^j},$$

or in the inverse manner,

31) $$\bar{\mu}^i(\bar{x}^1,\bar{x}^2,\bar{x}^3) = \mu^j(x^1,x^2,x^3)\frac{\partial \bar{x}^i}{\partial x^j},$$

the functions μ^i are called components of a contravariant vector, i.e., of a contravariant tensor of order one.

Compare (30) carefully with the transformation law (22) for the components of a covariant vector, or tensor of first order.

Observe that equations (3) giving the total differentials of the x^i functions may be written, with a rearrangement of terms, as

32) $$dx^i = d\bar{x}^j \frac{\partial x^i}{\partial \bar{x}^j}.$$

Compare equations (32) with (30) to see that the μ^i and the dx^i transform in precisely the same manner. Therefore, the differentials dx^i are components of a contravariant vector.

It is important to notice that although the transformation (1) is not linear, in general, it is accompanied by or induces a linear transformation on the differentials of the coordinates. Notice that equations (32) are linear in the differentials, but that the linear transformation varies from point to point in space because the coefficients $\partial x^i/\partial \bar{x}^j$ are functions of position.

Suppose, for instance, that the x^i are taken to be rectangular cartesian coordinates. Then, the differentials dx^i at a point $P(x^i)$ give the direction of a line in space joining $P(x^i)$ to $Q(x^i + dx^i)$. The differentials $d\bar{x}^i$ given by (32) determine the same direction from P to Q in the $\bar{x}^i$ system, for the direction of a line does not depend upon any particular coordinate system. The dx^i may be constants and therefore represent the same direction at all points of space. The $d\bar{x}^i$, however, depend upon $\partial \bar{x}^i/\partial x^k$

which cause $d\bar{x}^i$ to vary from point to point. From (32), notice that the inverse is

$$d\bar{x}^i = dx^p \frac{\partial \bar{x}^i}{\partial x^p}.$$

If dx^p are assigned as three constants, say $dx^p \equiv a^p$, then the $d\bar{x}^i$ are determined as soon as the transformation (1) of coordinates is known so that the $\partial \bar{x}^i / \partial x^p$ can be calculated for any point of the domain considered. Because the $d\bar{x}^i$ vary, they cannot be called direction numbers of a line in space.

5–3. A Quadratic Form from a Tensor Product. A tensor of arbitrary order will be introduced in the next section. However, it may be instructive at this point to define a certain type of fourth-order tensor and then to see an instance of this type and how it contracts to produce an invariant.

Definition: A set of functions $b_{ij}{}^{kl}$ which transform from x^i to $\bar{x}^i$ by equations (1) in the following manner

33) $$b_{ij}{}^{kl} = \bar{b}_{pq}{}^{rs} \frac{\partial \bar{x}^p}{\partial x^i} \frac{\partial \bar{x}^q}{\partial x^j} \frac{\partial x^k}{\partial \bar{x}^r} \frac{\partial x^l}{\partial \bar{x}^s},$$

are called components of a mixed tensor of the fourth order, covariant of order two, and contravariant of order two.

If a_{ij} are components of a covariant tensor, the mixed tensor of the fourth order with components given by the product $a_{ij} \, dx^k \, dx^l$ therefore has by (33) the transformation law

$$a_{ij} \, dx^k \, dx^l = \bar{a}_{pq} \frac{\partial \bar{x}^p}{\partial x^i} \frac{\partial \bar{x}^q}{\partial x^j} \, d\bar{x}^r \frac{\partial x^k}{\partial \bar{x}^r} \, d\bar{x}^s \frac{\partial x^l}{\partial \bar{x}^s},$$

or

34) $$(a_{ij} \, dx^k \, dx^l) = (\bar{a}_{pq} \, d\bar{x}^r \, d\bar{x}^s) \frac{\partial \bar{x}^p}{\partial x^i} \frac{\partial \bar{x}^q}{\partial x^j} \frac{\partial x^k}{\partial \bar{x}^r} \frac{\partial x^l}{\partial \bar{x}^s}.$$

This law of transformation should be compared with (4–17) where a restricted (linear) type of transformation of coordinates was used.

Now carry out the process of contraction on the tensor components in (34) by putting $k = i$ and $l = j$ to obtain the invariant form $a_{ij} \, dx^i \, dx^j$. In order to verify the invariance, notice that from (34)

35) $$a_{ij} \, dx^i \, dx^j = \bar{a}_{pq} \, d\bar{x}^r \, d\bar{x}^s \frac{\partial \bar{x}^p}{\partial x^i} \frac{\partial \bar{x}^q}{\partial x^j} \frac{\partial x^i}{\partial \bar{x}^r} \frac{\partial x^j}{\partial \bar{x}^s}$$

$$= \bar{a}_{pq} \, d\bar{x}^r \, d\bar{x}^s \, \delta_r{}^p \, \delta_s{}^q = \bar{a}_{pq} \, d\bar{x}^p \, d\bar{x}^q = \bar{a}_{ij} \, d\bar{x}^i \, d\bar{x}^j.$$

Hence, the quadratic differential form $a_{ij} \, dx^i \, dx^j$ is an invariant. This example illustrates a general principle. In a tensor expression, if all the indices sum out, the expression represents an invariant. On the other hand, if all indices except one sum out and the remaining one is covariant (subscript), the expression represents a covariant vector. If the single remaining index is a superscript, the expression represents a contravariant vector.

5–4. Definition of a General Tensor. From the structure of the law of transformation (33) for a mixed tensor of the fourth order one can extend the form to a tensor of higher order by writing additional indices of covariance and of contravariance. But instead of using, say, $i, j, k, l \cdots$ for covariant indices it will be advisable to write i_1 for i, i_2 for j, i_3 for k, and so on. (One may exhaust the letters of the alphabet but not numerical subscripts!) A similar remark holds for the superscripts.

For the sake of completeness it is well to exhibit the general tensor next.

Definition: Consider the set of functions $T^{l_1 \cdots l_q}_{k_1 \cdots k_p}$ in the coordinate system x^i and the set $\bar{T}^{j_1 \cdots j_q}_{i_1 \cdots i_p}$ in the $\bar{x}^i$ system, where each subscript and each superscript can range over 1, 2, 3 $(1, \cdots, n$ in n-space). The object of which the two sets are components in their respective coordinate systems is called a tensor if the sets are related by the law of transformation given by

$$36) \qquad \bar{T}^{j_1 \cdots j_q}_{i_1 \cdots i_p} = T^{l_1 \cdots l_q}_{k_1 \cdots k_p} \frac{\partial \bar{x}^{j_1}}{\partial x^{l_1}} \cdots \frac{\partial \bar{x}^{j_q}}{\partial x^{l_q}} \frac{\partial x^{k_1}}{\partial \bar{x}^{i_1}} \cdots \frac{\partial x^{k_p}}{\partial \bar{x}^{i_p}}.$$

The usual stipulation concerning the non-vanishing of the jacobian holds for the domain considered.

The tensor with components shown in (36) is covariant of order p and contravariant of order q. Observe that (36) is the same law of transformation as (33) if $p = 2$ and $q = 2$. Strictly speaking (36) defines an *absolute* tensor. The sets of T functions are components of a *relative* tensor if $|\partial x^i / \partial \bar{x}^j|^N$ appears as a factor in the right-hand member of (36). The number N is referred to as the *weight* of the tensor field. If $N = 0$, the tensor is absolute. If $N = 1$, the relative tensor field is called a *tensor density*.

5–5. Inner Product of Two Vectors. Consider the product of a covariant vector λ_i and a contravariant vector μ^j, i.e., $\lambda_i \mu^j$, which is a mixed tensor of the second order. A tensor of the second order obtained as the product of two vectors is sometimes called a *dyadic*. The law of transformation for a mixed tensor $b_i{}^j$ is [compare with (33)]

$$37) \qquad b_i{}^j = \bar{b}_p{}^q \frac{\partial \bar{x}^p}{\partial x^i} \frac{\partial x^j}{\partial \bar{x}^q}.$$

From the transformation laws (22) and (30), the tensor product $\lambda_i \mu^j$ transforms by

38) $$\lambda_i \mu^j = \bar{\lambda}_p \bar{\mu}^q \frac{\partial \bar{x}^p}{\partial x^i} \frac{\partial x^j}{\partial \bar{x}^q},$$

which is like (37), where $b_i{}^j \equiv \lambda_i \mu^j$.

If the tensor $\lambda_i \mu^j$ is contracted, the invariant $\lambda_i \mu^i$ results. This invariant, or scalar, is called the *inner* product, or the *scalar* product of the two vectors with components λ_i and μ^j.

Example. If the vector λ_i has components $(x^1, 2x^2, 3x^3)$ and the vector μ^j has components $(x^2, -2x^1, 3)$, find the invariant function $\lambda_i \mu^i$.

Solution: The sum $\lambda_i \mu^i$ means $\lambda_1 \mu^1 + \lambda_2 \mu^2 + \lambda_3 \mu^3$ which in this case is $x^1 x^2 - 4x^1 x^2 + 9x^3 \equiv 9x^3 - 3x^1 x^2$.

EXERCISES

1) Transform from orthogonal cartesian coordinates x^i to $\bar{x}^i$ by $x^1 = (\bar{x}^2)^2 + (\bar{x}^3)^2$, $x^2 = (\bar{x}^3)^2 + (\bar{x}^1)^2$, $x^3 = (\bar{x}^1)^2 + (\bar{x}^2)^2$. Use the definition of $\bar{a}_{kl}$ in (8) to find $\bar{a}_{11}$. Is the tensor $\bar{a}_{kl}$ symmetric?

2) Test a sufficient number of sets of values for $d\bar{x}^i$ in (18) to establish (19).

3) If the transformation is that of Exercise 1, and if a scalar field is determined by $\bar{f}(\bar{x}^1, \bar{x}^2, \bar{x}^3) \equiv (\bar{x}^1)^2 + (\bar{x}^2)^2 + (\bar{x}^3)^2$, find the function $f(x^1, x^2, x^3)$ into which $\bar{f}$ transforms. Calculate the components of the gradient of $\bar{f}$ directly, and then again by use of the inverse of equations (25).

4) Solve equations (21) for b_{ij}.

5) Given the scalar field $f \equiv (x^1)^2 + 2(x^2)(x^3) + (x^3)^2$, find the gradient vector field. Calculate the scalar product of this gradient with the contravariant vector μ^j with components $(x^3, -x^1, x^2)$.

6) Describe the kind of tensor which results after contracting $a_{ijk}{}^{pqr}$ with respect to i and p and then with respect to j and q.

7) If a_i is a covariant vector, show that $\partial a_i / \partial x^j$ is, in general, not a tensor.

8) A covariant vector has components (yz, zx, xy) in orthogonal cartesian coordinates. Find the components of the vector in cylindrical coordinates. (*Hint:* The coordinate transformation is $x = r \cos \theta$, $y = r \sin \theta$, $z = z$, or $x^1 = \bar{x}^1 \cos \bar{x}^2$, $x^2 = \bar{x}^1 \sin \bar{x}^2$, $x^3 = \bar{x}^3$.)

9) Show that if f and g denote scalar fields

$$\text{grad } (fg) = f \text{ grad } g + g \text{ grad } f$$

and

$$\text{grad } H(f) = H'(f) \text{ grad } f,$$

where $H(f)$ denotes a function of the scalar function f. These results may be written in the form

$$\nabla(fg) = f\nabla g + g\nabla f$$

and
$$\nabla H(f) = H'(f)\nabla f.$$

10) Show that under the transformation
$$2x^1 = (\bar{x}^1)^2 - (\bar{x}^2)^2, \qquad x^2 = \bar{x}^1\bar{x}^2, \qquad x^3 = \bar{x}^3$$
the curves $\bar{x}^2 =$ constant are parabolas. The $\bar{x}^i$ coordinates here may be styled parabolic cylindrical. Calculate ds^2 in the $\bar{x}$ system.

11) If the vector v_i transforms to $\bar{v}_i$ under $x^i \to \bar{x}^i$ and $\bar{v}_i$ transforms to $\bar{\bar{v}}_i$ under $\bar{x}^i \to \bar{\bar{x}}^i$, where the coordinate transformations are of the general type (1), show that the tensor law of transformation for a covariant vector is transitive (see Section 4-4).

5-6. Associate Tensors. Equations (3-22) stated that
$$39) \qquad a_j{}^i A_k{}^j = \delta_k{}^i a,$$
where a denoted the value of the determinant $|a_j{}^i|$ and $A_k{}^j$ was the cofactor of the element $a_j{}^k$. It will be useful to change the notation for the elements of the determinant from $a_j{}^i$ to a_{ij}, and to write $(1/a)A_k{}^j \equiv a^{kj}$, so that
$$40) \qquad a^{kj} = \frac{\text{cofactor of } a_{jk} \text{ in } |a_{ij}|}{|a_{ij}|}.$$

With the revised notation, equations (39) can be written in the form
$$41) \qquad a^{kj}a_{ij} = \delta_i{}^k.$$

The elements a_{ij} of the determinant $|a_{ij}|$ are to be identified in what follows with the coefficients of the fundamental quadratic form $ds^2 = a_{ij}\,dx^i\,dx^j$ of formula (5-10). From the definition of the $\bar{a}_{ij}$ in formula (5-8) it is apparent that $\bar{a}_{ij}$ (and therefore a_{ij}) is a symmetric tensor, i.e., $\bar{a}_{ij} = \bar{a}_{ji}$ for all i and j. It follows that a^{ij} are symmetric also, so that interchanging i and j in a_{ij} and a^{ij} is allowable.

One should note that equations (41) define the new quantities a^{kj}, for Cramer's rule can be applied to solve (41) for the "unknowns" a^{kj} in the following way. Multiply both sides of (41) by the cofactor A^{li} of a_{il} and sum on i to obtain
$$42) \qquad a^{kj}a_{ij}A^{li} = \delta_i{}^k A^{li} = A^{lk}.$$

Now replace $a_{ij}A^{li}$ by $a\delta_j{}^l$ by use of (39) to see that
$$43) \qquad a^{kj}a\delta_j{}^l = A^{lk},$$
or
$$44) \qquad a^{kl} = \frac{A^{lk}}{a},$$

which is equivalent to the statement in (40).

It is to be emphasized that although the a_{ij} are to be used here as the components of the fundamental metric tensor, the results obtained will hold for any symmetric covariant tensor of the second order with non-vanishing determinant $|a_{ij}|$.

It is now to be shown that the a^{ij} as defined in (41) are components of a contravariant tensor, i.e., that

$$45) \qquad \bar{a}^{ij} = a^{pq} \frac{\partial \bar{x}^i}{\partial x^p} \frac{\partial \bar{x}^j}{\partial x^q}.$$

Although (45) can be established otherwise, the following method is presented in order to illustrate some manipulations in tensor analysis.

Begin with the form of (41) for the transformed components, that is

$$46) \qquad \bar{a}^{pq} \bar{a}_{rq} = \delta_r{}^p,$$

and use (5-19) to obtain

$$47) \qquad \bar{a}^{pq} a_{lm} \frac{\partial x^l}{\partial \bar{x}^r} \frac{\partial x^m}{\partial \bar{x}^q} = \delta_r{}^p.$$

Multiply both sides of (47) by $\partial \bar{x}^r / \partial x^h$ to obtain

$$48) \qquad \bar{a}^{pq} a_{lm} \frac{\partial x^m}{\partial \bar{x}^q} \delta_h{}^l = \delta_r{}^p \frac{\partial \bar{x}^r}{\partial x^h} = \frac{\partial \bar{x}^p}{\partial x^h},$$

or

$$49) \qquad \bar{a}^{pq} a_{hm} \frac{\partial x^m}{\partial \bar{x}^q} = \frac{\partial \bar{x}^p}{\partial x^h}.$$

Multiply (49) by a^{hk} to arrive at

$$50) \qquad \bar{a}^{pq} a_{hm} a^{hk} \frac{\partial x^m}{\partial \bar{x}^q} = a^{hk} \frac{\partial \bar{x}^p}{\partial x^h}.$$

Now, after noticing that $a_{hm} a^{hk} = \delta_m{}^k$, and on multiplying by $\partial \bar{x}^r / \partial x^k$ and summing on k, there follows

$$51) \qquad \bar{a}^{pq} \delta_m{}^k \frac{\partial x^m}{\partial \bar{x}^q} \frac{\partial \bar{x}^r}{\partial x^k} = a^{hk} \frac{\partial \bar{x}^p}{\partial x^h} \frac{\partial \bar{x}^r}{\partial x^k},$$

or

$$52) \qquad \bar{a}^{pq} \frac{\partial x^m}{\partial \bar{x}^q} \frac{\partial \bar{x}^r}{\partial x^m} = a^{hk} \frac{\partial \bar{x}^p}{\partial x^h} \frac{\partial \bar{x}^r}{\partial x^k},$$

from which

$$53) \qquad \bar{a}^{pq} \delta_q{}^r = a^{hk} \frac{\partial \bar{x}^p}{\partial x^h} \frac{\partial \bar{x}^r}{\partial x^k},$$

or

$$54) \qquad \bar{a}^{pr} = a^{hk} \frac{\partial \bar{x}^p}{\partial x^h} \frac{\partial \bar{x}^r}{\partial x^k},$$

which, on changing dummy indices, becomes

54') $$\bar{a}^{ij} = a^{pq} \frac{\partial \bar{x}^i}{\partial x^p} \frac{\partial \bar{x}^j}{\partial x^q},$$

i.e., equations (45). Therefore, the a^{ij} are components of a contravariant tensor.

By the definition (41) of the quantities a^{kj} they are evidently symmetric if a_{kj} are symmetric. It has been seen that for any symmetric covariant tensor b_{ij} the equations $b^{kj}b_{ij} = \delta_i^k$ determine a set of quantities b^{kl} which are the components of a symmetric contravariant tensor of the second order. The tensors b_{ij} and b^{kl} may be called *associate* tensors. They are also referred to as *conjugate* tensors.

Observe that if b^i is a contravariant vector and a_{ij} is a covariant tensor, then $a_{ij}b^i = u_j$ is a covariant vector, and that if c_i is a covariant vector, then $a^{ij}c_i = v^j$ is a contravariant vector. The b^i and u_j are called associate vectors with respect to the tensor a_{ij}. Notice that the superscript on b^i is lowered by multiplying b^i by a_{ij} to obtain u_j. It will be seen later that b^i and u_j are the contravariant and covariant components of the same vector if the a_{ij} are the covariant components of the fundamental metric tensor. Similarly, c_i and v^j are related through a^{ij}. The subscript on c_i is raised by multiplying by a^{ij}.

6
Vector Analysis

6–1. Length of a Vector. Consider a vector **A** with contravariant components λ^i and form the tensor product $a_{ij}\lambda^k\lambda^l$, where a_{ij} is the fundamental covariant tensor of space in some coordinate system. Contract this fourth-order tensor twice to obtain the invariant $a_{ij}\lambda^i\lambda^j$. Because this expression is invariant under a transformation of coordinates, it must have some geometric meaning, which can be most readily discerned in orthogonal cartesian coordinates. Let, then, the a_{ij} be simply δ_{ij} so the invariant $a_{ij}\lambda^i\lambda^j$ becomes $\delta_{ij}\lambda^i\lambda^j = \lambda^i\lambda^i \equiv (\lambda^1)^2 + (\lambda^2)^2 + (\lambda^3)^2$, which is the square of the distance from the origin to the point with cartesian coordinates λ^i. Since the contravariant components λ^i of vector **A** are merely the coordinates of a point in orthogonal cartesian coordinates, the vector **A** may be visualized as the line segment from the origin to the point λ^i and its length l (which is also denoted by the notation $|\mathbf{A}|$) is given by

1) $$l^2 \equiv |\mathbf{A}|^2 = \delta_{ij}\lambda^i\lambda^j.$$

Because of the invariance of the quadratic form in (1), the square of the length of vector **A** in any other coordinate system $\bar{x}^i$ is given by

2) $$l^2 = |\mathbf{A}|^2 = \bar{a}_{ij}\bar{\lambda}^i\bar{\lambda}^j.$$

Remember that when λ^i are given and the expressions $\partial \bar{x}^i/\partial x^j$ are known from the transformation of coordinates $\bar{x}^i = \bar{x}^i(x^1,x^2,x^3)$, the components $\bar{\lambda}^j$ are uniquely determined by $\bar{\lambda}^i = \lambda^j\,(\partial \bar{x}^i/\partial x^j)$. The geometric meaning of the contravariant components of vector **A** in a general coordinate system will be explained in Section 6–4.

Consider next $u_j = a_{ij}\lambda^i$. If $a_{ij} = \delta_{ij}$, $u_j = \delta_{ij}\lambda^i = \lambda_j$. But from equation (1) it is seen that the invariant $l^2 = \delta_{ij}\lambda^i\lambda^j$ may now be written as $l^2 = \lambda_j\lambda^j$. Hence, the invariant $\lambda_i\lambda^i$ is the square of the length of the vector in any coordinate system. This invariant may be styled the inner product of a vector with its associate vector. The contravariant vector u^j associate to any vector u_i may be found from

3) $$a^{ij}u_i = u^j,$$

because, on expressing u_i as $a_{ik}u^k$, (3) gives

4) $$a^{ij}a_{ik}u^k = \delta_k^j u^k = u^j.$$

Similarly, the vector u_i associate to u^k is given by

5) $$u_i = a_{ik}u^k.$$

Notice that the square of the magnitude of any vector **B** with components u_i (or equivalently u^i) is given by either

6) $$|\mathbf{B}|^2 = a^{ij}u_i u_j = u^j u_j,$$

or

7) $$|\mathbf{B}|^2 = a_{ij}u^i u^j = u_j u^j.$$

Equations (6) and (7) hold for any coordinate system. In particular, if $a_{ij} = \delta_{ij}$, then $a^{ij} = \delta^{ij}$, and since δ_{ij} and δ^{ij} have the same values there is no distinction between contravariant and covariant components of a vector in the orthogonal cartesian system.

The inner product (or scalar product) of two vectors **A** (with components u_i or u^i) and **B** (with components v^i or v_i) is denoted by $\mathbf{A}\cdot\mathbf{B}$, so that

8) $$\mathbf{A}\cdot\mathbf{B} \equiv u_i v^i = a_{ij}u^j v^i = u^j v_j = a^{jk}u_k v_j.$$

It is evident that in the presence of the fundamental tensor a_{ij} (or a^{ij}) indices on vectorial components may be lowered or raised at will. Thus, in the metric geometry under study here, only the order of a tensor characterizes it, for the indices may be changed by use of the fundamental tensor. For instance, one index on the fourth-order tensor R_{ijkl} may be raised to obtain a different set of components of the tensor. Thus, one has

$$R^r{}_{jkl} = a^{ir}R_{ijkl}.$$

From (8) one should notice that if vector **B** coincides with **A**, the square of the magnitude of **A** is given by

9) $$\mathbf{A}\cdot\mathbf{A} = |\mathbf{A}|^2 = u_i u^i = a_{ij}u^i u^j = a^{ij}u_i u_j$$

in any coordinate system.

6–2. Angle Between Two Vectors; Orthogonal Vectors. A vector of length one is called a *unit* vector. It is readily seen that a vector **A** with components u_i can be reduced to a unit vector **U** by dividing each of the components u_i by $|\mathbf{A}|$. For, by formula (9), the square of the length of $\mathbf{U}$ is given by

10) $$\mathbf{U}\cdot\mathbf{U} \equiv |\mathbf{U}|^2 = a^{ij}\frac{u_i}{\sqrt{a^{pq}u_p u_q}}\frac{u_j}{\sqrt{a^{rs}u_r u_s}} = \frac{a^{ij}u_i u_j}{a^{kl}u_k u_l} = 1,$$

which means that $|\mathbf{U}| = 1$. The contravariant components of $\mathbf{A}$ could have been used in (10) to give

11) $$\mathbf{U}\cdot\mathbf{U} \equiv |\mathbf{U}|^2 = a_{ij}\frac{u^i}{\sqrt{a_{pq}u^p u^q}}\frac{u^j}{\sqrt{a_{rs}u^r u^s}} = \frac{a_{ij}u^i u^j}{a_{kl}u^k u^l} = 1.$$

The vector $\mathbf{A}$ is a unit vector itself if $a_{ij}u^i u^j = a^{ij}u_i u_j = 1$.

Consider two *unit* vectors $\mathbf{U}$ and $\mathbf{V}$ with components u_i and v^i, respectively. The relations $a^{ij}u_i u_j = 1$ and $a_{ij}v^i v^j = 1$ obtain. If the coordinates are orthogonal cartesian it is known that since $\mathbf{U}$ and $\mathbf{V}$ are unit vectors the components u_i and v^i are direction cosines of the directions of the vectors $\mathbf{U}$ and $\mathbf{V}$, and that if θ is the angle between $\mathbf{U}$ and $\mathbf{V}$

12) $$\mathbf{U}\cdot\mathbf{V} = \cos\theta = u_i v^i = \delta_{ij}u^i v^j.$$

In general coordinates the invariant form in (12) becomes

13) $$\cos\theta = a_{ij}u^i v^j.$$

From (13) it follows that a necessary and sufficient condition for two vectors to be orthogonal is that

14) $$\mathbf{U}\cdot\mathbf{V} \equiv a_{ij}u^i v^j = u_j v^j = 0,$$

or equivalently,

15) $$\mathbf{U}\cdot\mathbf{V} \equiv a^{ij}u_i v_j = u^j v_i = 0.$$

For any two non-unit vectors $\mathbf{P}$ and $\mathbf{Q}$ with components p_i and q_i in any coordinates the angle θ between them is given by

16) $$\cos\theta = \frac{\mathbf{P}}{|\mathbf{P}|}\cdot\frac{\mathbf{Q}}{|\mathbf{Q}|},$$

that is, by

17) $$\cos\theta = \frac{a_{ij}p^i q^j}{\sqrt{a_{rs}p^r p^s}\sqrt{a_{lm}q^l q^m}}.$$

From (17) it is seen that if the vectors with components p^i and q^i are orthogonal at every point of space, $a_{ij}p^i q^j = 0$ must be an identity in the x^i coordinates. If $a_{ij}p^i q^j = 0$ is not an identity but a conditional equation in x^i, this equation represents a locus at all points of which the p^i and q^i vectors are orthogonal.

Up to this point in this chapter the reader has probably been visualizing the space as three-dimensional. However, all formulas preceding (12) hold in case the indices take the range 1 to n. They may be taken as definitions in n-space. Formula (12), or more generally (17) will be used to *define* the measure of angle θ between vectors $\mathbf{P}$ and $\mathbf{Q}$ in any space, but this has meaning only in case $|\cos \theta| \leq 1$. It will be indicated in the following remarks that $|\cos \theta| \leq 1$ in (17) wherever the metric form $a_{ij}\, dx^i\, dx^j$ is *positive definite*. This means that $ds^2 \equiv a_{ij}\, dx^i\, dx^j$ is positive for all directions at every point in the space.

The vector $hp^i + kq^i$, where h and k are real parameters, is a member of the family of vectors determined by p^i and q^i. Assume that the form $a_{ij}\, dx^i\, dx^j$ is positive definite. Then the length l of vector $hp^i + kq^i$ given by

$$l^2 \equiv a_{ij}(hp^i + kq^i)(hp^j + kq^j)$$

can be zero only for non-real values of the ratio h/k. This means that the quadratic equation

$$(a_{ij}p^i p^j)\left(\frac{h}{k}\right)^2 + 2a_{ij}p^i q^j \left(\frac{h}{k}\right) + (a_{ij}q^i q^j) = 0$$

has complex roots. It follows that

$$(a_{ij}p^i q^j)^2 - (a_{ij}p^i p^j)(a_{kl}q^k q^l) < 0$$

or

$$\frac{(a_{ij}p^i q^j)^2}{(a_{ij}p^i p^j)(a_{kl}q^k q^l)} = \cos^2 \theta < 1.$$

If the direction of q^i is the same or opposite that of p^i, $\cos \theta = \pm 1$, so in all cases $|\cos \theta| \leq 1$. The reader should be aware that the inequality $|\cos \theta| \leq 1$ was proved entirely on the basis of the definiteness of the form $a_{ij}\, dx^i\, dx^j$ with the consequence that no geometrical notion is necessarily involved.

For a curve in three-space along which x^1 varies while x^2 and x^3 remain constant, the value of dx^1 is not zero while dx^2 and dx^3 are zero. Similarly, along the x^2-curve dx^2 is not zero while dx^1 and dx^3 are zero. Hence, from (17) the angle θ_{12} between the x^1- and x^2-curves is given by

18)
$$\cos \theta_{12} = \frac{a_{12}\, dx^1\, dx^2}{\sqrt{a_{11}\, dx^1\, dx^1}\, \sqrt{a_{22}\, dx^2\, dx^2}} = \frac{a_{12}}{\sqrt{a_{11}}\, \sqrt{a_{22}}}.$$

More generally, the angle θ_{ij} between the x^i- and x^j-curves ($i \neq j$) is readily seen to be given by

19)
$$\cos \theta_{ij} = \frac{a_{ij}\, dx^i\, dx^j}{\sqrt{a_{ii}\, dx^i\, dx^i}\, \sqrt{a_{jj}\, dx^j\, dx^j}}$$

$$= \frac{a_{ij}}{\sqrt{a_{ii}}\,\sqrt{a_{jj}}} \qquad (i, j \text{ not summed}),$$

which holds for any coordinate system.

From (19) one sees that if $a_{ij} = 0$ $(i \neq j)$ at any point of space, the x^i- and x^j-curves are orthogonal at that point. If the a_{ij} are zero for all i and j $(i \neq j)$ at every point of space, the three surfaces given by x^i-constant are called a triply orthogonal system. The spherical coordinates furnish an instance of this.

It is often necessary to find the component of one vector **Q** in the direction of a second vector **P**, that is, the scalar projection of **Q** on **P**.

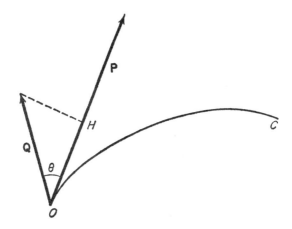

Fig. 2

From Fig. 2 it is clear that the projection OH is $|\mathbf{Q}| \cos \theta$. By (16)

20) $$OH = |\mathbf{Q}|\,\frac{\mathbf{P}}{|\mathbf{P}|} \cdot \frac{\mathbf{Q}}{|\mathbf{Q}|} = \mathbf{Q} \cdot \frac{\mathbf{P}}{|\mathbf{P}|}.$$

Note that if θ is obtuse, OH is negative.

An important application of (20) is the following. If a curve C: $x^i = x^i(s)$ through O is tangent to vector **P**, the unit vector $\mathbf{P}/|\mathbf{P}|$ has components dx^i/ds. If **Q** is taken to be the gradient $\partial\phi/\partial x^i$ of a scalar function ϕ at O, then the scalar projection of $\partial\phi/\partial x^i$ on the tangent at O is

21) $$\frac{\partial\phi}{\partial x^i}\,\frac{dx^i}{ds} = \frac{d\phi}{ds},$$

where $d\phi/ds$ is the directional derivative of ϕ along the curve C at O.

6-3. Some Applications. Before giving the geometric interpretation of covariant and contravariant components of a vector in general coordinates in space, it will be useful to see an example employing oblique cartesian coordinates in the plane. Refer the coordinates $x^i (i = 1, 2)$ of a point in the plane to a set of oblique axes inclined at an angle ω. Consider the triangle LMN in Fig. 3. Sides LM and MN are parallel to the axes.

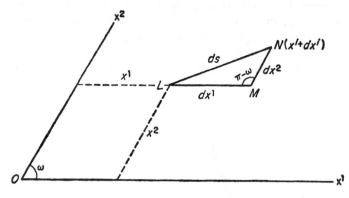

Fig. 3

If L and N have coordinates x^i and $x^i + dx^i$, respectively, the distance $ds = LN$ is obtained by the Law of Cosines from

22) $$ds^2 = (dx^1)^2 + (dx^2)^2 - 2(dx^1)(dx^2) \cos (\pi - \omega).$$

Hence, the fundamental quadratic differential form (5–10) is given by

23) $$ds^2 = dx^1\, dx^1 + dx^1\, dx^2 \cos \omega + dx^2\, dx^1 \cos \omega + dx^2\, dx^2 \equiv a_{ij}\, dx^i\, dx^j.$$

Therefore, the matrix of the components a_{ij} is

24) $$\begin{pmatrix} a_{11} & a_{12} \\ a_{21} & a_{22} \end{pmatrix} \equiv \begin{pmatrix} 1 & \cos \omega \\ \cos \omega & 1 \end{pmatrix},$$

and the determinant $a \equiv |a_{ij}|$ has the value given by

25) $$a = |a_{ij}| = \begin{vmatrix} 1 & \cos \omega \\ \cos \omega & 1 \end{vmatrix} = \sin^2 \omega.$$

Consider a point R with coordinates λ^i and the vector $\overrightarrow{OR}$ with components $OH = \lambda^1$, $OK = \lambda^2$. What are the covariant components λ_i of vector $\overrightarrow{OR}$? These are found, by using $\lambda_i = a_{ij}\lambda^j$ and the a_{ij} from (24), to be

26) $$\lambda_1 = a_{1j}\lambda^j = a_{11}\lambda^1 + a_{12}\lambda^2 = \lambda^1 + \lambda^2 \cos \omega,$$
$$\lambda_2 = a_{2j}\lambda^j = a_{21}\lambda^1 + a_{22}\lambda^2 = \lambda^1 \cos \omega + \lambda^2.$$

VECTOR ANALYSIS

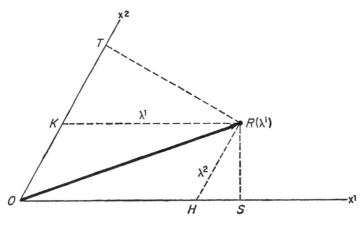

Fig. 4

By the geometry of Fig. 4, it is true that

27) $$\lambda_1 = \lambda^1 + \lambda^2 \cos \omega = OH + HS = OS,$$
$$\lambda_2 = \lambda^1 \cos \omega + \lambda^2 = KT + OK = OT.$$

This example exhibits the contravariant components OH and OK, and the covariant components OS and OT of the vector $\overrightarrow{OR}$. If the angle $\omega = 90°$, the covariant and contravariant components of $\overrightarrow{OR}$ are indistinguishable.

EXERCISES

1) Verify that the contravariant components a^{ij} of the metric tensor are $a^{11} = \csc^2 \omega$, $a^{12} = a^{21} = -\csc \omega \cot \omega$, $a^{22} = \csc^2 \omega$.

2) Verify that $a_{ij}a^{kj} = \delta_i{}^k$.

Consider next the problem of finding the angle θ between two given vectors $\overrightarrow{OR}$ and $\overrightarrow{OQ}$ with components λ^i and μ^i, respectively. By (17) θ is given by

28) $$\cos \theta = \frac{a_{ij}\lambda^i\mu^j}{\sqrt{a_{pq}\lambda^p\lambda^q}\sqrt{a_{rs}\mu^r\mu^s}}.$$

This formula for $\cos \theta$ may be verified by calculating it directly from Fig. 5 by trigonometry. Observe that, since θ is the difference between angles SOQ and SOR,

29) $$\cos \theta = \frac{OT}{OQ}\frac{OS}{OR} + \frac{TQ}{OQ}\frac{SR}{OR} = \frac{(OT)(OS) + (TQ)(SR)}{(OQ)(OR)}.$$

By use of $OS = \lambda_1$, $OT = \mu_1$, $TQ = \mu^2 \sin \omega$, $SR = \lambda^2 \sin \omega$, and $\mu_1 = \mu^1 + \mu^2 \cos \omega$, $\lambda_1 = \lambda^1 + \lambda^2 \cos \omega$, by (26), it is seen that (29) gives

30) $$\cos \theta = \frac{\mu_1 \lambda_1 + \mu^2 \lambda^2 \sin \omega}{|\overrightarrow{OR}||\overrightarrow{OQ}|} = \frac{\mu^1 \lambda^1 + \mu^2 \lambda^1 \cos \omega + \mu^1 \lambda^2 \cos \omega + \mu^2 \lambda^2}{|\overrightarrow{OR}||\overrightarrow{OQ}|},$$

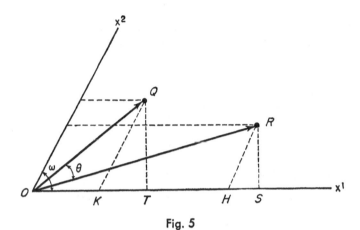

Fig. 5

or, because the numerator in the last member of (30) is $a_{ij}\lambda^i\mu^j$,

31) $$\cos \theta = \frac{a_{ij}\lambda^i\mu^j}{\sqrt{a_{pq}\lambda^p\lambda^q}\sqrt{a_{rs}\mu^r\mu^s}},$$

which verifies formula (28).

EXERCISES

1) Calculate the contravariant components a^{ij} of the metric tensor in spherical coordinates given by

$$x^1 = \bar{x}^1 \sin \bar{x}^2 \cos \bar{x}^3, \qquad x^2 = \bar{x}^1 \sin \bar{x}^2 \sin \bar{x}^3, \qquad x^3 = \bar{x}^1 \cos \bar{x}^2.$$

2) If the covariant components of a vector in spherical coordinates are $\lambda_i = (1,2,3)$ find the contravariant components λ^i.

3) Show that if H_i is a covariant vector, $H_{i,j}$ is not, in general, a tensor. (Remember that the notation $,j$ indicates a partial derivative with respect to x^j.)

4) Given a set of functions $H_{ij}(x^1,x^2,x^3)$ and the fact that $H_{ij}b^i{}_l = c_{jl}$, where $b^i{}_l$ and c_{jl} are known to be tensors, show by the appropriate transformation laws that H_{ij} are components of a tensor. (This is a particular instance of the *quotient* law for tensors.)

5) Given that $ds^2 = 3(dx^1)^2 + 3(dx^2)^2 + 6(dx^3)^2 - 2dx^1\,dx^2 - 4dx^1\,dx^3$ as an instance of $a_{ij}\,dx^i\,dx^j$ (5-10), calculate (a) the determinant $a \equiv |a_{ij}|$, (b) the values of a^{ij}, (c) the product of the determinants $|a_{ij}|$ and $|a^{ij}|$.

6) Show that if a_{ij} is skew-symmetric, $a_{ij}x^i x^j$ is zero.

7) If λ^i have the values (1,2,3) and μ^j the values $(-1,2,1)$, write the matrix of the components $c^{ij} \equiv \lambda^i \mu^j$.

8) If vectors **A** and **B** have contravariant components $(2,-1,3)$ and $(1,2,1)$ in an orthogonal cartesian system, find (a) the length of **A**, (b) the components of the unit vector in the direction of **B**, (c) the angle between **A** and **B**, and (d) the scalar projection of **A** on **B**.

9) Find the components of the gradient of the function $f \equiv x^2 x^3 + x^3 x^1 + x^1 x^2$, where x^i are orthogonal cartesian coordinates. Find the rate of change of the function with respect to arc length on the curve $x^1 = a\cos\theta$, $x^2 = a\sin\theta$, $x^3 = b\theta$, at the point where $\theta = \pi/4$.

10) In spherical coordinates the contravariant components of a vector are $\lambda^i = (1,1,1)$. Find the covariant components of the same vector.

11) Calculate the area of the parallelogram determined by the vectors $\overrightarrow{OR}$ and $\overrightarrow{OQ}$ in Fig. 5.

12) In a mechanical system with three degrees of freedom the virtual work is expressed by $H_i\,dq^i$, where q^i are generalized coordinates. Under a transformation of coordinates $q^i = q^i(\bar{q}^1, \bar{q}^2, \bar{q}^3)$, the virtual work (being an invariant) becomes $\bar{H}_i\,d\bar{q}^i$. Show that the quantities H_i are covariant components of a vector.

6–4. Geometric Meaning of Contravariant and Covariant Components of a Vector.

Consider, in a general coordinate system, the three coordinate curves through any point P of space, and the tangents to these curves at P. Because x^2 and x^3 remain constant along an x^1-curve, the direction of the tangent to the x^1-curve may be assigned components $(dx^1, 0, 0)$. Similarly, $(0, dx^2, 0)$ and $(0, 0, dx^3)$ are components of vectors from P tangent to the x^2- and x^3-curves, respectively. Let λ^i be the contravariant components of the vector from P to any other point R in space. From (17) the angle θ_1 between PR and the tangent to the x^1-curve is given by

32) $$\cos\theta_1 = \frac{a_{1j}\,dx^1\lambda^j}{\sqrt{a_{lm}\lambda^l\lambda^m}\,\sqrt{a_{11}}\,dx^1}.$$

From the observation that the length of the vector $(dx^1, 0, 0)$ is $\sqrt{a_{11}}\,dx^1$, it is seen from (32) that the scalar projection of this length upon the vector $\overrightarrow{PR}$ is

33) $$\sqrt{a_{11}}\,dx^1 \cos\theta_1 = \frac{a_{1j}\,dx^1\lambda^j}{\sqrt{a_{lm}\lambda^l\lambda^m}}.$$

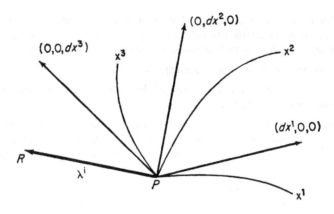

Fig. 6

Similar formulas hold with 1 replaced, in turn, by 2 and 3. Hence,

$$34) \qquad \sqrt{a_{ii}}\, dx^i \cos \theta_i = \frac{a_{ij}\, dx^i \lambda^j}{\sqrt{a_{lm} \lambda^l \lambda^m}} \qquad (i = 1, 2, 3)$$

are the three scalar projections of the tangent vectors upon the vector $\overrightarrow{PR}$. The sum of the three projections in (34) is

$$35) \qquad \sum_{i=1}^{3} \sqrt{a_{ii}}\, dx^i \cos \theta_i = \frac{a_{ij}\, dx^i \lambda^j}{\sqrt{a_{lm} \lambda^l \lambda^m}}.$$

Now each dx^i in (34) may be multiplied by a factor σ^i to make $\sigma^i\, dx^i$ (i not summed) equal to λ^i before the sum in (35) is effected. When this is done, (35) gives

$$36) \qquad \sum_{i=1}^{3} \sqrt{a_{ii}}\, \lambda^i \cos \theta_i = \frac{a_{ij} \lambda^i \lambda^j}{\sqrt{a_{ij} \lambda^i \lambda^j}} = \sqrt{a_{ij} \lambda^i \lambda^j}.$$

Hence, the geometrical interpretation, in any coordinate system, of the contravariant components λ^i of vector $\overrightarrow{PR}$ is the following:

> The vector $\overrightarrow{PR}$ with contravariant components λ^i is the diagonal of the parallelopiped built upon the three vectors from P tangent to the x^i-curves and of respective lengths $\sqrt{a_{ii}}\, \lambda^i$ (i not summed).

Notice that the parallelopiped is rectangular in the case of a triply orthogonal system of surfaces, but the projections of the diagonal upon the tangent lines to the coordinate curves are, respectively, $\lambda^i (i = 1, 2, 3)$

only in case the coordinate system is orthogonal cartesian, that is, in case $a_{ij} \equiv \delta_{ij}$.

The geometrical interpretation of covariant components will now be studied. In formula (6-9) it was seen that the square of the length of a vector with components λ^i and λ_i can be written either as $a_{ij}\lambda^i\lambda^j$ or $a^{ij}\lambda_i\lambda_j$. Also from (17) the cosine of the angle between two vectors **A** and **B** with components λ^i (or λ_i) and μ^i (or μ_i), respectively, can be expressed by either $|\mathbf{A}|\,|\mathbf{B}|\cos\theta = a_{ij}\lambda^i\mu^j$ or $|\mathbf{A}|\,|\mathbf{B}|\cos\theta = a^{ij}\lambda_i\mu_j$. Now, as in (34), let θ_i be the angle between any vector $\overrightarrow{PR}$ at P with components λ^i (or λ_i) and the tangent line to a coordinate x^i-curve at P. The dx^i can be divided out of equation (34) to give

37) $\qquad \sqrt{a_{lm}\lambda^l\lambda^m}\,\sqrt{a_{ii}}\cos\theta_i = a_{ij}\lambda^j = \lambda_i \qquad (i = 1, 2, 3),$

from which

38) $\qquad \sqrt{a_{lm}\lambda^l\lambda^m}\cos\theta_i = \dfrac{\lambda_i}{\sqrt{a_{ii}}} \qquad (i = 1, 2, 3).$

The left member of (38) is the scalar projection of the vector $\overrightarrow{PR}$ upon the tangent to the x^i-curve at P. Hence, equation (38) gives the geometric interpretation of the covariant components of any vector as follows:

> Each covariant component λ_i of a vector $\overrightarrow{PR}$ in any coordinate system is equal to $\sqrt{a_{ii}}$ times the orthogonal projection of the length of the vector upon the tangent to the x^i-curve at P.

It is important to realize that λ_i and λ^i are merely different kinds of description of the same vector, and that λ_i and λ^i are related by the fundamental metric tensor a_{ij}, namely, by $\lambda_i = a_{ij}\lambda^j$ and $\lambda^j = a^{ij}\lambda_i$. There is no fundamental metric tensor in non-Riemannian geometry, so in that study the λ_i and λ^i must be considered as separate entities. (See Section 5-1 or Section 8-7 for the definition of a Riemannian geometry.)

EXERCISES

1) In spherical coordinates the contravariant components of a vector are $\lambda^i = (1,1,1)$. The covariant components λ_i were found in Exercise 10 of Section 6-3. Sketch a figure and illustrate by it the covariant and contravariant components of the vector.

2) The covariant components of a vector in cylindrical coordinates are $\lambda_i = (2,1,3)$. Find the contravariant components.

3) Show that the angle θ_{23} between the x^2- and x^3-coordinate curves is given by $\cos \theta_{23} = a_{23}/(\sqrt{a_{22}} \sqrt{a_{33}})$, and, by cyclic permutation, deduce the other two formulas for $\cos \theta_{31}$, and $\cos \theta_{12}$.

4) Show that the coordinate surfaces $x^i = $ constant are mutually orthogonal if and only if $a_{ij} = 0$, $i \neq j$.

5) Given a vector with components p_i in orthogonal cartesian coordinates, find the covariant components of this vector in cylindrical coordinates, where $x^1 = \bar{x}^1 \cos \bar{x}^2$, $x^2 = \bar{x}^1 \sin \bar{x}^2$, $x^3 = \bar{x}^3$. Now show that the lengths of the projections of the vector upon the tangents to the $\bar{x}^1, \bar{x}^2, \bar{x}^3$- "curves" are given by

$$\bar{p}_1 = p_1 \cos \bar{x}^2 + p_2 \sin \bar{x}^2, \qquad \frac{\bar{p}_2}{\bar{x}^1} = -p_1 \sin \bar{x}^2 + p_2 \cos \bar{x}^2, \qquad \bar{p}_3 = p_3.$$

6) Solve Exercise 5 by making use of the contravariant components of the vector in the cylindrical system.

7) The gradient of a scalar $\phi(x^1, x^2, x^3)$ has been defined as the vector with components $\phi_{,i}$ in an orthogonal cartesian system. Find the covariant components of this vector in the cylindrical system and then show that the scalar projections of the vector upon the coordinate curve tangents in the cylindrical system are

$$\frac{\partial \phi}{\partial \bar{x}^1}, \qquad \frac{1}{\bar{x}^1} \frac{\partial \phi}{\partial \bar{x}^2}, \qquad \frac{\partial \phi}{\partial \bar{x}^3}.$$

8) Repeat Exercise 5, using spherical coordinates given by

$$x^1 = \bar{x}^1 \sin \bar{x}^2 \cos \bar{x}^3, \qquad x^2 = \bar{x}^1 \sin \bar{x}^2 \sin \bar{x}^3, \qquad x^3 = \bar{x}^1 \cos \bar{x}^2$$

to obtain the covariant components in the spherical coordinates for the vector with components p_i in orthogonal cartesian coordinates. Then show that the lengths of the projections of the vector upon the tangents to the $\bar{x}^1$-, $\bar{x}^2$-, and $\bar{x}^3$-"curves" are given by

$$\bar{p}_1 = p_1 \sin \bar{x}^2 \cos \bar{x}^3 + p_2 \sin \bar{x}^2 \sin \bar{x}^3 + p_3 \cos \bar{x}^2,$$

$$\frac{\bar{p}_2}{\bar{x}^1} = p_1 \cos \bar{x}^2 \cos \bar{x}^3 + p_2 \cos \bar{x}^2 \sin \bar{x}^3 - p_3 \sin \bar{x}^2,$$

$$\frac{\bar{p}_3}{\bar{x}^1 \sin \bar{x}^2} = -p_1 \sin \bar{x}^3 + p_2 \cos \bar{x}^3.$$

9) Refer to the proof of $|\cos \theta| \leq 1$ in Section 6–2 and deduce that for two unit vectors u_i and v^i in rectangular cartesian coordinates in n-space, the Cauchy-Schwarz inequality

$$(u_i v^i)(u_j v^j) \leq (\delta_{ij} u^i u^j)(\delta_{kl} v^k v^l)$$

holds.

6–5. Alternative (Reciprocal) Geometrical Interpretation of Contravariant and Covariant Components of a Vector.

Instead of the tangent lines to the x^i-coordinate curves, one may use the normal lines to the surfaces x^i = constant, and consider the projections of the vector $\overrightarrow{PR}$ upon the three normals at P. In order to obtain the formulas for the desired projections, it will be helpful to consider first the general surface represented by $f(x^1,x^2,x^3)$ = constant. The total differential $f_{,i}\,dx^i = 0$ shows that the vector $\nabla f \equiv \operatorname{grad} f$ with covariant components $f_{,i}$ is normal to the surface. The length $|\nabla f|$ of the gradient is given by

39) $$|\nabla f|^2 = a^{ij} f_{,i} f_{,j},$$

and the unit normal to the surface is $\nabla f/|\nabla f|$.

In particular, for the coordinate surface x^1 = constant, it is seen from (39) that

40) $$|\nabla x^1|^2 = a^{ij} x^1_{,i} x^1_{,j} = a^{11},$$

and that for any coordinate surface x^i = constant it is true that

41) $$|\nabla x^i|^2 = a^{ii} \qquad (i = 1, \text{ or } 2, \text{ or } 3).$$

Now if ϕ_i is the angle between $\overrightarrow{PR}$ and the normal to the surface x^i = constant, the scalar projection of vector $\overrightarrow{PR}$ upon the normal to the surface x^1 = constant is equal to

42) $$|\overrightarrow{PR}|\cos\phi_1 = \overrightarrow{PR}\cdot\frac{\nabla f}{|\nabla f|} = \frac{\lambda^i f_{,i}}{|\nabla f|} = \frac{\lambda^i x^1_{,i}}{|\nabla x^1|} = \frac{\lambda^1}{\sqrt{a^{11}}}.$$

Note that $|\nabla x^1| = \sqrt{a^{11}}$ from (40). By projecting upon the other two surface normals in a similar manner, it follows that the three scalar projections of vector $\overrightarrow{PR}$ upon the normals to the surfaces x^i = constant are given by

43) $$|\overrightarrow{PR}|\cos\phi_i = \frac{\lambda^i}{\sqrt{a^{ii}}} \qquad (i = 1, \text{ or } 2, \text{ or } 3).$$

The geometrical interpretation of the contravariant components of a vector may therefore be expressed as follows:

> Each contravariant component λ^i of a vector $\overrightarrow{PR}$ in any coordinate system is equal to $\sqrt{a^{ii}}$ times the orthogonal projection of the length of the vector upon the normal to the surface x^i = constant through the point P. (Compare this with the preceding interpretation of covariant components λ_i.)

Next, write equation (43) in the form

44) $$\sqrt{a^{ii}}\cos\phi_i = \frac{\lambda^i}{|\overrightarrow{PR}|},$$

multiply both sides by λ_i and sum on i to obtain

45) $$\sum_{i=1}^{3} \lambda_i \sqrt{a^{ii}} \cos \phi_i = \frac{\lambda^i \lambda_i}{|\overrightarrow{PR}|} = \frac{|\overrightarrow{PR}|^2}{|\overrightarrow{PR}|} = |\overrightarrow{PR}|.$$

Equation (45) has the following interpretation:

> The vector $\overrightarrow{PR}$ with covariant components λ_i is the diagonal of the parallelopiped built upon the three vectors from P normal to the x^i = constant surfaces and of respective lengths $\sqrt{a^{ii}}\,\lambda_i$ (i not summed). (Compare this with the analogous statement relative to contravariant components λ^i.)

A set of three vectors along the tangents to the respective x^i-curves and a set of three vectors normal to the respective surfaces x^i = constant are called *reciprocal* sets of vectors. Any vector at P may be referred to one or the other of these two sets of base vectors.

EXERCISES

1) Show that at any point on the curve of intersection of the surfaces $f(x^1,x^2,x^3) = 0$, $g(x^1,x^2,x^3) = 0$, they are inclined at an angle θ given by

$$\cos \theta = \frac{\nabla f \cdot \nabla g}{|\nabla f|\,|\nabla g|}.$$

Deduce that $\nabla f \cdot \nabla g \equiv 0$ is a necessary and sufficient condition for the surfaces to be orthogonal.

2) Show that the surfaces x^i = constant and x^j = constant are orthogonal if and only if $a^{ij} = 0$.

3) Given the transformation $\bar{x}^1 = x^1 - 3x^2 + 2x^3$, $\bar{x}^2 = 2x^1 - x^2 + x^3$, $\bar{x}^3 = 3x^1 - 4x^2 + 5x^3$ from orthogonal cartesian coordinates x^i to oblique cartesian coordinates $\bar{x}^i$, and the vector **V** with components $(1,1,1)$ in x^i coordinates, (a) find the covariant and contravariant components of **V** in the $\bar{x}^i$ system, (b) verify that $\bar{\lambda}_p = \bar{a}_{pq}\bar{\lambda}^q$, and (c) find the scalar projection of **V** onto the $\bar{x}^1$-"curve." *Ans.* (c): $-13\sqrt{3}/15$.

4) Consider the extension of Fig. 4 to three-space referred to a system of oblique cartesian coordinates. Denote the angles x^2Ox^3, x^3Ox^1, x^1Ox^2 by λ, μ, ν, respectively. Show that

$$(a_{ij}) \equiv \begin{pmatrix} 1 & \cos \nu & \cos \mu \\ \cos \nu & 1 & \cos \lambda \\ \cos \mu & \cos \lambda & 1 \end{pmatrix}.$$

Draw lines through the end point R of vector $\overrightarrow{OR}$ parallel to the coordinate axes to find the contravariant components λ^i of $\overrightarrow{OR}$. Drop perpendiculars to the axes to find the covariant components λ_i of $\overrightarrow{OR}$. If A_1 is the foot of the perpendicular from R to the x^1-axis, then by projection, the segment

$$OA_1 = \lambda^1 + \lambda^2 \cos \nu + \lambda^3 \cos \mu.$$

Verify that $\lambda_i = a_{ij}\lambda^j$, for $i = 1, 2, 3$.

7

Vector Algebra

A vector has been defined as a tensor of first order, and the behavior of vector components under a change of coordinates has been investigated. Some of the essential facts concerning the algebra of vectors in three-space will be set forth in this chapter. Orthogonal cartesian coordinates will be employed throughout.

7-1. Base Vectors. The components of a vector have been emphasized heretofore. It is useful to think of the vector as a line segment with three attributes—length, direction, and sense of direction. If the components of a vector are λ^i, the vector may be realized as the segment *from* the origin *to* the point with coordinates λ^i. If the vector is moved parallel to itself to any other position, the vector in the new position has the same components λ^i as before. That is, the scalar projections λ^i of the vector on the x^i-coordinate axes are unchanged. Two vectors will be considered equal if they have the same components, i.e., vectors **A** and **B** having components a^i and b^i are equal if and only if $a^i = b^i$ for all i. Notice that the scalar projections a^i and b^i are directed segments. The vector $-\mathbf{A}$ is defined as the vector with components $-a^i$. A scalar multiple of **A**, written $s\mathbf{A}$, has components sa^i. Multiplying a vector by a scalar stretches or contracts the vector according as the scalar is greater than or less than 1. The sum $\mathbf{A} + \mathbf{B}$ of two vectors has components $a^i + b^i$. The sum $\mathbf{A} + (-\mathbf{A})$ has components $a^i - a^i = 0$. Thus, the vector **0** has all components 0. One should distinguish between the zero vector **0** and the zero scalar 0.

If vectors **A** and **B** are in the x^1x^2-plane each of **A** and **B** has two components and i takes the range 1, 2. In this case it is evident from the geometry of Fig. 7 that the vector sum $\mathbf{A} + \mathbf{B}$ with components $a^i + b^i$ is a vector from the origin to the point $(a^i + b^i)$, and that it is therefore the diagonal of the parallelogram built upon **A** and **B**. This extends readily to three-space. Given three non-coplanar vectors **A**, **B**, **C** with components

a^i, b^i, c^i, the sum vector **A** + **B** + **C** is represented by the diagonal vector (from the origin) of the parallelopiped built upon the vectors **A**, **B**, **C**.

In the applications of vectors to physical problems, three classifications of vectors arise—free vectors, sliding vectors, and fixed vectors. If a vector **V** is determined only by its components v^i, then the vector is free

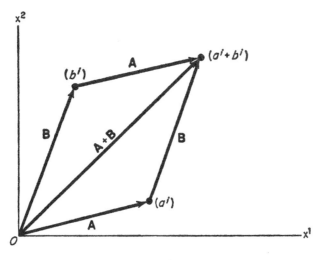

Fig. 7

in the sense that any vector in space with the same length, direction, and sense of direction has the same components. A vector may represent a force which must act along a single specified line. The point of application of the force along the line may be immaterial. If so, the vector representing the force may slide along the line. If the point of application of the force is required to be fixed, the vector representing it must be fixed. Consider all vectors in this chapter as free vectors.

It is convenient to introduce a set of unit base vectors along the x^i-axes. These are customarily denoted by **i**, **j**, **k**, as indicated in Fig. 8; but in order to use the tensor notation, these unit vectors may be denoted by $\mathbf{e}_h$ or by $\mathbf{e}^h$. It will be recalled that in orthogonal cartesian coordinates the contravariant and covariant components coincide. Note that $\mathbf{i} = \mathbf{e}_1 = \mathbf{e}^1$, $\mathbf{j} = \mathbf{e}_2 = \mathbf{e}^2$, $\mathbf{k} = \mathbf{e}_3 = \mathbf{e}^3$, and that the length of each of these vectors is unity. The vector **V** with components v^i may be expressed as $\mathbf{V} = v^1\mathbf{e}_1 + v^2\mathbf{e}_2 + v^3\mathbf{e}_3 = v^i\mathbf{e}_i$, or equally well by $\mathbf{V} = v_1\mathbf{e}^1 + v_2\mathbf{e}^2 + v_3\mathbf{e}^3 = v_i\mathbf{e}^i$, if the covariant notation v_i is desired. It is important to keep in mind that in the vector sum $v_i\mathbf{e}^i$ the v_i are scalars and the $\mathbf{e}^i$ are vectors. Note that the sum of $\mathbf{V}(= v_i\mathbf{e}^i)$ and $W(= w_i\mathbf{e}^i)$ is $\mathbf{V} + \mathbf{W} = v_i\mathbf{e}^i + w_i\mathbf{e}^i = (v_i + w_i)\mathbf{e}^i$.

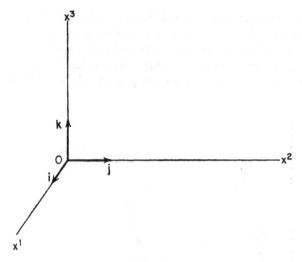

Fig. 8

7-2. Products of Vectors. The operations of taking the sum and difference of two vectors, and of multiplying a vector by a scalar, have been defined. Multiplication of vectors is now in order. Several kinds of products of vectors will be defined.

In Section 5-5 it was seen that if the first-order tensors with components λ_i and μ^j are multiplied, a second-order tensor, called a dyadic, with components $\lambda_i \mu^j$ results. The dyadic has nine components in three-space. Further analysis of tensors of the second and higher orders will appear later, so nothing further will be done at this point concerning the algebra of dyadics.

DOT PRODUCT. The *dot* product or scalar product of two first-order tensors with components λ_i and μ^j has been defined as the contraction of the dyadic $\lambda_i \mu^j$, i.e., the sum $\lambda_i \mu^i$, which is a scalar invariant. The notation $\mathbf{V} \cdot \mathbf{W} \equiv v_i w^i$, for instance, will be employed. It was shown in Section 6-2 that the cosine of the angle θ between $\mathbf{V}$ and $\mathbf{W}$ is given by

1) $$\cos \theta = \frac{a_{ij} v^i w^j}{\sqrt{a_{pq} v^p v^q} \sqrt{a_{rs} w^r w^s}},$$

which may be written as

2) $$\cos \theta = \frac{\delta_{ij} v^i w^j}{|\mathbf{V}| |\mathbf{W}|} = \frac{v_j w^j}{|\mathbf{V}| |\mathbf{W}|}.$$

(Remember that $a_{ij} \equiv \delta_{ij}$ in orthogonal cartesian coordinates.) It follows from $\mathbf{V} \cdot \mathbf{W} = v_i w^i$ and (2) that

3) $$\mathbf{V} \cdot \mathbf{W} = |\mathbf{V}| |\mathbf{W}| \cos \theta.$$

Recall that if **W** is a unit vector, **V**·**W** gives the scalar projection of **V** in the direction of **W**. Observe that the dot product of two vectors is a scalar.

Cross Product. The *cross* product (also called *vector* product and *outer* product) of **V** and **W** is denoted by **V** × **W** and is defined as follows: The vectors **V** and **W** with components v^i and w^i give rise to a third vector **H** with components $h_i = e_{ijk}v^j w^k$, which are the three second-order determinants formed from the matrix

4) $$\begin{pmatrix} v^1 & v^2 & v^3 \\ w^1 & w^2 & w^3 \end{pmatrix},$$

i.e.,

5) $$e_{1jk}v^j w^k = v^2 w^3 - v^3 w^2, \qquad e_{2jk}v^j w^k = v^3 w^1 - v^1 w^3,$$
$$e_{3jk}v^j w^k = v^1 w^2 - v^2 w^1.$$

From (3) it is evident that two vectors **V** and **W** are perpendicular if and only if **V**·**W** = 0. Consider the dot product of **V** and **H**, i.e.,

6) $$\mathbf{V} \cdot \mathbf{H} = \mathbf{V} \cdot (\mathbf{V} \times \mathbf{W}) = v^i e_{ijk} v^j w^k = e_{ijk} v^i v^j w^k.$$

The determinant in the last member of (6) is zero (two rows identical), and therefore vector **H** is perpendicular to **V**. Similarly, **W**·(**V** × **W**) = $w^i e_{ijk} v^j w^k = e_{ijk} w^i v^j w^k = 0$, so vector **H** is perpendicular to **W**. The direction of **H** is taken so that **V**, **W**, **H** form a right-handed triple. This means that if the fingers of the right-hand point in the direction of rotation from **V** to **W**, the thumb will point in the direction of **H** ≡ (**V** × **W**). It remains to determine the length of **H**. This is given by

7) $$|\mathbf{H}|^2 = |\mathbf{V} \times \mathbf{W}|^2 = \delta^{ij} h_i h_j = \delta^{ij} e_{ipq} v^p w^q e_{jrs} v^r w^s = \sum_{i=1}^{3} (e_{ilm} v^l w^m)^2$$
$$= \begin{vmatrix} v^2 & v^3 \\ w^2 & w^3 \end{vmatrix}^2 + \begin{vmatrix} v^3 & v^1 \\ w^3 & w^1 \end{vmatrix}^2 + \begin{vmatrix} v^1 & v^2 \\ w^1 & w^2 \end{vmatrix}^2,$$

which, by the Lagrange identity (2–27′), is equal to

$$(v^i v_i)(w^j w_j) - (v^i w_i)^2 = |\mathbf{V}|^2 |\mathbf{W}|^2 - |\mathbf{V}|^2 |\mathbf{W}|^2 \cos^2 \theta = |\mathbf{V}|^2 |\mathbf{W}|^2 \sin^2 \theta,$$

where θ is the angle of rotation from **V** to **W**. It follows that the length of **H** is given by

8) $$|\mathbf{H}| \equiv |\mathbf{V} \times \mathbf{W}| = |\mathbf{V}| \, |\mathbf{W}| \sin \theta.$$

Thus, the number representing the length of **V** × **W** is the number representing the area of the parallelogram built upon the vectors **V** and **W**. One concludes that the cross product of two vectors is a vector with three properties. It is perpendicular to the plane of the two given vectors, it is equal in length to the magnitude of the area of the parallelogram determined by the two vectors, and it is directed in a right-handed manner relative to the given vectors. One should observe that $\mathbf{H} \equiv \mathbf{V} \times \mathbf{W}$ can be written as $\mathbf{H} = h_i \mathbf{e}^i = (e_{ijk} v^j w^k) \mathbf{e}^i$ which can be expressed in the form

$$8') \qquad \mathbf{H} \equiv \mathbf{V} \times \mathbf{W} = \begin{vmatrix} \mathbf{e}^1 & \mathbf{e}^2 & \mathbf{e}^3 \\ v^1 & v^2 & v^3 \\ w^1 & w^2 & w^3 \end{vmatrix}.$$

It is important to have a closer look at the vector $\mathbf{H}$, the components of which may be expressed by either $h_i = e_{ijk} v^j w^k$ or $h^i = e^{ijk} v_j w_k$. It will be shown that if $\mathbf{V}$ and $\mathbf{W}$ are absolute vectors [see the remarks following (5–36)], then the vector $\mathbf{H}$ with components h_i transforms as a relative covariant vector of weight -1, while the vector $\mathbf{H}$ with components h^i transforms as a relative contravariant vector of weight 1. In order to show this, it is necessary to determine the transformation law for the tensor components e_{ijk} and e^{ijk}. Recall from (2–30) and Exercise 9 of Section 2–3 that (with a slight change of notation)

$$e^{ijk} a_i{}^1 a_j{}^2 a_k{}^3 = |a_m{}^l| \equiv a$$

and that the same determinant $|a_m{}^l|$ can be expanded by

$$|a_m{}^l| = e_{ijk} a_1{}^i a_2{}^j a_3{}^k,$$

where e_{ijk} take on the same values as e^{ijk}. If the subscripts 1, 2, 3 are replaced by p, q, r, it is readily verified that

$$e_{ijk} a_p{}^i a_q{}^j a_r{}^k = a e_{pqr}.$$

If now the elements $a_j{}^i$ in the determinant are replaced by $\partial x^i / \partial \bar{x}^j$, the last equality reads as follows:

$$e_{ijk} \frac{\partial x^i}{\partial \bar{x}^p} \frac{\partial x^j}{\partial \bar{x}^q} \frac{\partial x^k}{\partial \bar{x}^r} = \left| \frac{\partial x^l}{\partial \bar{x}^m} \right| e_{pqr},$$

which may be written as

$$\bar{e}_{pqr} = \left| \frac{\partial x^l}{\partial \bar{x}^m} \right|^{-1} e_{ijk} \frac{\partial x^i}{\partial \bar{x}^p} \frac{\partial x^j}{\partial \bar{x}^q} \frac{\partial x^k}{\partial \bar{x}^r},$$

and this shows that e_{ijk} transform as a covariant tensor of weight -1. Since $\mathbf{V}$ and $\mathbf{W}$ are absolute vectors, the h_i therefore transform as a relative covariant vector of weight -1. The proof concerning the manner in which the h^i transform is effected in a similar fashion.

As a final statement on the cross product, it is emphasized that whereas the inner product $v_i w^i$ of two vectors is meaningful if i is summed from 1 to n, the cross product of two vectors is a vector only in case $n = 3$. Multiplying $v^j w^k$ by e_{ijk} and summing on j and k produces three components of a vector if the range of the indices is 1, 2, 3. If the indices range over $1, \cdots, n (n > 3)$, the quantities $c^{jk} \equiv v^j w^k - v^k w^j$ are seen to be components of a skew-symmetric tensor of the second order.

BOX PRODUCT. A geometrical interpretation can be given for the scalar product $\mathbf{A} \cdot (\mathbf{B} \times \mathbf{C})$ of three vectors. Let the three vectors $\mathbf{A}, \mathbf{B}, \mathbf{C}$ be drawn from the origin O as a common point. Take first the case in which $\mathbf{A}$ lies in the plane of $\mathbf{B}$ and $\mathbf{C}$. The consequence is that $\mathbf{A} \cdot \mathbf{B} \times \mathbf{C}$ is zero, because the vector $\mathbf{B} \times \mathbf{C}$ is perpendicular to the plane of $\mathbf{B}$ and $\mathbf{C}$, and therefore to $\mathbf{A}$. If $\mathbf{A}, \mathbf{B}, \mathbf{C}$ do not have a common origin, the condition $\mathbf{A} \cdot \mathbf{B} \times \mathbf{C} = 0$ means that the three vectors are all parallel to a plane.

Next, if $\mathbf{A}, \mathbf{B}, \mathbf{C}$ are not coplanar, the expression

$$9) \qquad \mathbf{A} \cdot \frac{\mathbf{B} \times \mathbf{C}}{|\mathbf{B} \times \mathbf{C}|}$$

gives the scalar projection of $\mathbf{A}$ onto the direction of $\mathbf{B} \times \mathbf{C}$, and this projection is the altitude h of the parallelopiped of which the vectors $\mathbf{A}, \mathbf{B}, \mathbf{C}$ are coterminous edges at the origin O (Fig. 9). Because the area

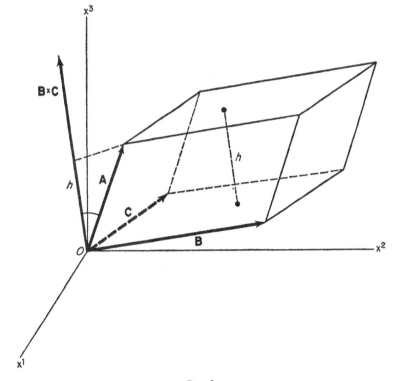

Fig. 9

of the base parallelogram on **B** and **C** is $|\mathbf{B} \times \mathbf{C}|$, it is evident from (9) that the volume of the parallelopiped on **A, B, C** is given by $\mathbf{A} \cdot (\mathbf{B} \times \mathbf{C})$. If **A, B, C** are coplanar, the volume is zero. If the components of **A, B, C** are a^i, b^i, c^i, the scalar value of $\mathbf{A} \cdot \mathbf{B} \times \mathbf{C}$ is $a^i e_{ijk} b^j c^k$ or $e_{ijk} a^i b^j c^k$. The product $\mathbf{A} \cdot \mathbf{B} \times \mathbf{C}$ is called the *triple scalar* product and also the *box* product.

The product $(\mathbf{A} \cdot \mathbf{B})\mathbf{C}$ is clearly a scalar $\mathbf{A} \cdot \mathbf{B}$ times vector **C**. The components of the resulting vector may be written as $(a_i b^i) c^k$.

TRIPLE VECTOR PRODUCT. A final type of (triple) product remains, namely, $\mathbf{A} \times (\mathbf{B} \times \mathbf{C})$, which is called the *triple vector* product, or triple cross product. If d_i denote the components of $\mathbf{D} \equiv (\mathbf{B} \times \mathbf{C})$, then $d_i = e_{ijk} b^j c^k$. The components of $\mathbf{A} \times \mathbf{D}$ are $e^{ilm} a_l d_m$, so that $\mathbf{A} \times (\mathbf{B} \times \mathbf{C})$ has components

10) $$e^{ilm} a_l e_{mpq} b^p c^q.$$

In order to facilitate operations with the e_{ijk} and e^{ijk} symbols, a generalized Kronecker delta is introduced. Define

11) $$\begin{aligned} \delta_{kl}^{ij} &= 1 \quad \text{if} \quad i = k, \quad j = l \quad (i \neq j), \\ \delta_{kl}^{ij} &= -1 \quad \text{if} \quad i = l, \quad j = k \quad (i \neq j), \\ \delta_{kl}^{ij} &= 0 \quad \text{for any other set of indices.} \end{aligned}$$

For instance, $\delta_{12}^{12} = 1$, $\delta_{23}^{23} = 1$, $\delta_{32}^{23} = -1$, $\delta_{31}^{13} = -1$, $\delta_{31}^{12} = 0$, $\delta_{12}^{23} = 0$. By evaluating e_{ijk}, e^{ijk}, and δ_{kl}^{ij} for various combinations of indices, it can be shown that

12) $$\begin{aligned} \delta_{kj}^{ij} &= -\delta_{jk}^{ij} = 2\delta_k^i, \\ \delta_{kl}^{ij} &= \delta_k^i \delta_l^j - \delta_l^i \delta_k^j, \\ e^{ijk} e_{klm} &= e^{ijk} e_{lmk} = \delta_{lm}^{ij}. \end{aligned}$$

By use, first of the third line in (12) and then the second line, the expression in (10) may be reduced as follows.

13) $$\begin{aligned} e^{ilm} e_{mpq} a_l b^p c^q &= \delta_{pq}^{il} a_l b^p c^q = (\delta_p^i \delta_q^l - \delta_q^i \delta_p^l) a_l b^p c^q \\ &= \delta_p^i a_q b^p c^q - \delta_p^l a_l b^p c^i = (a_q c^q) b^i - (a_p b^p) c^i. \end{aligned}$$

Hence, the triple cross product $\mathbf{A} \times (\mathbf{B} \times \mathbf{C})$ takes the form

14) $$\mathbf{A} \times (\mathbf{B} \times \mathbf{C}) = (\mathbf{A} \cdot \mathbf{C})\mathbf{B} - (\mathbf{A} \cdot \mathbf{B})\mathbf{C}.$$

Note that this triple product is a vector.

VECTOR IDENTITIES. Various vector identities can be proved by use of the product formulas in this section. For instance, consider $(\mathbf{A} \times \mathbf{B}) \times (\mathbf{C} \times \mathbf{D})$. Replace $(\mathbf{A} \times \mathbf{B})$ by a single vector **P** and then apply (14) to obtain

$$\mathbf{P} \times (\mathbf{C} \times \mathbf{D}) = (\mathbf{P} \cdot \mathbf{D})\mathbf{C} - (\mathbf{P} \cdot \mathbf{C})\mathbf{D}.$$

Replacement of **P** by **A** × **B** yields

15) $\quad (\mathbf{A} \times \mathbf{B}) \times (\mathbf{C} \times \mathbf{D}) = (\mathbf{A} \times \mathbf{B} \cdot \mathbf{D})\mathbf{C} - (\mathbf{A} \times \mathbf{B} \cdot \mathbf{C})\mathbf{D}.$

Observe that in (15) the scalar $(\mathbf{A} \times \mathbf{B}) \cdot \mathbf{D}$ may be written without parentheses because $\mathbf{A} \times (\mathbf{B} \cdot \mathbf{D})$ has no meaning. Notice also that

$$\mathbf{A} \times \mathbf{B} \cdot \mathbf{D} = \mathbf{B} \cdot \mathbf{D} \times \mathbf{A} = \mathbf{D} \times \mathbf{A} \cdot \mathbf{B} = \mathbf{A} \cdot \mathbf{B} \times \mathbf{D} = \mathbf{B} \times \mathbf{D} \cdot \mathbf{A} = \mathbf{D} \cdot \mathbf{A} \times \mathbf{B}$$

because all of these scalars represent the volume of the same parallelopiped. Of course, a change of orientation will change the sign. Interchanging **B** and **D**, say, in $\mathbf{A} \times \mathbf{B} \cdot \mathbf{D}$ means that two rows in the determinant $e_{ijk}a^i b^j d^k$ are interchanged. Finally, observe that $\mathbf{A} \times (\mathbf{B} \times \mathbf{C}) = -(\mathbf{B} \times \mathbf{C}) \times \mathbf{A}$.

The dot and cross products of pairs of unit vectors along the axes are easily calculated from the definitions to be

16) $\quad \begin{array}{ll} \mathbf{j} \cdot \mathbf{k} = \mathbf{k} \cdot \mathbf{i} = \mathbf{i} \cdot \mathbf{j} = 0, & \mathbf{i} \cdot \mathbf{i} = \mathbf{j} \cdot \mathbf{j} = \mathbf{k} \cdot \mathbf{k} = 1, \\ \mathbf{j} \times \mathbf{k} = \mathbf{i}, \mathbf{k} \times \mathbf{i} = \mathbf{j}, \mathbf{i} \times \mathbf{j} = \mathbf{k}, & \mathbf{i} \times \mathbf{i} = \mathbf{j} \times \mathbf{j} = \mathbf{k} \times \mathbf{k} = 0. \end{array}$

By use of $\mathbf{e}_i$ and $\mathbf{e}^i$, the results in (16) may take the respective forms

17) $\quad\quad\quad\quad\quad\quad\quad \mathbf{e}_i \cdot \mathbf{e}^j = \delta_i{}^j,$

$\mathbf{e}_i \times \mathbf{e}^j = \mathbf{e}^k$ for any even permutation of 1, 2, 3; $\mathbf{e}_i \times \mathbf{e}^j = -\mathbf{e}^k$ for an odd permutation of 1, 2, 3; and $\mathbf{e}_i \times \mathbf{e}^j = 0$ if $i = j$. Hence, $\mathbf{e}_i \times \mathbf{e}^j = e_{ijk}\mathbf{e}^k$.

If **V** and **W** are expressed by $v^i \mathbf{e}_i$ and $w_j \mathbf{e}^j$, then $\mathbf{V} \cdot \mathbf{W} = (v^i \mathbf{e}_i) \cdot (w_j \mathbf{e}^j) = v^i w_j (\mathbf{e}_i \cdot \mathbf{e}^j) = v^i w_j \delta_i{}^j = v^i w_i$.

Notice also that

$$\mathbf{V} \times \mathbf{W} = v^i \mathbf{e}_i \times w_j \mathbf{e}^j = v^i w_j (\mathbf{e}_i \times \mathbf{e}^j) = v^i w_j e_{ijk} \mathbf{e}^k$$
$$= (v^2 w_3 - v^3 w_2)\mathbf{e}^1 + (v^3 w_1 - v^1 w_3)\mathbf{e}^2 + (v^1 w_2 - v^2 w_1)\mathbf{e}^3.$$

7-3. Linear Dependence. If vectors $\mathbf{V}_1$ and $\mathbf{V}_2$ are parallel, there exists a number m such that $\mathbf{V}_1 = m\mathbf{V}_2$, or $\mathbf{V}_1 - m\mathbf{V}_2 = 0$, or $c_1 \mathbf{V}_1 + c_2 \mathbf{V}_2 = 0$, where $c_2/c_1 = -m$. Conversely, if $c_1 \mathbf{V}_1 + c_2 \mathbf{V}_2 = 0$, then $\mathbf{V}_1 = -(c_2/c_1)\mathbf{V}_2$, so $\mathbf{V}_1$ and $\mathbf{V}_2$ are parallel. Vectors $\mathbf{V}_1$ and $\mathbf{V}_2$ are said to be *linearly dependent* if there exist scalars c_1 and c_2 such that $c_1 \mathbf{V}_1 + c_2 \mathbf{V}_2 = 0$, where c_1 and c_2 are not both zero.

Consider three vectors $\mathbf{V}_i (i = 1, 2, 3)$ with common origin O, and with $\mathbf{V}_3$ lying in the plane of $\mathbf{V}_1$ and $\mathbf{V}_2$. Through the end point P of $\mathbf{V}_3$ draw lines parallel to the other two vectors to form a parallelogram. In Fig. 10, $\overrightarrow{OH} + \overrightarrow{HP} = \mathbf{V}_3$, but there are scalars a_1 and a_2 such that $\overrightarrow{OH} = a_1 \mathbf{V}_1$ and $\overrightarrow{OK} = \overrightarrow{HP} = a_2 \mathbf{V}_2$. Hence, $\mathbf{V}_3 = a_1 \mathbf{V}_1 + a_2 \mathbf{V}_2$ so $\mathbf{V}_3$ is a linear combination of $\mathbf{V}_1$ and $\mathbf{V}_2$. Multiplication by a suitable constant allows the last equation to be written as $c_i \mathbf{V}_i = 0$, where the c_i are not all zero. Clearly, any three vectors in a plane are linearly dependent, i.e., one of the three can always be expressed as a linear combination of the other two. If the three vectors do not lie in a plane and are not parallel to the same plane, they are linearly independent.

Take next the case of four vectors $V_i (i = 1, \cdots, 4)$ in space. Let them have a common origin O, and let V_1, V_2, V_3 be linearly independent. By drawing lines through the end point P of V_4 (Fig. 11) parallel to the

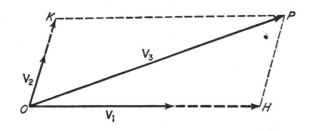

Fig. 10

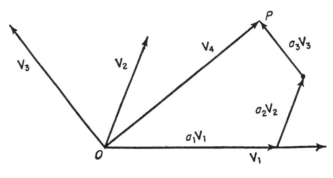

Fig. 11

other three vectors, it is seen that scalars exist such that $V_4 = a_1 V_1 + a_2 V_2 + a_3 V_3$, which may be written in the form $c_i V_i = 0 (i = 1, \cdots, 4)$. This means that four vectors in three-space are linearly dependent. In general, n vectors are linearly dependent if there exist scalars $c_1, \cdots, c_n$ not all zero such that $c_i V_i = 0 (i = 1, \cdots, n)$. It can be shown that $n + 1$ vectors in n-space are linearly dependent.

7–4. Vector Equation of a Line. Consider first the case in which the line l contains the origin and the fixed point C with coordinates c^i. The vector $\mathbf{X}$ from O to a variable point x^i on l, and the vector $\overrightarrow{OC}$ are linearly dependent. Hence, there exists a scalar σ such that $\mathbf{X} = \sigma \overrightarrow{OC}$. This is the vector equation of the line l where σ is the parameter. In coordinate form the equations of l are $x^i = \sigma c^i$. A unit vector on $\vec{OC}$ may be used. It has components $c^i/\sqrt{c_i c^i}$. The equations of l are now $x^i = tc^i/\sqrt{c_i c^i}$, where t is the distance from O to $P(x^i)$.

Consider next the case in which l does not contain the origin, but passes through a fixed point $C(c^i)$ in the direction of a unit vector $\mathbf{U}$ with components u^i. Let $\mathbf{X}$ denote the variable vector from O to $P(x^i)$ on l

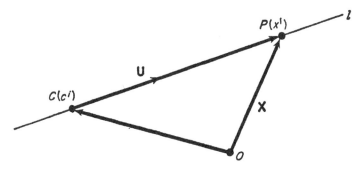

Fig. 12

(Fig. 12). Then $\vec{CP} = t\mathbf{U}$, where t is the length of $\vec{CP}$, and the vector equation of line l is

18) $$\mathbf{X} = \vec{OC} + t\mathbf{U}.$$

As the parameter t varies, point $P(x^i)$ generates the line l. In component form (18) becomes

19) $$x^i = c^i + tu^i.$$

It should be noticed that the components of vector $\vec{CP}$ are $x^i - c^i$, and that

20) $$u^i = \frac{x^i - c^i}{|\vec{CP}|} = \frac{x^i - c^i}{\sqrt{\sum_{i=1}^{3}(x^i - c^i)^2}}.$$

If line l is determined by the points $A(a^i)$ and $B(b^i)$, the variable vector $\mathbf{X}$ from O to P on l may be expressed by

21) $$\mathbf{X} = \vec{OA} + \vec{AP}.$$

Notice in Fig. 13 that $\vec{AB} = -\vec{OA} + \vec{OB} = \vec{OB} - \vec{OA}$ and that $\vec{AP}$ is a scalar times $\vec{AB}$. Hence,

22) $$\mathbf{X} = \vec{OA} + \sigma(\vec{OB} - \vec{OA}) = (1 - \sigma)\vec{OA} + (\sigma)\vec{OB}.$$

Observe that the sum of the scalar multiples of $\vec{OA}$ and $\vec{OB}$ is 1, so that $\mathbf{X} = \alpha \vec{OA} + \beta \vec{OB}$, where $\alpha + \beta = 1$.

If σ in (22) is such that $0 < \sigma < 1$, point P is on the segment AB. If $\sigma > 1$, P is on the extension of AB, while if $\sigma < 0$, P is on the extension of BA.

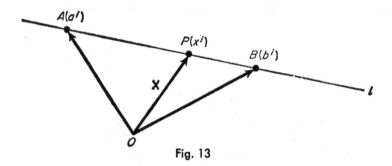

Fig. 13

It is not difficult to show that the equation of the plane through the end points of three non-coplanar vectors $\vec{OA}$, $\vec{OB}$, $\vec{OC}$ is given by

23) $\qquad \mathbf{X} = \alpha \vec{OA} + \beta \vec{OB} + \gamma \vec{OC}, \qquad \alpha + \beta + \gamma = 1.$

By using an obvious change in notation, the vector equation

24) $\qquad \mathbf{X} = \sum_{i=1}^{n} \alpha_i \vec{OA}_i, \qquad \sum_{i=1}^{n} \alpha_i = 1,$

is seen to represent a hyperplane in an n-dimensional linear space.

7–5. Applications in Mechanics. The moment of a force F with respect to a point P has magnitude equal to the magnitude of the force times the perpendicular distance from P to the line of action of the force. Let the vector $\mathbf{AB}$ represent the force (Fig. 14). Draw the vector $\vec{PA}$. The magnitude of the moment of F about P is $|\vec{PA}||\vec{AB}| \sin \theta$. But this is the magnitude of the cross product $\mathbf{H} = \vec{PA} \times \vec{AB}$. The force F tends to produce rotation about a line through P perpendicular to the plane of P and F. The cross product vector determines the direction of this axis of rotation. Clearly then the cross product $\vec{PA} \times \vec{AB}$ is useful in describing the moment of a force about a point.

Another example of the use of the cross product arises in the rotation of a rigid body about an axis l, the angular velocity ω being constant. A vector $\mathbf{\Omega}$, of magnitude ω, along the axis l describes the angular velocity.

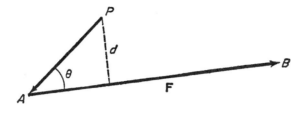

Fig. 14

If the fingers of the right hand point in the direction of rotation of the body, the thumb points in the direction of Ω on l (Fig. 15). Take a point O on l as origin for Ω. Let $\mathbf{R}$ be the position vector of any point P in the body, and θ the angle between $\overrightarrow{OP}$ and the axis l. Because the radius of

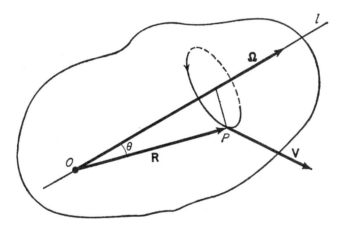

Fig. 15

the circle in which P moves is $|\mathbf{R}| \sin \theta$, the speed of P in its path is $|\mathbf{V}| = |\Omega|\,|\mathbf{R}|\,|\sin \theta| = |\Omega \times \mathbf{R}|$, where $\mathbf{V}$ is the velocity vector of P. Notice that the vector $\Omega \times \mathbf{R}$ gives the direction of motion of P as well as the magnitude of the velocity. With O as the origin of coordinates, let P have coordinates x^i and let Ω have components ω^i. Then

$$\mathbf{V} = \Omega \times \mathbf{R} = \begin{vmatrix} \mathbf{e}^1 & \mathbf{e}^2 & \mathbf{e}^3 \\ \omega^1 & \omega^2 & \omega^3 \\ x^1 & x^2 & x^3 \end{vmatrix} = e_{ijk}\omega^j x^k \mathbf{e}^i.$$

The velocity components of P are then $v_i = e_{ijk}\omega^j x^k$.

7–6. Vector Methods in Geometry.
The purpose of this section is to give a brief indication of the usefulness of vectors in solving problems of elementary geometry.

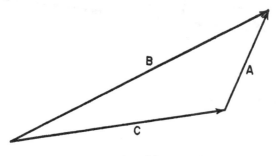

Fig. 16

One necessary fact is that if three vectors form a triangle, as, for instance **A, B, C** in Fig. 16, then the sum of the vectors is zero. In Fig. 16, **C + A = B** or **C + A + (−B) = 0**. The concept of linear dependence and the vector equations of lines and planes are also essential tools. Two examples are provided to illustrate the use of vector methods.

Example 1. A line is drawn from vertex A of parallelogram $ABCD$ to the midpoint M of side BC. Prove that AM and the diagonal BD trisect each other.

First Solution: Let I be the intersection of AM and BD (Fig. 17). Assign the vectors $\vec{AB} = \mathbf{a}$, $\vec{AD} = \mathbf{b}$, with the consequence that $\vec{BD} = \mathbf{b} - \mathbf{a}$, and $\vec{AM} = \mathbf{a} + \tfrac{1}{2}\mathbf{b}$. Let $\vec{AI} = \sigma \vec{AM}$. Hence, $\vec{AI} = \sigma(\mathbf{a} + \tfrac{1}{2}\mathbf{b})$. Let $\vec{BI} = \tau \vec{BD}$, so that $\vec{BI} = \tau(\mathbf{b} - \mathbf{a})$. Now, by the fact that the sum of the vectors around triangle ABI is zero, it follows that $\vec{AB} + \vec{BI} = \vec{AI}$, or

$$\mathbf{a} + \tau(\mathbf{b} - \mathbf{a}) = \sigma(\mathbf{a} + \tfrac{1}{2}\mathbf{b})$$

from which

$$(1 - \sigma - \tau)\mathbf{a} + \left(\tau - \frac{\sigma}{2}\right)\mathbf{b} = \mathbf{0}.$$

This condition can be satisfied for independent vectors **a** and **b** only if

$$1 - \sigma - \tau = 0, \qquad \tau - \frac{\sigma}{2} = 0$$

from which

$$\sigma = \tfrac{2}{3}, \qquad \tau = \tfrac{1}{3}.$$

But σ is the ratio of AI to AM, and τ is the ratio of BI to BD, so the proof is complete.

Second Solution: By use of (22) the vector equation of BD (with A as origin) is $\mathbf{X} = (1 - t)\mathbf{a} + t\mathbf{b}$, and the equation of AM (with B as origin) is $\mathbf{Y} =$

$(1 - r)(-\mathbf{a}) + r(\tfrac{1}{2}\mathbf{b})$. If P and Q, the end points of $\mathbf{X}$ and $\mathbf{Y}$, coincide, then $\mathbf{X} = \mathbf{a} + \mathbf{Y}$, which yields

$$(1 - r - t)\mathbf{a} + \left(t - \frac{r}{2}\right)\mathbf{b} = 0.$$

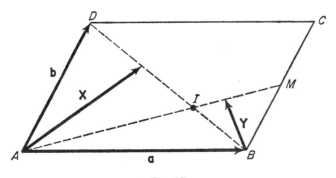

Fig. 17

Therefore, $r = \tfrac{2}{3}$ and $t = \tfrac{1}{3}$. Use of these values in $\mathbf{X}$ and $\mathbf{Y}$ gives

$$\mathbf{X} = \tfrac{2}{3}\mathbf{a} + \tfrac{1}{3}\mathbf{b}, \qquad \mathbf{Y} = \tfrac{1}{3}(-\mathbf{a}) + \tfrac{1}{3}\mathbf{b}$$

as the vectors from A and B to the point I.

One important element in the strategy used in solving a problem in the plane is the choice of two vectors of the configuration for base vectors in terms of which all other vectors may be expressed. For problems in which it is desired to maintain some degree of symmetry in the analysis, a choice of origin and base vectors outside the figure may be advisable. This is illustrated in the following example.

Example 2. Prove that the midpoints of the sides of any space quadrilateral $ABCD$ are vertices of a parallelogram.

Solution: Choose an arbitrary origin O and vectors $\mathbf{a}, \mathbf{b}, \mathbf{c}, \mathbf{d}$ from O to A, B, C, D. The vector from O to the midpoint M_1 of AB is

$$\overrightarrow{OM_1} = \tfrac{1}{2}(\mathbf{a} + \mathbf{b}),$$

and

$$\overrightarrow{OM_3} = \tfrac{1}{2}(\mathbf{c} + \mathbf{d})$$

is the vector from O to M_3, the midpoint of CD. The vectors from O to the midpoints M_2, M_4 of BC and DA are

$$\overrightarrow{OM_2} = \tfrac{1}{2}(\mathbf{b} + \mathbf{c}), \qquad \overrightarrow{OM_4} = \tfrac{1}{2}(\mathbf{d} + \mathbf{a}).$$

Now vector

$$\overrightarrow{M_1M_2} = \overrightarrow{OM_2} - \overrightarrow{OM_1} = \tfrac{1}{2}(\mathbf{c} - \mathbf{a}),$$

and
$$M_4\vec{M_3} = \vec{OM_3} - \vec{OM_4} = \tfrac{1}{2}(c - a).$$

Therefore, the points $M_1, \cdots, M_4$ determine a parallelogram.

EXERCISES

1) A line is determined by the points $C(1,2,-3)$ and $D(2,4,7)$. Find its vector equation.

2) Find the angle between the line CD in Exercise 1 and the line determined by the points $E(-8,2,1)$ and $F(2,1,0)$.

3) Find the scalar projection of the vector $2\mathbf{i} - \mathbf{j} + 3\mathbf{k}$ in the direction from the origin to the point $(1,2,3)$.

4) Given the three points $P_1(2,2,1)$, $P_2(0,1,0)$, $P_3(1,-1,2)$, find (a) the components of vector $\vec{P_3P_1}$, (b) the length of $\vec{P_3P_1}$, (c) $\vec{P_3P_1}\cdot\vec{P_3P_2}$, (d) the cosine of the angle $P_1P_3P_2$, and (e) the area of the triangle $P_1P_2P_3$.

5) Write the vector equation of the plane through the point $A(1,2,3)$ and perpendicular to the direction $(2:-1:2)$. [*Hint:* Express the condition that the vector from A to any point $P(x^i)$ in the plane be perpendicular to any vector with the direction $(2:-1:2)$.]

6) Use the result of Exercise 5 to obtain the equation of the plane in the form
$$2(x^1 - 1) - (x^2 - 2) + 2(x^3 - 3) = 0.$$

7) A line through the origin passes through the points $A(1,2,-3)$ and $B(4,8,-12)$. Find the vector from the origin to (a) the midpoint M of the segment AB, (b) the point R such that $AR/RB = \tfrac{2}{3}$.

8) Find the vector from the origin to (a) the midpoint M of the segment from $A(2,-4,5)$ to $B(3,4,6)$, and (b) the point R on AB such that $AR/RB = \tfrac{2}{3}$.

9) For any vectors **a**, **b**, **c** show that
$$\mathbf{a} \times (\mathbf{b} + \mathbf{c}) = \mathbf{a} \times \mathbf{b} + \mathbf{a} \times \mathbf{c}.$$

10) Find the vector equation of the plane through the point $(1,2,3)$ and parallel to $\mathbf{i} + \mathbf{j} - \mathbf{k}$ and $2\mathbf{i} - \mathbf{j} + 3\mathbf{k}$.

11) Write the equation of the plane in Exercise 10 in rectangular coordinate form.

12) If $\mathbf{U} = \mathbf{i} + \mathbf{j} + \mathbf{k}$, $\mathbf{V} = 2\mathbf{i} - \mathbf{j} + 3\mathbf{k}$, $\mathbf{W} = \mathbf{i} + 3\mathbf{j} - \mathbf{k}$, calculate (a) $\mathbf{U}\cdot\mathbf{V} \times \mathbf{W}$, (b) $\mathbf{U} \times (\mathbf{V} \times \mathbf{W})$, (c) $(\mathbf{U} \times \mathbf{V}) \times \mathbf{W}$, (d) $(\mathbf{U} - \mathbf{V})\cdot(\mathbf{V} - \mathbf{W}) \times (\mathbf{W} - \mathbf{U})$.

13) Use formula (23) to write the vector equation of the plane determined by the points $A(1,2,6)$, $B(-1,2,4)$, $C(2,6,2)$.

14) Develop equation (23) in Section 7-4.

15) Find the volume of the parallelopiped with coterminous edges OA, OB, OC where A, B, C are given in Exercise 13.

16) Prove the identity $\mathbf{a} \times (\mathbf{b} \times \mathbf{c}) + \mathbf{b} \times (\mathbf{c} \times \mathbf{a}) + \mathbf{c} \times (\mathbf{a} \times \mathbf{b}) = \mathbf{0}$.

17) Use a vector method to develop the Law of Cosines. [*Hint:* Let the vectors on the three sides be $\mathbf{a}$, $\mathbf{b}$, $\mathbf{a} - \mathbf{b}$. Calculate $|\mathbf{a} - \mathbf{b}|^2 = (\mathbf{a} - \mathbf{b}) \cdot (\mathbf{a} - \mathbf{b})$.]

18) Show that $(\mathbf{A} \times \mathbf{B}) \times (\mathbf{C} \times \mathbf{D}) = (\mathbf{C} \times \mathbf{D} \cdot \mathbf{A})\mathbf{B} - (\mathbf{C} \times \mathbf{D} \cdot \mathbf{B})\mathbf{A}$. [*Hint:* Let $\mathbf{P} = \mathbf{C} \times \mathbf{D}$ instead of $\mathbf{A} \times \mathbf{B}$ in the development of formula (15).]

19) Use the notation $\mathbf{A} \times \mathbf{B} \cdot \mathbf{C} = [\mathbf{ABC}]$ and deduce from formula (15) and the result in Exercise 18 that

$$\mathbf{D} = \frac{[\mathbf{DBC}]}{[\mathbf{ABC}]} \mathbf{A} + \frac{[\mathbf{ADC}]}{[\mathbf{ABC}]} \mathbf{B} + \frac{[\mathbf{ABD}]}{[\mathbf{ABC}]} \mathbf{C},$$

which shows that any four vectors in three-space are linearly dependent.

20) By use of Exercise 19 express the vector $\mathbf{i} + \mathbf{j} + \mathbf{k}$ as a linear combination of the vectors $2\mathbf{i} - \mathbf{j}$, $\mathbf{i} + 2\mathbf{j} - \mathbf{k}$, $\mathbf{i} + 4\mathbf{j} + 5\mathbf{k}$.

21) Show that $(\mathbf{A} \times \mathbf{B}) \cdot (\mathbf{C} \times \mathbf{D}) \equiv (\mathbf{A} \cdot \mathbf{C})(\mathbf{B} \cdot \mathbf{D}) - (\mathbf{A} \cdot \mathbf{D})(\mathbf{B} \cdot \mathbf{C})$.

22) A force of 10 lb. acts along the line $x^1 = 1 + 2\sigma$, $x^2 = 2 + 6\sigma$, $x^3 = 3 + 3\sigma$. Find the moment of the force about the origin. *Ans.* $10\sqrt{157}/7$.

23) The faces of a cube are the planes $x^1 = 0$, $x^1 = 5$, $x^2 = 0$, $x^2 = 5$, $x^3 = 0$, $x^3 = 5$. It is made to spin about the diagonal from $(0,0,0)$ to $(5,5,5)$. Find the velocity of the point $P(1,2,3)$ in the cube if the constant angular velocity of the cube is 10 radians per sec. *Ans.* $|\mathbf{V}| = 10\sqrt{2}$ ft./sec.

24) A solid cylinder is rotating about the line $x^1 = 1 + 2\sigma$, $x^2 = 2 + 6\sigma$, $x^3 = 3 + 3\sigma$ with a constant angular velocity of 10 radians per sec. Find the velocity of the particle of the body which is at the point $P(2,4,4)$.

25) Prove that the point of intersection of the three medians of a triangle trisects each of them.

26) Use the vector method to prove that the bisectors of the angles of a triangle are concurrent.

8

Differentiation of Vectors

8-1. Vector Functions of a Scalar Variable. In (2-3) a curve C in three-space was defined by the equations $x^i = x^i(s)$, where s is the arc length measured from some fixed point on the curve. The position vector $\mathbf{R}$ from the origin O to any point P with parameter s on C is $\mathbf{R} = x^i(s)\mathbf{e}_i$. As the scalar parameter s varies, the position vector $\mathbf{R}$ varies, so $\mathbf{R}$ is a vector function of s.

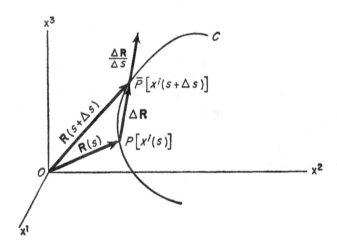

Fig. 18

From the triangle $OP\bar{P}$ in Fig. 18 it is evident that $\mathbf{R}(s) + \overrightarrow{P\bar{P}} = \mathbf{R}(s + \Delta s)$, or

1) $\qquad \overrightarrow{P\bar{P}} \equiv \Delta \mathbf{R} = \mathbf{R}(s + \Delta s) - \mathbf{R}(s).$

If both sides of (1) are multiplied by the scalar $1/\Delta s$, a vector $\Delta \mathbf{R}/\Delta s$ is obtained. Vectors $\Delta\mathbf{R}$ and $\Delta\mathbf{R}/\Delta s$ are linearly dependent. Consider the limit as Δs tends to zero, i.e.,

2) $$\lim_{\Delta s \to 0} \frac{\Delta \mathbf{R}}{\Delta s} = \lim_{\Delta s \to 0} \frac{\mathbf{R}(s + \Delta s) - \mathbf{R}(s)}{\Delta s}.$$

When this limit exists, it is called the derivative of the vector $\mathbf{R}$ with respect to s and is denoted by $d\mathbf{R}/ds$, or by $\mathbf{R}'$. Clearly, the vector $\mathbf{R}'$ is tangent to the curve C at P. Because $\mathbf{R} = x^i(s)\mathbf{e}_i$, $\mathbf{R}'$ is given by $x^{i\prime}(s)\mathbf{e}_i$, so that the components of $\mathbf{R}'$ are the direction cosines $x^{i\prime}(s)$ of the tangent to C at P (see 2–17), so $\mathbf{R}'$ is therefore a unit vector. As P moves on C the arc length s is a function of the time t. Differentiation of $\mathbf{R}$ with respect to t yields

3) $$\frac{d\mathbf{R}}{dt} = \mathbf{R}'\frac{ds}{dt} = x^{i\prime}(s)\frac{ds}{dt}\mathbf{e}_i,$$

which shows that the vector representing the velocity $\mathbf{V}$ of P is the speed $v \equiv ds/dt$ in the path, times the unit vector tangent to C at P, i.e., $\mathbf{V} = d\mathbf{R}/dt$. Notice that

$$|\mathbf{V}| = |d\mathbf{R}/dt| = |\mathbf{R}'ds/dt| = |\mathbf{R}'|\,|ds/dt| = |ds/dt|.$$

The curve C may be defined in terms of the parameter t by

4) $$\mathbf{R} = x^i(t)\mathbf{e}_i.$$

From (4)

5) $$\mathbf{V} = \frac{d\mathbf{R}}{dt} = \frac{dx^i}{dt}\mathbf{e}_i,$$

which exhibits the components dx^i/dt of the velocity vector. A second differentiation gives the acceleration vector $\mathbf{A}$, i.e.,

6) $$\mathbf{A} = \frac{d\mathbf{V}}{dt} = \frac{d^2\mathbf{R}}{dt^2} = \frac{d^2x^i}{dt^2}\mathbf{e}_i$$

so that the components of $\mathbf{A}$ parallel to the coordinate axes are d^2x^i/dt^2.

In mechanics it is sometimes inconvenient to use the coordinate components d^2x^i/dt^2 of the acceleration vector. To derive another set of components, differentiate equation (3) to obtain

7) $$\mathbf{A} = \frac{d^2\mathbf{R}}{dt^2} = \frac{ds}{dt}x^{i\prime\prime}(s)\frac{ds}{dt}\mathbf{e}_i + \frac{d^2s}{dt^2}x^{i\prime}(s)\mathbf{e}_i.$$

Recall that $\alpha^i, \beta^i, \gamma^i$ are the direction cosines of the tangent, principal normal, and binormal to C at P, and that $x^{i\prime}(s) = \alpha^i$, $x^{i\prime\prime}(s) = \sqrt{x^{k\prime\prime}x^{k\prime\prime}}\,\beta^i$. Notice that $\alpha^i\mathbf{e}_i, \beta^i\mathbf{e}_i, \gamma^i\mathbf{e}_i$ are unit vectors along the tangent, principal normal, and binormal. Equation (7) can now be written as

$$8) \quad \mathbf{A} = \frac{d^2s}{dt^2} \alpha^i \mathbf{e}_i + \left(\frac{ds}{dt}\right)^2 \sqrt{x^{k\prime\prime} x^{k\prime\prime}} \, \beta^i \mathbf{e}_i + 0 \, \gamma^i \mathbf{e}_i,$$

which shows that the acceleration vector **A** always lies in the osculating plane to C at P. The components a_t and a_n of **A** may be calculated from the equations $x^i = x^i(s)$ of C from the formulas

$$9) \quad a_t = \frac{d^2s}{dt^2}, \qquad a_n = \sqrt{x^{k\prime\prime} x^{k\prime\prime}} \left(\frac{ds}{dt}\right)^2,$$

where, of course, a_t and a_n are the components of **A** along the tangent and principal normal, respectively. The expression $\sqrt{x^{i\prime\prime} x^{i\prime\prime}}$ will appear in the next section as the *first curvature* κ of C at P. It is usually denoted by $1/\rho$, and ρ is called the *radius* of first curvature. With this notation equations (9) become

$$10) \quad a_t = \frac{dv}{dt}, \qquad a_n = \frac{v^2}{\rho}.$$

8-2. Frenet Formulas for Space Curves. The curvature of a plane curve is defined as follows. Consider the angle $\Delta\phi$ between the two tangents to the curve at points P and $\bar{P}$ which are separated by arc length Δs along the curve. The ratio $\Delta\phi/\Delta s$ is the average amount of turning of the tangent per unit length of arc. The limit of this ratio as Δs tends to zero is defined as the *curvature* of the curve at P.

For a curve C in space given by $x^i = x^i(s)$, there are two independent curvatures. The first curvature is defined in a manner similar to that for a plane curve. Consider the tangent to C at P and the tangent at P' distant Δs along C from P. If $\Delta\phi$ is the angle between these tangents, the *first curvature* of C at P is the limit of $\Delta\phi/\Delta s$ as P' approaches P. This limit, denoted by κ, is a measure of the instantaneous rate of change (with respect to arc length) of the orientation of the normal plane to C at P. An analytical expression for the first curvature is obtained by differentiating

$$\mathbf{T} = \frac{d\mathbf{R}}{ds} = x^{i\prime} \mathbf{e}_i$$

to obtain

$$11) \quad \frac{d\mathbf{T}}{ds} = x^{i\prime\prime} \mathbf{e}_i.$$

The magnitude of $d\mathbf{T}/ds$, that is $\sqrt{x^{i\prime\prime} x^{i\prime\prime}}$, is the value of κ. The radius ρ of first curvature is defined by $\rho = (x^{i\prime\prime} x^{i\prime\prime})^{-1/2}$. From (11) it follows that

$$12) \quad \frac{d\mathbf{T}}{ds} = \kappa \mathbf{N},$$

where **N** is the unit vector along the principal normal to C at P. A unit vector **B** along the binormal at P is defined by $\mathbf{B} = \mathbf{T} \times \mathbf{N}$, from which differentiation gives

13) $$\frac{d\mathbf{B}}{ds} = \frac{d\mathbf{T}}{ds} \times \mathbf{N} + \mathbf{T} \times \frac{d\mathbf{N}}{ds} = \kappa(\mathbf{N} \times \mathbf{N}) + \mathbf{T} \times \frac{d\mathbf{N}}{ds} = \mathbf{T} \times \frac{d\mathbf{N}}{ds}.$$

Since **B** is a unit vector $\mathbf{B} \cdot \mathbf{B} = 1$ and therefore $\mathbf{B} \cdot (d\mathbf{B}/ds) = 0$. By taking the dot product of both sides of (13) with **T** it follows that

14) $$\mathbf{T} \cdot \frac{d\mathbf{B}}{ds} = 0.$$

Hence, $d\mathbf{B}/ds$ is perpendicular to both **T** and **B** and is therefore some vector along the principal normal. It is written as

15) $$\frac{d\mathbf{B}}{ds} = -\tau \mathbf{N},$$

where τ is called the *torsion* or *second curvature* of the curve C at P. The torsion is seen to be a measure of the rate of change of orientation of the osculating plane. The reciprocal of τ may be called the *radius* of torsion.

In order to find the rate of change of direction of the principal normal (with respect to arc length), differentiate $\mathbf{N} = \mathbf{B} \times \mathbf{T}$ to obtain

$$\frac{d\mathbf{N}}{ds} = \mathbf{B} \times \frac{d\mathbf{T}}{ds} + \frac{d\mathbf{B}}{ds} \times \mathbf{T},$$

which, by use of (12) and (15), can be given the form

16) $$\frac{d\mathbf{N}}{ds} = -\kappa \mathbf{T} + \tau \mathbf{B}.$$

The formulas

17) $$\frac{d\mathbf{T}}{ds} = \kappa \mathbf{N}, \qquad \frac{d\mathbf{N}}{ds} = -\kappa \mathbf{T} + \tau \mathbf{B}, \qquad \frac{d\mathbf{B}}{ds} = -\tau \mathbf{N}$$

from (12), (15), (16) are called the Frenet-Serret formulas. They are fundamental in the differential geometry of space curves. The quantities in (17) can be calculated in terms of the functions $x^i(s)$ defining the curve C, where the x^i are referred to an orthogonal cartesian frame of reference. In Section 10–6 the Frenet-Serret formulas appear in arbitrary space coordinates. These formulas are referred to in the literature both as Frenet and as Frenet-Serret formulas.

As an indication of the use of formulas (17), consider the following

Example. Find the line about which the normal plane to a space curve at $P(x^i)$ tends to turn as the point P moves along the curve. This line is called the *instantaneous axis* of rotation of the plane or the *characteristic* line of the plane.

Solution: The position vector from the origin O to a point $P(x^i)$ on the curve C is $\mathbf{R}$. Let $\mathbf{Q}$ be the vector from O to any point Q in the normal plane. The vector $\overrightarrow{PQ}$ in the normal plane is then $\mathbf{Q} - \mathbf{R}$, and since $\mathbf{T}$ is the unit tangent vector, one has for the vector equation of the normal plane

$$(\mathbf{Q} - \mathbf{R}) \cdot \mathbf{T} = 0,$$

where $\mathbf{R}$ and $\mathbf{T}$ are functions of arc length s. At the point $s + \Delta s$, the normal plane has equation

$$(\mathbf{Q} - \mathbf{R} - \Delta \mathbf{R}) \cdot (\mathbf{T} + \Delta \mathbf{T}) = 0.$$

The linear combination

$$(\mathbf{Q} - \mathbf{R}) \cdot \mathbf{T} - \mathbf{T} \cdot \Delta \mathbf{R} - \Delta \mathbf{R} \cdot \Delta \mathbf{T} = 0,$$

obtained by subtracting the equation of one normal plane from that of its neighbor, represents a plane through the intersection of the neighboring normal planes. On dividing by Δs and allowing Δs to go to zero, one obtains in the limit

$$(\mathbf{Q} - \mathbf{R}) \cdot \frac{d\mathbf{T}}{ds} = \mathbf{T} \cdot \frac{d\mathbf{R}}{ds} = \mathbf{T} \cdot \mathbf{T} = 1,$$

or, by use of the first of (17),

$$(\mathbf{Q} - \mathbf{R}) \cdot \kappa \mathbf{N} = 1$$

or

$$(\mathbf{Q} - \mathbf{R}) \cdot \mathbf{N} = \rho.$$

The instantaneous axis is therefore the intersection of the planes

$$(\mathbf{Q} - \mathbf{R}) \cdot \mathbf{T} = 0, \quad (\mathbf{Q} - \mathbf{R}) \cdot \mathbf{N} = \rho$$

which is the line with vector equation

$$\mathbf{Q} = \mathbf{R} + \rho \mathbf{N} + t \mathbf{B},$$

where t is a parameter. The locus of this characteristic line as P moves along C is an interesting surface.

In a similar way one shows that the instantaneous axis of rotation of the osculating plane to C at P is the intersection of the planes

$$(\mathbf{Q} - \mathbf{R}) \cdot \mathbf{B} = 0, \qquad (\mathbf{Q} - \mathbf{R}) \cdot \frac{d\mathbf{B}}{ds} + \left(-\frac{d\mathbf{R}}{ds}\right) \cdot \mathbf{B} = 0,$$

the second of which is obtained by differentiating the first. By use of $\mathbf{T} \cdot \mathbf{B} = 0$ and the third formula in (17), these equations reduce to

$$(\mathbf{Q} - \mathbf{R}) \cdot \mathbf{B} = 0, \qquad (\mathbf{Q} - \mathbf{R}) \cdot \mathbf{N} = 0.$$

The line sought is therefore the tangent line to C at P with vector equation

$$\mathbf{Q} = \mathbf{R} + t\mathbf{T}.$$

EXERCISE

Find the instantaneous axis of rotation of the rectifying plane of a space curve. Give the vector equation of the line. *Ans.* $\mathbf{Q} = \mathbf{R} + (\tau\mathbf{T} + \kappa\mathbf{B})t$.

8-3. Application in Mechanics. The choice of the components in (6) for the acceleration vector, or those in (10), depends upon the nature of the problem in hand. From (6) the length of the vector representing acceleration at P is

$$|\mathbf{A}| = \sqrt{\frac{d^2x^i}{dt^2}\frac{d^2x^i}{dt^2}}.$$

On the other hand, if the components in (10) are used, the magnitude of the acceleration at P is $|\mathbf{A}| = \sqrt{a_t^2 + a_n^2}$.

The components of the acceleration vector are useful in applying Newton's second law $\mathbf{F} = m\mathbf{A}$, for the motion of a particle, where $\mathbf{F}$ is a vector describing the force acting on the particle P, the scalar m denotes the mass of the particle, and $\mathbf{A}$ is the acceleration vector. The quantities involved must be expressed in appropriate units. If no force acts, $\mathbf{F}$ is the zero vector, and the d^2x^i/dt^2 components of $\mathbf{A}$ are convenient. From $\mathbf{A} = \mathbf{0}$, i.e., $d^2x^i/dt^2 = 0$, one finds on integrating that $dx^i/dt = c^i$, and consequently

18) $$x^i = c^i t + d^i,$$

where c^i and d^i are arbitrary constants. Equations (18) represent the straight line of the motion. If the position vector and velocity vector are specified at time $t = 0$, say, the constants c^i and d^i can be evaluated.

It should be observed that from $\mathbf{A} \equiv d^2\mathbf{R}/dt^2 = \mathbf{0}$, it follows that $\mathbf{V} = d\mathbf{R}/dt = \mathbf{C}$, and $\mathbf{R} = \mathbf{C}t + \mathbf{D}$, where $\mathbf{C}$ and $\mathbf{D}$ are constant vectors, so that (18) may be obtained by using vectors instead of their components.

If the components in (10) are employed for the problem in which the force vector $\mathbf{F} = \mathbf{0}$, the vanishing of the components, i.e., $a_t \equiv dv/dt = 0$ and $a_n \equiv v^2/\rho = 0$, shows that $v = $ constant and $1/\rho = 0$. The latter condition means that the locus of the particle is a straight line (zero curvature).

If the problem involves a force in the direction of the tangent to a curve, use of the components of acceleration in (10) is indicated.

Additional applications in mechanics are furnished in some of the exercises to follow. A third type of resolution of the velocity and acceleration vectors is considered in the next section for the case of a particle moving on a plane curve.

8-4. Motion in a Plane. Three different modes of resolution of the acceleration vector are useful in the study of the motion of a particle along a plane curve C. Again, the nature of the problem dictates the choice of components for the most convenient solution.

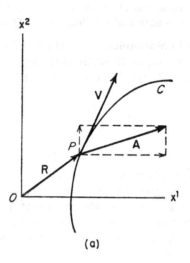

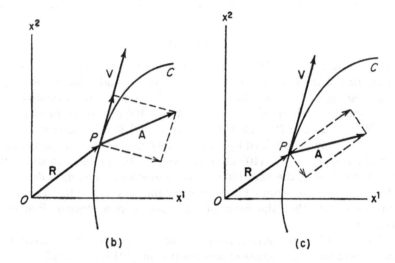

Fig. 19

In Fig. 19a the components of the acceleration vector **A** are parallel to the coordinate axes and are given by d^2x^i/dt^2 ($i = 1, 2$). Because the curve is plane, the plane in which C lies is the osculating plane of C, so the acceleration components a_t and a_n are along the tangent and normal to the plane curve and are given by dv/dt and v^2/ρ, where now $1/\rho$ is the curvature of C at P. (See Fig. 19b.) In Fig. 19c, the vector **A** is projected upon the radial line of the position vector **R** to P, and upon the normal to the radial line at P. Call these components a_r and a_θ, respectively. In order to compute a_r and a_θ, let (r, θ) be the polar coordinates of P, and let **K** be a unit vector perpendicular to the unit vector $\mathbf{H} \equiv \mathbf{R}/|\mathbf{R}|$ along the radius vector. One has

19) $$\mathbf{H} \equiv \mathbf{R}/|\mathbf{R}| = \cos\theta\, \mathbf{e}_1 + \sin\theta\, \mathbf{e}_2,$$

and

20) $$\mathbf{K} = -\sin\theta\, \mathbf{e}_1 + \cos\theta\, \mathbf{e}_2.$$

Notice that vector **K** is obtained from **H** by replacing θ in (19) by $\theta + \pi/2$ to obtain (20). Now on the curve C, r and θ are functions of the time t, so by differentiation of (19) with respect to t (letting the angular speed $d\theta/dt$ be denoted by ω) one finds

21) $$\frac{d\mathbf{H}}{dt} = (-\sin\theta)\frac{d\theta}{dt}\mathbf{e}_1 + (\cos\theta)\frac{d\theta}{dt}\mathbf{e}_2 = \omega\mathbf{K}.$$

Because the velocity vector **V** is $d\mathbf{R}/dt$, and $\mathbf{R} = r\mathbf{H}$, one sees from (21) that

22) $$\mathbf{V} = r\frac{d\mathbf{H}}{dt} + \mathbf{H}\frac{dr}{dt} = \frac{dr}{dt}\mathbf{H} + r\omega\mathbf{K},$$

which exhibits the components dr/dt and $r\omega$ along and perpendicular to the radius vector to P. Hence, the speed in the path can be found from

23) $$|\mathbf{V}| \equiv v = \sqrt{\left(\frac{dr}{dt}\right)^2 + r^2\left(\frac{d\theta}{dt}\right)^2}.$$

On differentiating equation (22), one finds the acceleration vector

24) $$\mathbf{A} = \frac{d\mathbf{V}}{dt} = \frac{d^2r}{dt^2}\mathbf{H} + \frac{dr}{dt}\frac{d\mathbf{H}}{dt} + \frac{dr}{dt}\omega\mathbf{K} + r\frac{d\omega}{dt}\mathbf{K} + r\omega\frac{d\mathbf{K}}{dt}.$$

Differentiation of **K** in (20) shows that

25) $$\frac{d\mathbf{K}}{dt} = (-\cos\theta\, \mathbf{e}_1 - \sin\theta\, \mathbf{e}_2)\omega = -\omega\mathbf{H}.$$

Therefore, by use of (21) and (25), equation (24) takes the form

26) $$\mathbf{A} = \left(\frac{d^2r}{dt^2} - r\omega^2\right)\mathbf{H} + \left(r\frac{d\omega}{dt} + 2\omega\frac{dr}{dt}\right)\mathbf{K},$$

from which the components of the acceleration vector along and perpendicular to the radius vector are given by

27) $$a_r = \frac{d^2r}{dt^2} - r\omega^2, \qquad a_\theta = r\frac{d\omega}{dt} + 2\omega\frac{dr}{dt}.$$

An indication of the use of formulas (27) is provided by the following example.

Example. Investigate the equations of motion for a particle of mass m which moves in a force field with components f_1 and f_2 along and perpendicular to the radius vector. In particular, consider the case of a central force $mf(r)$, and, finally, treat the case of the inverse square law, in which $f(r) = -k^2/r^2$.

Solution: From (27) the equations of motion are

28) $$m\left(\frac{d^2r}{dt^2} - r\omega^2\right) = f_1, \qquad m\left(r\frac{d\omega}{dt} + 2\omega\frac{dr}{dt}\right) = f_2.$$

The second of these can be written as

$$\frac{d}{dt}(mr^2\omega) = rf_2$$

or

$$\frac{d}{dt}(mvr) = rf_2,$$

which means that the rate of change of the moment of momentum of the particle about the origin is given by the moment of the force f_2 about the origin.

Equations (28) are difficult to solve even for simple functions f_1 and f_2 of r and θ. However, if the force is always toward the center (the origin) f_1 may be written as $mf(r)$, $f_2 \equiv 0$, and integration of the second equation in (28) gives

$$mr^2\omega \equiv mr^2\frac{d\theta}{dt} = \text{constant},$$

or, by choosing the constant as mh,

$$r^2\frac{d\theta}{dt} = h.$$

If the last equation is used to eliminate ω from the first of (28), there results

$$\frac{d^2r}{dt^2} - \frac{h^2}{r^3} = f(r),$$

and multiplication of this equation by $2(dr/dt)$ yields

$$\frac{d}{dt}\left(\frac{dr}{dt}\right)^2 - \frac{2h^2}{r^3}\frac{dr}{dt} = 2f(r)\frac{dr}{dt},$$

which can be integrated to obtain

29) $$\left(\frac{dr}{dt}\right)^2 + \frac{h^2}{r^2} = 2\int f(r)\,dr + \text{constant}.$$

Equation (29) determines r as a function of t, and with this the equation $r^2\, d\theta = h\, dt$ can be used to find θ as a function of t. The orbit of the particle is then determined with r and θ as functions of a single parameter. It can be shown that if $f(r)$ is taken as $-k^2 r$ in (29), the orbit of the particle is an ellipse of the form $q^2 x^2 + p^2 y^2 = p^2 q^2$.

To treat the final case, let $f(r) = -k^2/r^2$. Equations (28) then become

$$\text{30)} \qquad \frac{d^2 r}{dt^2} - r\omega^2 = -\frac{k^2}{r^2}, \qquad r^2 \frac{d\theta}{dt} = h.$$

The first of these assumes a simpler form under the change of dependent variable $r = 1/u$. One has

$$\frac{dr}{dt} = \frac{dr}{du}\frac{du}{dt} = \left(-\frac{1}{u^2}\right)\frac{du}{dt} = -\frac{1}{u^2}\frac{du}{d\theta}\frac{d\theta}{dt} = -\frac{1}{u^2}\frac{du}{d\theta}\left(\frac{h}{r^2}\right) = -h\frac{du}{d\theta},$$

and

$$\frac{d^2 r}{dt^2} = \frac{d}{dt}\left(-h\frac{du}{d\theta}\right) = -h\frac{d}{d\theta}\left(\frac{du}{d\theta}\right)\frac{d\theta}{dt} = -h\frac{d^2 u}{d\theta^2}\left(\frac{h}{r^2}\right) = -h^2 u^2 \frac{d^2 u}{d\theta^2},$$

so that the first of (30) becomes (on using $\omega = hu^2$)

$$\text{31)} \qquad \frac{d^2 u}{d\theta^2} + u = \left(\frac{k}{h}\right)^2.$$

The solution of (31) is easily found to be

$$u = \left(\frac{k}{h}\right)^2 + A\cos\theta + B\sin\theta,$$

or, in another form,

$$\text{32)} \qquad \frac{1}{r} = \frac{1}{a} + \lambda \cos(\theta - \alpha) \qquad (a = h^2 k^{-2})$$

with λ and α as the arbitrary constants.

The equation

$$r = \frac{l}{1 + e\cos(\theta - \alpha)}$$

represents a conic section with focus at the origin, semilatus rectum l, eccentricity e, and α as the angle from the polar axis to an axis of symmetry of the conic. Solve (32) for r to obtain

$$\text{33)} \qquad r = \frac{a}{1 + \lambda a \cos(\theta - \alpha)},$$

from which it is seen that the orbit of the particle is a conic with eccentricity λa, and length of latus rectum $2a$. The constants λ and α are determined by initial conditions.

It is interesting to observe that the following three laws of Kepler can be deduced from the result of using the inverse square law of Newton.

I. The planets describe ellipses with the sun at one focus. This follows from (33) for it can be shown that the eccentricity λa is less than one. Note from (33) that if $\lambda a \geq 1$, r would become infinite for some θ so the motion would not be periodic.

II. The radius vector from a planet to the sun sweeps over equal areas in equal intervals of time. This follows by integrating $r^2 \, d\theta = h \, dt$ and using $\theta = \theta_1$ when $t = t_1$, and $\theta = \theta_2$ when $t = t_2$.

III. The squares of the periods of the planets are proportional to the cubes of their mean distances from the sun. If p and q are the lengths of the semiaxes of the ellipse, the area is πpq, $q = p\sqrt{1 - e^2}$, and the semi-latus rectum $l = p(1 - e^2) = h^2 k^{-2}$. Integrate $r^2 \, d\theta = h \, dt$ over the entire ellipse to obtain

$$2\pi pq = hP,$$

where P is the period. It follows that

$$P^2 = \frac{4\pi^2 p^2 q^2}{h^2} = \frac{4\pi^2 p^4 (1 - e^2)}{k^2 p (1 - e^2)} = C p^3,$$

where C is a constant. From this the third law is seen to hold.

8–5. Law of Transformation for Velocity Components. In formulas (22) and (27) the components of the velocity and acceleration vectors were derived for the polar coordinate system. In Chapter 5 it was shown that if any tensor of first order has known components in one coordinate system and if the transformation to new coordinates is known, the components of the tensor (vector) in the new coordinate system can be determined. It is proposed to find the components of the velocity vector **V** in polar coordinates from the components dx^1/dt, dx^2/dt in rectangular cartesian coordinates. The transformation is

34) $\qquad x^1 = \bar{x}^1 \cos \bar{x}^2, \qquad x^2 = \bar{x}^1 \sin \bar{x}^2.$

Differentiation gives

35) $\qquad \dfrac{dx^i}{dt} = \dfrac{d\bar{x}^k}{dt} \dfrac{\partial x^i}{\partial \bar{x}^k}$

so dx^i/dt evidently transform in the contravariant manner, namely, like

36) $\qquad \lambda^i = \bar{\lambda}^k \dfrac{\partial x^i}{\partial \bar{x}^k}.$

The partial derivatives $\partial x^i / \partial \bar{x}^k$ are found from (34) to be

37) $\qquad \begin{aligned} \dfrac{\partial x^1}{\partial \bar{x}^1} &= \cos \bar{x}^2, & \dfrac{\partial x^1}{\partial \bar{x}^2} &= -\bar{x}^1 \sin \bar{x}^2, \\[4pt] \dfrac{\partial x^2}{\partial \bar{x}^1} &= \sin \bar{x}^2, & \dfrac{\partial x^2}{\partial \bar{x}^2} &= \bar{x}^1 \cos \bar{x}^2. \end{aligned}$

Use of (37) in (36) gives the transformation

38)
$$\frac{dx^1}{dt} = \frac{d\bar{x}^1}{dt}\cos\bar{x}^2 - \frac{d\bar{x}^2}{dt}\bar{x}^1\sin\bar{x}^2,$$
$$\frac{dx^2}{dt} = \frac{d\bar{x}^1}{dt}\sin\bar{x}^2 + \frac{d\bar{x}^2}{dt}\bar{x}^1\cos\bar{x}^2,$$

which could have been obtained by direct differentiation in (34). Solution of (38) for $d\bar{x}^1/dt$ and $d\bar{x}^2/dt$ yields

39)
$$\frac{d\bar{x}^1}{dt} = \frac{dx^1}{dt}\cos\bar{x}^2 + \frac{dx^2}{dt}\sin\bar{x}^2,$$
$$\frac{d\bar{x}^2}{dt} = \frac{1}{\bar{x}^1}\left(-\frac{dx^1}{dt}\sin\bar{x}^2 + \frac{dx^2}{dt}\cos\bar{x}^2\right).$$

Equations (39) give the contravariant components of the velocity vector **V** in the polar coordinate system. It remains to find the projections of the vector **V** upon the radial line and perpendicular to it. This is accomplished by use of (6–43), that is, by

40)
$$|\mathbf{V}|\cos\phi_i = \frac{\bar{\lambda}^i}{\sqrt{\bar{a}^{ii}}}.$$

(Note that the radial line is normal to the curve $\bar{x}^1$ = constant, i.e., r = constant, and the line perpendicular to the radial line is normal to the "curve" $\bar{x}^2$ = constant, i.e., the straight line θ = constant.) For polar coordinates $ds^2 = dr^2 + r^2\,d\theta^2 = (d\bar{x}^1)^2 + (\bar{x}^1)^2(d\bar{x}^2)^2$, from which $\bar{a}_{11} = 1$, $\bar{a}_{12} = \bar{a}_{21} = 0$, $\bar{a}_{22} = (\bar{x}^1)^2$, and $\bar{a}^{11} = 1$, $\bar{a}^{12} = \bar{a}^{21} = 0$, $\bar{a}^{22} = (\bar{x}^1)^{-2}$. Therefore, from (40), the projection of the velocity vector **V** upon the radial line is

$$\frac{\bar{\lambda}^1}{\sqrt{\bar{a}^{11}}} = \frac{d\bar{x}^1}{dt} = \frac{dr}{dt},$$

and the projection of **V** perpendicular to the radial line is

$$\frac{\bar{\lambda}^2}{\sqrt{\bar{a}^{22}}} = \bar{x}^1\bar{\lambda}^2 = \bar{x}^1\frac{d\bar{x}^2}{dt} = r\frac{d\theta}{dt}.$$

These formulas agree with those obtained for v_r and v_θ in (22).

The question arises: Can the components of the acceleration vector in polar coordinates (or any other coordinates) be found by the method used above for the velocity vector? The answer is in the negative unless the coordinate transformation is linear. The change from rectangular to polar coordinates is not linear. The fundamental reason why one cannot use the tensor law of transformation to obtain the components of the acceleration vector in any coordinate system starting from the components d^2x^i/dt^2 in the orthogonal cartesian system is the following: Although dx^i/dt are components of a tensor of the first order, the derivatives of dx^i/dt are not components of a tensor, and therefore one may not apply a tensor law of transformation. Covariant differentiation is introduced in the next chapter, and it will be seen that the covariant derivative of a tensor of any order is again a tensor but of one order higher in the covariant indices. The acceleration vector will be treated later (Section 10-7) by means of the covariant derivative of the velocity vector.

8-6. Vector Functions of Two Scalar Parameters. As an extension of $\mathbf{R} = x^i(s)\mathbf{e}_i$ for the position vector of a point on a curve, one may consider $\mathbf{R} = x^i(u^1,u^2)\mathbf{e}_i$ as the position vector for a point P on a surface with

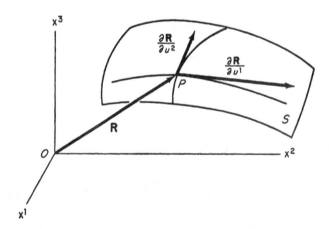

Fig. 20

Gauss equations $x^i = x^i(u^1,u^2)$. At a point P of the curve $u^2 = $ constant on the surface S (Fig. 20), a tangent vector is given by

41) $$\frac{\partial \mathbf{R}}{\partial u^1} = \frac{\partial x^i}{\partial u^1}\mathbf{e}_i,$$

as a vector in three-space, and at the same point a tangent vector to the curve $u^1 = $ constant has the form

42) $$\frac{\partial \mathbf{R}}{\partial u^2} = \frac{\partial x^i}{\partial u^2}\mathbf{e}_i.$$

Unit vectors $\mathbf{T}_1$ and $\mathbf{T}_2$ are now desired along the coordinate curves. It is necessary to divide equation (41), for instance, by the square root of the sum of the squares of the components $\partial x^i/\partial u^1$. In order to simplify the notation, make the following definitions

43) $\quad g_{11} \equiv \dfrac{\partial x^i}{\partial u^1} \dfrac{\partial x^i}{\partial u^1}, \qquad g_{12} = g_{21} \equiv \dfrac{\partial x^i}{\partial u^1} \dfrac{\partial x^i}{\partial u^2}, \qquad g_{22} \equiv \dfrac{\partial x^i}{\partial u^2} \dfrac{\partial x^i}{\partial u^2}.$

(Some authors use $E = g_{11}$, $F = g_{12} = g_{21}$, $G = g_{22}$.) The unit tangent vectors to the $u^1 =$ constant and $u^2 =$ constant curves through P are given by

44) $\quad \mathbf{T}_\alpha = \dfrac{1}{\sqrt{g_{\alpha\alpha}}} \dfrac{\partial \mathbf{R}}{\partial u^\alpha} = \dfrac{1}{\sqrt{g_{\alpha\alpha}}} \dfrac{\partial x^i}{\partial u^\alpha} \mathbf{e}_i \qquad (\alpha = 1, 2).$

The angle ω between $\mathbf{T}_1$ and $\mathbf{T}_2$ is given by $\mathbf{T}_1 \cdot \mathbf{T}_2$, that is,

45) $\quad \cos \omega = \dfrac{1}{\sqrt{g_{11}}} \dfrac{1}{\sqrt{g_{22}}} \dfrac{\partial x^i}{\partial u^1} \dfrac{\partial x^j}{\partial u^2} \mathbf{e}_i \cdot \mathbf{e}_j = \dfrac{1}{\sqrt{g_{11}g_{22}}} \dfrac{\partial x^i}{\partial u^1} \dfrac{\partial x^i}{\partial u^2} = \dfrac{g_{12}}{\sqrt{g_{11}g_{22}}}.$

Observe that the coordinate curves on the surface are everywhere orthogonal if and only if g_{12} is identically zero.

8–7. Riemannian Metric. In order to study either metric geometry or the dynamics of a particle moving on a surface, a measure of distance is needed. Length of arc of a curve on a surface will be obtained by thinking of the curve as immersed in three-space. Let ds be the differential of an arc joining two points $x^i(u^1, u^2)$ and $x^i(u^1 + du^1, u^2 + du^2)$ on the surface S. Because the arc is immersed in three-space $ds^2 = \delta_{ij}\, dx^i\, dx^j$. Calculation of dx^i from $x^i = x^i(u^1, u^2)$ gives

46) $\quad dx^i = \dfrac{\partial x^i}{\partial u^1} du^1 + \dfrac{\partial x^i}{\partial u^2} du^2 = \dfrac{\partial x^i}{\partial u^\alpha} du^\alpha,$

and

47) $\quad (ds)^2 = \delta_{ij}\, dx^i\, dx^j = \delta_{ij} \dfrac{\partial x^i}{\partial u^\alpha} du^\alpha \dfrac{\partial x^j}{\partial u^\beta} du^\beta = \delta_{ij} \dfrac{\partial x^i}{\partial u^\alpha} \dfrac{\partial x^j}{\partial u^\beta} du^\alpha\, du^\beta.$

As in (43), let

48) $\quad g_{\alpha\beta} \equiv \delta_{ij} \dfrac{\partial x^i}{\partial u^\alpha} \dfrac{\partial x^j}{\partial u^\beta}.$

The differential of arc on the surface is then expressed by (47) as

49) $\quad (ds)^2 = g_{\alpha\beta}\, du^\alpha\, du^\beta,$

where the Greek letters have the range 1, 2.

A metric defined as in (49) by a quadratic differential form is called a Riemannian metric, the resulting geometry is called Riemannian geometry, and the space to which such a metric has been assigned is called a Riemannian space. Consequently a surface having the metric (49) is a two-dimensional Riemannian space, and the geometry of the surface is a Riemannian geometry. A Riemannian geometry is, in general, an instance of a non-euclidean geometry. The tensor calculus is useful in the study of curved surfaces, and is indispensable in the study of Riemannian spaces of higher than two dimensions.

If a surface curve C is described by u^2 = constant, then $du^2 = 0$, and (49) gives $ds = \sqrt{g_{11}}\, du^1$. Similarly, on a u^2-curve where u^1 = constant, $du^1 = 0$, and $ds = \sqrt{g_{22}}\, du^2$. Thus, the derivatives of the surface parameters u^α with respect to arc length are given by

$$50)\qquad \frac{du^\alpha}{ds} = \frac{1}{\sqrt{g_{\alpha\alpha}}} \qquad (\alpha \text{ not summed}).$$

If the curve C is not a coordinate curve on S, it may be represented by $u^\alpha = u^\alpha(\sigma)$, where σ is a general parameter. By using the differentials

$$51)\qquad du^\alpha = \frac{du^\alpha}{d\sigma}\, d\sigma,$$

the form for $(ds)^2$ along the curve C becomes

$$52)\qquad (ds)^2 = g_{\alpha\beta}\, \frac{du^\alpha}{d\sigma}\, \frac{du^\beta}{d\sigma}\, (d\sigma)^2.$$

If $\sigma = t$, where t represents time, the velocity of a particle moving along the curve C is expressed from (52) by

$$53)\qquad v = \frac{ds}{dt} = \left(g_{\alpha\beta}\, \frac{du^\alpha}{dt}\, \frac{du^\beta}{dt}\right)^{\!\!1/2},$$

where the positive root is used if s increases as t increases.

It should be realized that the arc element formula (52) may be arrived at by using the position vector $\mathbf{R} = x^i(u^1, u^2)\mathbf{e}_i$, where $u^\alpha = u^\alpha(\sigma)$ on curve C. Differentiation of $\mathbf{R}$ with respect to s yields

$$54)\qquad \frac{d\mathbf{R}}{ds} = \frac{\partial x^i}{\partial u^\alpha}\, \frac{du^\alpha}{d\sigma}\, \frac{d\sigma}{ds}\, \mathbf{e}_i.$$

Because $d\mathbf{R}/ds$ is a unit vector, one has from (54) that

$$55)\qquad \frac{d\mathbf{R}}{ds} \cdot \frac{d\mathbf{R}}{ds} = 1 = \left(\frac{\partial x^i}{\partial u^\alpha}\, \frac{du^\alpha}{d\sigma}\, \frac{d\sigma}{ds}\right)\!\left(\frac{\partial x^j}{\partial u^\beta}\, \frac{du^\beta}{d\sigma}\, \frac{d\sigma}{ds}\right) \mathbf{e}_i \cdot \mathbf{e}_j$$

$$= \delta_{ij}\, \frac{\partial x^i}{\partial u^\alpha}\, \frac{\partial x^j}{\partial u^\beta}\, \frac{du^\alpha}{d\sigma}\, \frac{du^\beta}{d\sigma}\, \left(\frac{d\sigma}{ds}\right)^{\!2}.$$

It follows from (55), on using (48), that

56) $$\left(\frac{ds}{d\sigma}\right)^2 = g_{\alpha\beta}\frac{du^\alpha}{d\sigma}\frac{du^\beta}{d\sigma},$$

as in (52).

8–8. Extrinsic and Intrinsic Geometry. It is important to realize that there are two types of geometry of a surface—*extrinsic* and *intrinsic*. The former has to do with geometric properties relative to the space of immersion of the surface. Just as a curve immersed in a plane, and a tortuous curve in three-space have distinguishing extrinsic geometric properties, so do surfaces immersed in three-space have distinguishing extrinsic geometric properties which are different for surfaces immersed in four-space. On the other hand, the intrinsic geometry refers to properties of the surface itself without regard for the space of immersion. Whether a surface is in a three- or four-dimensional space, the intrinsic geometry of the surface is the same.

As an instance of an extrinsic property, notice that the arc element ds was imposed upon the surface by virtue of the fact that the surface is in three-space. On the other hand, the formula (45) for the angle between the parametric curves at a point of the surface is intrinsic. Of course, it is true that $\cos \omega$ was obtained by thinking of the surface curves as curves in three-space, but the same formula can be obtained in an intrinsic manner once the arc length ds is established. To do this, consider the formula (6–17) for the cosine of the angle between two directions p^i and q^i in a general coordinate system, namely

57) $$\cos \theta = \frac{a_{ij}p^i q^j}{\sqrt{a_{rs}p^r p^s}\sqrt{a_{lm}q^l q^m}}.$$

Let θ be the angle between the curves $C_1: u_1^\alpha = u_1^\alpha(\sigma)$ and $C_2: u_2^\alpha = u_2^\alpha(\sigma)$ through point P on S. The directions of the surface curves are given by du_1^α and du_2^α, and the lengths of the vectors with these components are $\sqrt{g_{\alpha\beta}\,du_1^\alpha\,du_1^\beta}$ and $\sqrt{g_{\alpha\beta}\,du_2^\alpha\,du_2^\beta}$, where the fundamental metric tensor $g_{\alpha\beta}$ on the surface is employed. From (57), the angle θ between curves C_1 and C_2 on the surface is given by

58) $$\cos \theta = \frac{g_{\alpha\beta}\,du_1^\alpha\,du_2^\beta}{\sqrt{g_{\gamma\delta}\,du_1^\gamma\,du_1^\delta}\sqrt{g_{\rho\tau}\,du_2^\rho\,du_2^\tau}}.$$

Now, as a particular case, let θ be ω, the angle between the parametric curves. In this case C_1 becomes the u^1-curve and C_2 the u^2-curve, so that $du_1^1 \equiv du^1, du_1^2 = 0;\ du_2^1 = 0, du_2^2 \equiv du^2$. Placing these values in (58) yields

59) $$\cos \omega = \frac{g_{12}\,du^1\,du^2}{\sqrt{g_{11}\,du^1\,du^1}\sqrt{g_{22}\,du^2\,du^2}} = \frac{g_{12}}{\sqrt{g_{11}g_{22}}},$$

in agreement with (45). Any geometric property of the surface which can be expressed in terms of the fundamental tensor $g_{\alpha\beta}$ (including partial derivatives of $g_{\alpha\beta}$) is an intrinsic property.

By means of the tensor analysis an extensive literature has been developed not only for a surface immersed in three- and higher-dimensional linear spaces but also for a curved space of n dimensions (call it V_n) immersed either in a curved space V_m or in a linear space of sufficiently high dimensionality. Denote a linear space of k-dimensions by S_k. One writes $V_2 \subset S_3$ to indicate that an ordinary surface is immersed in a linear space of three dimensions. A question arises. What is the least dimensionality of a linear space in order that a V_n may be immersed in it? It is possible to show that

$$60) \qquad V_n \subset S_{\frac{n(n+1)}{2}}.$$

The symbol $\subset$ can be read "is immersible in," or "belongs to," or "is a subset of."

Although no attempt is made here toward an actual proof of the statement in (60), it may be of interest to outline a few suggestive steps. Let $u^1, \cdots, u^n$ be the variables (coordinates) in V_n, and let $x^1, \cdots, x^k$ be the variables in the ambient space S_k. Then $x^i = x^i(u^1, \cdots, u^n)$ ($i = 1, \cdots, k$) are the equations of the V_n which is imbedded in the S_k, and the number k of these equations is to be determined. The element of arc ds is given by $ds^2 = a_{ij} \, dx^i \, dx^j$ in S_k and by $ds^2 = g_{\alpha\beta} \, du^\alpha \, du^\beta$ in V_n. If these are to be equal, one has

$$61) \qquad a_{ij} \, dx^i \, dx^j \equiv a_{ij} \frac{\partial x^i}{\partial u^\alpha} \frac{\partial x^j}{\partial u^\beta} du^\alpha \, du^\beta \equiv g_{\alpha\beta} \, du^\alpha \, du^\beta,$$

from which it follows that the functions $x^i(u^1, \cdots, u^n)$ are solutions of the partial differential equations

$$62) \qquad a_{ij} \frac{\partial x^i}{\partial u^\alpha} \frac{\partial x^j}{\partial u^\beta} = g_{\alpha\beta}.$$

The number of equations in (62) is $n(n + 1)/2$, so $k \geq n(n + 1)/2$. Thus, in general, a curved space V_n can be immersed in a linear space of not less than $n(n + 1)/2$ dimensions. By putting $n = 2$, it is comforting to find that $V_2 \subset S_3$. Similarly, a V_3 can be immersed in a space S_6, while a curved space V_4 of four dimensions can be immersed in a linear space of ten dimensions. For a particular V_n, a smaller number k for the dimensionality of S_k may suffice. For instance, if V_n is euclidean n-space itself, then k may be taken as n. If $n + p$ is the least possible value of k, V_n is said to be of class p. The class of a euclidean space is zero. The class of a V_n cannot be greater than $(n/2)(n+1) - n = n(n-1)/2$. If $n = 2$, the class is one or zero.

8–9. Surface Normal and Tangent Plane. Another instance of an extrinsic geometric construct for a surface V_2 in S_3 is the surface normal at a point P on the surface. (For a V_2 in S_4 there are infinitely many surface normals at a point.) The vectors $\partial \mathbf{R}/\partial u^1$ and $\partial \mathbf{R}/\partial u^2$ in formulas (41) and (42) determine the tangent plane to the surface at P. The normal N to the surface at P is normal to the tangent plane at P, and the direction of N is given by

$$63) \qquad \frac{\partial \mathbf{R}}{\partial u^1} \times \frac{\partial \mathbf{R}}{\partial u^2}.$$

The vector (63) defines also the sense of direction of the surface normal.

From (41) and (42) it is seen that

$$64) \quad \frac{\partial \mathbf{R}}{\partial u^1} \times \frac{\partial \mathbf{R}}{\partial u^2} = \frac{\partial x^i}{\partial u^1} \frac{\partial x^j}{\partial u^2} \mathbf{e}_i \times \mathbf{e}_j = e_{ijk} \frac{\partial x^i}{\partial u^1} \frac{\partial x^j}{\partial u^2} \mathbf{e}^k = e_{ijk} \frac{\partial x^j}{\partial u^1} \frac{\partial x^k}{\partial u^2} \mathbf{e}^i.$$

If $(X^i - x^i)$ are the components of any vector $\mathbf{H}$ through $P(x^i)$ in the tangent plane, it follows that

$$65) \qquad \mathbf{H} \cdot \frac{\partial \mathbf{R}}{\partial u^1} \times \frac{\partial \mathbf{R}}{\partial u^2} = 0,$$

which can be written as

$$66) \qquad e_{ijk}(X^i - x^i) \frac{\partial x^j}{\partial u^1} \frac{\partial x^k}{\partial u^2} = 0,$$

or, finally, as

$$67) \qquad \begin{vmatrix} X^1 - x^1 & X^2 - x^2 & X^3 - x^3 \\ \dfrac{\partial x^1}{\partial u^1} & \dfrac{\partial x^2}{\partial u^1} & \dfrac{\partial x^3}{\partial u^1} \\ \dfrac{\partial x^1}{\partial u^2} & \dfrac{\partial x^2}{\partial u^2} & \dfrac{\partial x^3}{\partial u^2} \end{vmatrix} = 0.$$

This is the equation of the tangent plane to the surface $x^i = x^i(u^1, u^2)$ at the point $P(x^i)$. [Compare (2–39).]

8–10. Local and Global Geometry. Besides the classification of geometric properties of a configuration into extrinsic and intrinsic categories, as in Section 8–8, there is another important classification. Geometric properties which require knowledge of the configuration as a whole are referred to as *global* properties, or *integral* geometry, or geometry "*in the large.*" For instance, the problem of finding the points of intersection of a line and a conic in a plane is a problem of integral geometry, for it requires a knowledge of the figure as a whole. On the other hand, if attention is directed only to the geometry in some neighborhood of an element of the configuration (say, near a point or a line), then the geometry is referred to as *local*, or as geometry *"in the small."* Differential geometry employs the differential calculus to investigate geometric properties of a configuration in some neighborhood of an element. Instances of this are found in the study of tangent lines at nearby points on a space curve, and in the curvatures of a space curve at a point. Differential geometry is therefore a local geometry. It considers configurations in the small. Another instance of this kind appears in Chapter 12 where the *total* curvature of a surface at a point is introduced. Clearly, differential geometry did not exist prior to the invention of the calculus (Newton 1642–1727, Leibniz 1646–1716).

Integral or global geometry (which may or may not employ the integral calculus) was the sort of geometry developed prior to Newton. This category may be subdivided into synthetic and analytic geometry. In the former, geometric properties are deduced without use of a coordinate system, while in the latter a frame of reference is selected and coordinates are attached to geometrical elements (points, lines, planes). Clearly, prior to the time of Descartes (1596–1650)—at least before his *La Géométrie* (1637)—all geometry could be classified as synthetic with the global aspect.

EXERCISES

1) A curve is described by the position vector $\mathbf{R} = (t)\mathbf{e}_1 + (t^2)\mathbf{e}_2 + (t^3)\mathbf{e}_3$. Find the length of the acceleration vector at time $t = 1$.

2) For the curve in Exercise 1 find the cosine of the angle between the tangents at $t = 1$ and $t = 2$.

3) If vectors $\mathbf{U}$ and $\mathbf{V}$ are functions of t, show that

(a) $\dfrac{d}{dt}(\mathbf{U}\cdot\mathbf{V}) = \mathbf{U}\cdot\dfrac{d\mathbf{V}}{dt} + \dfrac{d\mathbf{U}}{dt}\cdot\mathbf{V}$, (b) $\dfrac{d}{dt}(\mathbf{U}\times\mathbf{V}) = \mathbf{U}\times\dfrac{d\mathbf{V}}{dt} + \dfrac{d\mathbf{U}}{dt}\times\mathbf{V}$.

[*Hint:* (a) Write $\mathbf{U}\cdot\mathbf{V} = u_i v^i$, (b) Differentiate a determinant.]

4) If $\mathbf{T}(s)$ is a unit vector, show that $d\mathbf{T}/ds$ is a vector perpendicular to $\mathbf{T}(s)$.

5) (a) The formula $\kappa \equiv 1/\rho = (x^{i\prime\prime} x^{i\prime\prime})^{1/2}$ gives the first curvature of the curve $x^i = x^i(s)$ with arc length s as parameter. Find the formula for κ if the curve is given by $x^i = x^i(\theta)$, where θ is an arbitrary parameter. (b) Use the result to find the first curvature of the helix

$$x^1 = a\cos\theta, \qquad x^2 = a\sin\theta, \qquad x^3 = b\theta$$

at the point where $\theta = \pi/4$. (c) Calculate $\alpha^i, \beta^i, \gamma^i$ for the helix and use one of the Frenet formulas to find the second curvature τ. Ans. $\kappa = a/(a^2 + b^2)$, $\tau = b/(a^2 + b^2)$.

6) Illustrate (8–35) as relating contravariant components of the velocity vector in orthogonal cartesian and spherical coordinates.

7) Write the velocity vector as $\mathbf{V} = |\mathbf{V}|\,\mathbf{T} = v\mathbf{T}$, and differentiate with respect to t to show that the acceleration vector $\mathbf{A}$ at a point on the locus of a particle in space can be written as

$$\mathbf{A} = \frac{dv}{dt}\mathbf{T} + v^2\kappa\mathbf{N}.$$

8) A particle moves on a circle of radius a with constant angular speed ω. Show that the acceleration vector is given by $a\omega^2$ in magnitude and is always directed toward the center of the circle.

9) If a particle moves on the curve given by $\mathbf{R} = (a\cos nt)\mathbf{e}_1 + (b\sin nt)\mathbf{e}_2$ show that the acceleration vector is given by $-n^2\mathbf{R}$ and is therefore always directed toward the origin.

10) A particle of mass m moves through a resisting medium. The resistance $\mathcal{R}$ is a function of the speed and is directed along the tangent to the curve of the motion. If ϕ is the angle which the velocity vector makes with the x^1-axis, show that the differential equations of motion of the particle are

$$m\frac{dv}{dt} = -mg\sin\phi - \mathcal{R}, \qquad m\frac{v^2}{\rho} = mg\cos\phi.$$

11) If a force vector is $\mathbf{F} = q^i\mathbf{e}_i$ in an orthogonal cartesian system x^i and the transformation $x^1 = \bar{x}^1 + \bar{x}^2 - \bar{x}^3$, $x^2 = \bar{x}^1 - \bar{x}^2 + \bar{x}^3$, $x^3 = -\bar{x}^1 + \bar{x}^2 + \bar{x}^3$ is effected, what are the components of the force vector $\mathbf{F}$ in the $\bar{x}^i$ system? Ans. $Q^1 = \frac{1}{2}(q^1 + q^2)$, $Q^2 = \frac{1}{2}(q^1 + q^3)$, $Q^3 = \frac{1}{2}(q^2 + q^3)$.

12) Show that if the curve $x^3 = f(x^1)$, $x^2 = 0$ is revolved about the x^3-axis, the surface of revolution generated has the parametric equations

$$x^1 = u^1\cos u^2, \qquad x^2 = u^1\sin u^2, \qquad x^3 = f(u^1).$$

Describe the coordinate curves $u^1 = $ constant, $u^2 = $ constant. Find the expression $ds^2 = g_{\alpha\beta}\,du^\alpha\,du^\beta$ for this surface.

13) For the helicoid given by

$$x^1 = u^1\cos u^2, \qquad x^2 = u^1\sin u^2, \qquad x^3 = bu^2,$$

show that the parametric curves are circular helices and straight lines. Find the element of arc ds for this surface.

14) Find the component of grad ϕ along the normal to the surface $x^1 = (u^1)^2 + (u^2)^2$, $x^2 = (u^1)^2 + u^1u^2 + (u^2)^2$, $x^3 = (u^1)^3 + (u^2)^3$, at the point $u^1 = 1$, $u^2 = 0$, where $\phi \equiv (x^1)^2 + (x^2)^2 + (x^3)^2$. Ans. $-2/\sqrt{13}$.

15) Show that the surface $x^i = a_1{}^i u^1 + a_2{}^i u^2 + a_3{}^i u^1 u^2$ ($i = 1, 2, 3$) is a hyperbolic paraboloid if $|a_j{}^i| \neq 0$, and a pair of planes if $|a_j{}^i| = 0$. If $|a_j{}^i| \neq 0$,

find the equation of the tangent plane to the surface at the point where $u^1 = u^2 = 0$. Ans. $A_i{}^3 X^i = 0$, ($A_i{}^3 \equiv$ cofactor of $a_3{}^i$ in $|a_j{}^i|$).

16) For the surface defined by
$$x^1 = \frac{u^1 + u^2}{1 + u^1 u^2}, \qquad x^2 = \frac{u^1 - u^2}{1 + u^1 u^2}, \qquad x^3 = \frac{1 - u^1 u^2}{1 + u^1 u^2},$$
what is the curve $u^1 = u^2$? Describe the curve $u^1 u^2 = 2$.

17) Show that the angle ω between the curves $u^1 =$ constant and $u^2 =$ constant on the surface in Exercise 16 is given by $\cos \omega = (u^1 - u^2)^2 / [1 + (u^1)^2][1 + (u^2)^2]$. Deduce that at all points of the curve $u^1 = u^2$, the coordinate curves meet at right angles.

18) On the surface $x^i = x^i(u^1, u^2)$ the family of curves given by $\phi(u^1, u^2) =$ constant has the differential equation $\dfrac{\partial \phi}{\partial u^\alpha} du^\alpha = 0$, which may be written as $\lambda_\alpha du^\alpha = 0$. Show that the differential equation of the family of orthogonal trajectories to the curves $\phi =$ constant on the surface is
$$(\lambda_2 g_{1\alpha} - \lambda_1 g_{2\alpha}) du^\alpha = 0.$$

9
Differentiation of Tensors

9-1. Equivalence of Forms; Christoffel Symbols. In the preceding chapter ordinary and partial derivatives of vectors were a subject of study. It was seen in (5-25) that the partial derivatives $\partial \phi / \partial x^i$ of a scalar point function ϕ (tensor of zero order), constitute the components of a gradient vector (tensor of order one), but that the partial derivatives $\partial^2 \phi / \partial x^i \, \partial x^j$ are not components of a tensor under a general transformation of coordinates. A generalized type of differentiation called *covariant* differentiation will be introduced. By use of it the covariant derivative of a tensor always produces a tensor of one order higher in the covariant indices. In 1869, Christoffel wrote a paper[1] on the equivalence of quadratic differential forms in which he introduced certain symbols which are now called Christoffel symbols, and which appear in the definition of the covariant derivative.

In order to see how these symbols arise in a natural manner and how they are related under a general transformation of coordinates, the equivalence of forms will be considered. The treatment of the equivalence problem will not be exhaustive but will proceed only far enough to yield the desired results.

One may consider a quadratic differential form $a_{ij} \, dx^i \, dx^j$ in a certain coordinate system x^i and another form $\bar{a}_{ij} \, d\bar{x}^i \, d\bar{x}^j$ in $\bar{x}^i$ coordinates. Then one may seek the transformation $x^i = x^i(\bar{x}^1, \cdots, \bar{x}^n)$ under which the first form goes into the second. Here the dimensionality of the space is n. If a transformation can be found, the forms are said to be *equivalent*. Another point of view is provided by considering $a_{ij} \, dx^i \, dx^j$ as the fundamental metric for a space, and $\bar{a}_{ij} \, d\bar{x}^i \, d\bar{x}^j$ the metric of another space. If a mapping $x^i = x^i(\bar{x}^1, \cdots, \bar{x}^n)$ can be found such that $a_{ij} \, dx^i \, dx^j$ transforms into $\bar{a}_{ij} \, d\bar{x}^i \, d\bar{x}^j$, the forms are said to be equivalent and the two

[1] E. B. Christoffel, Über die Transformation der homogenen Differentialausdrücke zweiten Grades. *Jour. fur reine und angew. Math.*, vol. 70, 1869, pp. 46–70.

spaces are said to be *applicable* to each other. The measure of arc length ds in the first space is equal to the arc length $d\bar{s}$ in the second. [Here the space V_n with $a_{ij}\,dx^i\,dx^j$ as metric may be considered as immersed in a linear space of dimensionality $n(n+1)/2$.] An example is the applicability of two surfaces. For instance, a cylindrical surface may be rolled on a plane. A curve on the surface has the same arc length as the curve into which it is developed on the plane. This means that a transformation of coordinates exists under which the formula for ds^2 on the surface transforms into the formula for $d\bar{s}^2$ on the plane. Any two surfaces with the same fundamental quadratic form for the metric are applicable. It should be mentioned that applicability may hold only locally. If a curve is drawn in ink on a closed cylinder rolling on a plane, the curve on the cylinder maps into an image curve on the plane. This image repeats itself during each complete revolution of the cylinder. A portion of the cylinder maps into a portion of the plane in a one-to-one manner.

It has been seen that the law of transformation for the covariant components of a tensor is

1) $$\bar{a}_{pq} = a_{ij}\frac{\partial x^i}{\partial \bar{x}^p}\frac{\partial x^j}{\partial \bar{x}^q}.$$

The desired transformation (or mapping) functions x^i of $\bar{x}^i$ must satisfy the partial differential equations in (1). Because $a_{ij} = a_{ji}$ and $\bar{a}_{pq} = \bar{a}_{qp}$, there are $n(n+1)/2$ equations in (1). Since only n functions $x^i(\bar{x}^1, \cdots, \bar{x}^n)$ are to be found, the a_{ij} and $\bar{a}_{pq}$ must satisfy some relations not containing the partial derivatives $\partial x^i/\partial \bar{x}^j$. It will be of interest to calculate the derivative of $\bar{a}_{pq}$ in (1) with respect to $\bar{x}^r$ to obtain

2) $$\frac{\partial \bar{a}_{pq}}{\partial \bar{x}^r} = \frac{\partial a_{ij}}{\partial x^k}\frac{\partial x^i}{\partial \bar{x}^p}\frac{\partial x^j}{\partial \bar{x}^q}\frac{\partial x^k}{\partial \bar{x}^r} + a_{ij}\left(\frac{\partial^2 x^i}{\partial \bar{x}^p\,\partial \bar{x}^r}\frac{\partial x^j}{\partial \bar{x}^q} + \frac{\partial x^i}{\partial \bar{x}^p}\frac{\partial^2 x^j}{\partial \bar{x}^q\,\partial \bar{x}^r}\right).$$

By cyclic permutation of p, q, r and of i, j, k in (2) one arrives at

3) $$\frac{\partial \bar{a}_{qr}}{\partial \bar{x}^p} = \frac{\partial a_{jk}}{\partial x^i}\frac{\partial x^j}{\partial \bar{x}^q}\frac{\partial x^k}{\partial \bar{x}^r}\frac{\partial x^i}{\partial \bar{x}^p} + a_{jk}\left(\frac{\partial^2 x^j}{\partial \bar{x}^q\,\partial \bar{x}^p}\frac{\partial x^k}{\partial \bar{x}^r} + \frac{\partial x^j}{\partial \bar{x}^q}\frac{\partial^2 x^k}{\partial \bar{x}^r\,\partial \bar{x}^p}\right),$$

and a similar cyclic permutation in (3) yields

4) $$\frac{\partial \bar{a}_{rp}}{\partial \bar{x}^q} = \frac{\partial a_{ki}}{\partial x^j}\frac{\partial x^k}{\partial \bar{x}^r}\frac{\partial x^i}{\partial \bar{x}^p}\frac{\partial x^j}{\partial \bar{x}^q} + a_{ki}\left(\frac{\partial^2 x^k}{\partial \bar{x}^r\,\partial \bar{x}^q}\frac{\partial x^i}{\partial \bar{x}^p} + \frac{\partial x^k}{\partial \bar{x}^r}\frac{\partial^2 x^i}{\partial \bar{x}^p\,\partial \bar{x}^q}\right).$$

By a suitable change of dummy indices, the first term in the parentheses in (2) and the second term in the parentheses in (3) are equal. The second term of the parentheses in (2) and the first term of the parentheses in (4) are equal. The same is true for the remaining pair in parentheses. Now subtract equation (4) from the sum of equations (2) and (3) and divide

Sec. 9-1]   DIFFERENTIATION OF TENSORS   111

by two to obtain

5) $$\frac{1}{2}\left(\frac{\partial \bar{a}_{pq}}{\partial \bar{x}^r} + \frac{\partial \bar{a}_{qr}}{\partial \bar{x}^p} - \frac{\partial \bar{a}_{rp}}{\partial \bar{x}^q}\right)$$
$$= \frac{1}{2}\left(\frac{\partial a_{ij}}{\partial x^k} + \frac{\partial a_{jk}}{\partial x^i} - \frac{\partial a_{ki}}{\partial x^j}\right)\frac{\partial x^i}{\partial \bar{x}^p}\frac{\partial x^j}{\partial \bar{x}^q}\frac{\partial x^k}{\partial \bar{x}^r} + a_{jk}\frac{\partial^2 x^k}{\partial \bar{x}^p \partial \bar{x}^r}\frac{\partial x^j}{\partial \bar{x}^q}.$$

In order to simplify equation (5), some convenient symbols are introduced. The Christoffel symbol of the first kind is defined by

6) $$\Gamma_{ki;j} \equiv \frac{1}{2}\left(\frac{\partial a_{ij}}{\partial x^k} + \frac{\partial a_{jk}}{\partial x^i} - \frac{\partial a_{ki}}{\partial x^j}\right)$$

with respect to the a_{ij} tensor, and therefore by

7) $$\bar{\Gamma}_{rp;q} \equiv \frac{1}{2}\left(\frac{\partial \bar{a}_{pq}}{\partial \bar{x}^r} + \frac{\partial \bar{a}_{qr}}{\partial \bar{x}^p} - \frac{\partial \bar{a}_{rp}}{\partial \bar{x}^q}\right)$$

with respect to the $\bar{a}_{pq}$ components of the same tensor. Use of (6) and (7) in (5) gives

8) $$\bar{\Gamma}_{rp;q} = \Gamma_{ki;j}\frac{\partial x^i}{\partial \bar{x}^p}\frac{\partial x^j}{\partial \bar{x}^q}\frac{\partial x^k}{\partial \bar{x}^r} + a_{jk}\frac{\partial^2 x^k}{\partial \bar{x}^p \partial \bar{x}^r}\frac{\partial x^j}{\partial \bar{x}^q}.$$

From the transformation law for a contravariant tensor, that is, from

9) $$\bar{a}^{pq} = a^{ij}\frac{\partial \bar{x}^p}{\partial x^i}\frac{\partial \bar{x}^q}{\partial x^j},$$

it follows that

10) $$\bar{a}^{pq}\frac{\partial x^k}{\partial \bar{x}^q} = a^{ij}\frac{\partial \bar{x}^p}{\partial x^i}\frac{\partial \bar{x}^q}{\partial x^j}\frac{\partial x^k}{\partial \bar{x}^q} = a^{ij}\frac{\partial \bar{x}^p}{\partial x^i}\delta_j{}^k = a^{ik}\frac{\partial \bar{x}^p}{\partial x^i},$$

or, by a change of indices, that

11) $$\bar{a}^{qt}\frac{\partial x^j}{\partial \bar{x}^t} = a^{mj}\frac{\partial \bar{x}^q}{\partial x^m}.$$

Next multiply the left-hand member of (8) by $\bar{a}^{qt}(\partial x^j/\partial \bar{x}^t)$ and the right-hand member by $a^{mj}(\partial \bar{x}^q/\partial x^m)$ to arrive at

12) $$\bar{a}^{qt}\bar{\Gamma}_{rp;q}\frac{\partial x^j}{\partial \bar{x}^t} = a^{mj}\Gamma_{ki;j}\frac{\partial \bar{x}^q}{\partial x^m}\frac{\partial x^i}{\partial \bar{x}^p}\frac{\partial x^j}{\partial \bar{x}^q}\frac{\partial x^k}{\partial \bar{x}^r} + a^{mj}a_{lk}\frac{\partial^2 x^k}{\partial \bar{x}^p \partial \bar{x}^r}\frac{\partial x^l}{\partial \bar{x}^q}\frac{\partial \bar{x}^q}{\partial x^m}.$$

The Christoffel symbol of the second kind is defined by

13) $$\Gamma_{ij}{}^k \equiv a^{kl}\Gamma_{ij;l}.$$

By use of (13), and the fact that

$$\frac{\partial x^l}{\partial \bar{x}^q}\frac{\partial \bar{x}^q}{\partial x^m} = \delta_m{}^l,$$

(12) becomes

14) $$\bar{\Gamma}_{rp}{}^t \frac{\partial x^j}{\partial \bar{x}^t} = \Gamma_{ki}{}^j \frac{\partial x^i}{\partial \bar{x}^p} \frac{\partial x^k}{\partial \bar{x}^r} + a^{mj} a_{lk} \delta_m{}^l \frac{\partial^2 x^k}{\partial \bar{x}^p \partial \bar{x}^r}.$$

The Kronecker delta in (14) leaves the survivor $a^{mj}a_{mk}$, which is, in turn, $\delta_k{}^j$. Hence, (14) leads to

15) $$\bar{\Gamma}_{rp}{}^t \frac{\partial x^j}{\partial \bar{x}^t} = \Gamma_{ki}{}^j \frac{\partial x^i}{\partial \bar{x}^p} \frac{\partial x^k}{\partial \bar{x}^r} + \frac{\partial^2 x^j}{\partial \bar{x}^p \partial \bar{x}^r},$$

which is the important law of transformation of the Christoffel symbols of the second kind.

Two observations should be made here. First, from the definition of the Christoffel symbols, it is seen that $\Gamma_{ij}{}^k = \Gamma_{ji}{}^k$; that is, the symbols are symmetric in the lower indices. Second, the quantities $\Gamma_{ij}{}^k$ are not components of a tensor for a general change of coordinates. They are, however, components of a tensor for a linear transformation of coordinates, because for such a transformation one has $\partial^2 x^j / \partial \bar{x}^p \partial \bar{x}^q \equiv 0$, and (15) in this case is merely an instance of the transformation law

$$\bar{b}_{rp}{}^t \frac{\partial x^j}{\partial \bar{x}^t} = b_{ki}{}^j \frac{\partial x^i}{\partial \bar{x}^p} \frac{\partial x^k}{\partial \bar{x}^r},$$

which, by multiplication by $\partial \bar{x}^q / \partial x^j$, is seen to be equivalent to

$$\bar{b}_{rp}{}^q = b_{ki}{}^j \frac{\partial x^i}{\partial \bar{x}^p} \frac{\partial x^k}{\partial \bar{x}^r} \frac{\partial \bar{x}^q}{\partial x^j}.$$

Multiplication of (15) by $\partial \bar{x}^q / \partial x^j$ leads to the useful explicit form for $\bar{\Gamma}_{rp}{}^q$, that is,

16) $$\bar{\Gamma}_{rp}{}^q = \Gamma_{ki}{}^j \frac{\partial x^i}{\partial \bar{x}^p} \frac{\partial x^k}{\partial \bar{x}^r} \frac{\partial \bar{x}^q}{\partial x^j} + \frac{\partial \bar{x}^q}{\partial x^j} \frac{\partial^2 x^j}{\partial \bar{x}^p \partial \bar{x}^r}.$$

9–2. The Riemann-Christoffel Tensor. It should be realized that the theory of quadratic differential forms is completely equivalent to that of symmetric covariant tensors of the second order. For to any symmetric tensor b_{ij} of the second order there corresponds the quadratic form $b_{ij} dx^i dx^j$. It will be recalled that b_{ij} is symmetric if $b_{ij} = b_{ji}$.

On demanding the equivalence of two forms $a_{ij} dx^i dx^j$ and $\bar{a}_{pq} d\bar{x}^p d\bar{x}^q$, one is brought to the conditions expressed by the partial differential equations in (1). For such a system of equations, the integrability conditions should be investigated. It would be expected to demand of the solution functions $x^i(\bar{x}^1, \cdots, \bar{x}^n)$ of equations (1) that

17) $$\frac{\partial^2 x^j}{\partial \bar{x}^p \partial \bar{x}^r} = \frac{\partial^2 x^j}{\partial \bar{x}^r \partial \bar{x}^p},$$

but this leads to no condition on the coefficients of equations (1) because the single equation obtained by differentiating (1) is symmetric in p and r. Also, it is evidently impossible to eliminate the second derivatives from equations (1) and (15), so it is clear that there is no tensor whose components are functions of the first derivatives of the a_{ij}, or of these derivatives and the a_{ij} components themselves. It will be seen, however, that by going on to the third derivatives, a tensor arises which involves the second derivatives of the a_{ij}. The integrability condition of equations (1) and (15) involving the third derivatives requires that

18) $$\frac{\partial^3 x^i}{\partial \bar{x}^p \, \partial \bar{x}^q \, \partial \bar{x}^r} = \frac{\partial^3 x^i}{\partial \bar{x}^p \, \partial \bar{x}^r \, \partial \bar{x}^q}.$$

Let equation (15) be written as

19) $$\frac{\partial^2 x^j}{\partial \bar{x}^p \, \partial \bar{x}^r} + \Gamma_{ki}{}^j \frac{\partial x^i}{\partial \bar{x}^p} \frac{\partial x^k}{\partial \bar{x}^r} = \bar{\Gamma}_{pr}{}^t \frac{\partial x^j}{\partial \bar{x}^t},$$

and write the same equation with r replaced by q to have

20) $$\frac{\partial^2 x^j}{\partial \bar{x}^p \, \partial \bar{x}^q} + \Gamma_{ki}{}^j \frac{\partial x^i}{\partial \bar{x}^p} \frac{\partial x^k}{\partial \bar{x}^q} = \bar{\Gamma}_{pq}{}^t \frac{\partial x^j}{\partial \bar{x}^t}.$$

Now, in order to satisfy (18), differentiate (19) with respect to $\bar{x}^q$, and (20) with respect to $\bar{x}^r$ to obtain

21) $$\frac{\partial^3 x^j}{\partial \bar{x}^p \, \partial \bar{x}^q \, \partial \bar{x}^r} + \frac{\partial \Gamma_{ki}{}^j}{\partial x^l} \frac{\partial x^i}{\partial \bar{x}^p} \frac{\partial x^k}{\partial \bar{x}^r} \frac{\partial x^l}{\partial \bar{x}^q} + \Gamma_{ki}{}^j \frac{\partial^2 x^i}{\partial \bar{x}^p \, \partial \bar{x}^q} \frac{\partial x^k}{\partial \bar{x}^r}$$
$$+ \Gamma_{ki}{}^j \frac{\partial x^i}{\partial \bar{x}^p} \frac{\partial^2 x^k}{\partial \bar{x}^r \, \partial \bar{x}^q} = \frac{\partial \bar{\Gamma}_{pr}{}^t}{\partial \bar{x}^q} \frac{\partial x^j}{\partial \bar{x}^t} + \bar{\Gamma}_{pr}{}^t \frac{\partial^2 x^j}{\partial \bar{x}^t \, \partial \bar{x}^q},$$

and

22) $$\frac{\partial^3 x^j}{\partial \bar{x}^p \, \partial \bar{x}^q \, \partial \bar{x}^r} + \frac{\partial \Gamma_{ki}{}^j}{\partial x^l} \frac{\partial x^i}{\partial \bar{x}^p} \frac{\partial x^k}{\partial \bar{x}^q} \frac{\partial x^l}{\partial \bar{x}^r} + \Gamma_{ki}{}^j \frac{\partial^2 x^i}{\partial \bar{x}^p \, \partial \bar{x}^r} \frac{\partial x^k}{\partial \bar{x}^q}$$
$$+ \Gamma_{ki}{}^j \frac{\partial x^i}{\partial \bar{x}^p} \frac{\partial^2 x^k}{\partial \bar{x}^q \, \partial \bar{x}^r} = \frac{\partial \bar{\Gamma}_{pq}{}^t}{\partial \bar{x}^r} \frac{\partial x^j}{\partial \bar{x}^t} + \bar{\Gamma}_{pq}{}^t \frac{\partial^2 x^j}{\partial \bar{x}^t \, \partial \bar{x}^r}.$$

The first terms of the last two equations are equal by (18), and the fourth terms are also equal. Subtraction of (22) from (21) therefore yields

23) $$\left(\frac{\partial \Gamma_{li}{}^j}{\partial x^k} - \frac{\partial \Gamma_{ki}{}^j}{\partial x^l} \right) \frac{\partial x^i}{\partial \bar{x}^p} \frac{\partial x^k}{\partial \bar{x}^q} \frac{\partial x^l}{\partial \bar{x}^r} + \Gamma_{ki}{}^j \left(\frac{\partial^2 x^i}{\partial \bar{x}^p \, \partial \bar{x}^q} \frac{\partial x^k}{\partial \bar{x}^r} - \frac{\partial^2 x^i}{\partial \bar{x}^p \, \partial \bar{x}^r} \frac{\partial x^k}{\partial \bar{x}^q} \right)$$
$$= \left(\frac{\partial \bar{\Gamma}_{pr}{}^t}{\partial \bar{x}^q} - \frac{\partial \bar{\Gamma}_{pq}{}^t}{\partial \bar{x}^r} \right) \frac{\partial x^j}{\partial \bar{x}^t} + \bar{\Gamma}_{pr}{}^t \frac{\partial^2 x^j}{\partial \bar{x}^t \, \partial \bar{x}^q} - \bar{\Gamma}_{pq}{}^t \frac{\partial^2 x^j}{\partial \bar{x}^t \, \partial \bar{x}^r}.$$

The second derivatives can be eliminated from (23) by use of (19) or (20) to obtain

$$24)\quad \left(\frac{\partial \Gamma_{li}^{j}}{\partial x^{k}} - \frac{\partial \Gamma_{ki}^{j}}{\partial x^{l}}\right)\frac{\partial x^{i}}{\partial \bar{x}^{p}}\frac{\partial x^{k}}{\partial \bar{x}^{q}}\frac{\partial x^{l}}{\partial \bar{x}^{r}} + \Gamma_{ki}^{j}\frac{\partial x^{k}}{\partial \bar{x}^{r}}\left(\bar{\Gamma}_{pq}^{t}\frac{\partial x^{i}}{\partial \bar{x}^{t}} - \Gamma_{ls}^{i}\frac{\partial x^{s}}{\partial \bar{x}^{p}}\frac{\partial x^{l}}{\partial \bar{x}^{q}}\right)$$

$$- \Gamma_{ki}^{j}\frac{\partial x^{k}}{\partial \bar{x}^{q}}\left(\bar{\Gamma}_{pr}^{t}\frac{\partial x^{i}}{\partial \bar{x}^{t}} - \Gamma_{ls}^{i}\frac{\partial x^{s}}{\partial \bar{x}^{p}}\frac{\partial x^{l}}{\partial \bar{x}^{r}}\right)$$

$$= \left(\frac{\partial \bar{\Gamma}_{pr}^{t}}{\partial \bar{x}^{q}} - \frac{\partial \bar{\Gamma}_{pq}^{t}}{\partial \bar{x}^{r}}\right)\frac{\partial x^{j}}{\partial \bar{x}^{t}} + \bar{\Gamma}_{pr}^{t}\left(\bar{\Gamma}_{tq}^{s}\frac{\partial x^{j}}{\partial \bar{x}^{s}} - \Gamma_{ki}^{j}\frac{\partial x^{i}}{\partial \bar{x}^{t}}\frac{\partial x^{k}}{\partial \bar{x}^{q}}\right)$$

$$- \bar{\Gamma}_{pq}^{t}\left(\bar{\Gamma}_{tr}^{s}\frac{\partial x^{j}}{\partial \bar{x}^{s}} - \Gamma_{ki}^{j}\frac{\partial x^{i}}{\partial \bar{x}^{t}}\frac{\partial x^{k}}{\partial \bar{x}^{r}}\right).$$

The terms involving a product of Γ and $\bar{\Gamma}$ symbols in (24) cancel. After a rearrangement of dummy indices, (24) becomes

$$25)\quad \left(\frac{\partial \Gamma_{li}^{j}}{\partial x^{k}} - \frac{\partial \Gamma_{ki}^{j}}{\partial x^{l}} - \Gamma_{lm}^{j}\Gamma_{ki}^{m} + \Gamma_{km}^{j}\Gamma_{li}^{m}\right)\frac{\partial x^{i}}{\partial \bar{x}^{p}}\frac{\partial x^{k}}{\partial \bar{x}^{q}}\frac{\partial x^{l}}{\partial \bar{x}^{r}}$$

$$= \left(\frac{\partial \bar{\Gamma}_{pr}^{t}}{\partial \bar{x}^{q}} - \frac{\partial \bar{\Gamma}_{pq}^{t}}{\partial \bar{x}^{r}} + \bar{\Gamma}_{pr}^{s}\bar{\Gamma}_{sq}^{t} - \bar{\Gamma}_{pq}^{s}\bar{\Gamma}_{sr}^{t}\right)\frac{\partial x^{j}}{\partial \bar{x}^{t}}.$$

Next define

$$26)\quad R^{j}{}_{ikl} \equiv \frac{\partial \Gamma_{li}^{j}}{\partial x^{k}} - \frac{\partial \Gamma_{ki}^{j}}{\partial x^{l}} - \Gamma_{ki}^{m}\Gamma_{lm}^{j} + \Gamma_{li}^{m}\Gamma_{km}^{j},$$

by which (25) takes the form

$$27)\quad R^{j}{}_{ikl}\frac{\partial x^{i}}{\partial \bar{x}^{p}}\frac{\partial x^{k}}{\partial \bar{x}^{r}}\frac{\partial x^{l}}{\partial \bar{x}^{q}} = \bar{R}^{t}{}_{pqr}\frac{\partial x^{j}}{\partial \bar{x}^{t}}.$$

Multiplication of (27) by $\partial \bar{x}^{s}/\partial x^{j}$ and use of the fact that

$$\frac{\partial x^{j}}{\partial \bar{x}^{t}}\frac{\partial \bar{x}^{s}}{\partial x^{j}} = \delta_{t}^{s}$$

lead to

$$28)\quad \bar{R}^{s}{}_{pqr} = R^{j}{}_{ikl}\frac{\partial x^{i}}{\partial \bar{x}^{p}}\frac{\partial x^{k}}{\partial \bar{x}^{q}}\frac{\partial x^{l}}{\partial \bar{x}^{r}}\frac{\partial \bar{x}^{s}}{\partial x^{j}},$$

which reveals that $R^{j}{}_{ikl}$ as defined in (26) are components of a tensor of the fourth order. It is called the Riemann-Christoffel tensor. If it is desired to have all indices covariant, the contravariant index may be lowered by multiplying $R^{j}{}_{ikl}$ by the tensor of the fundamental metric to give

$$29)\quad R_{hikl} = a_{hj}R^{j}{}_{ikl}.$$

By analogy with the Christoffel symbols, R_{hikl} are called the Riemann symbols of the first kind, and $R^{h}{}_{ikl}$ Riemann symbols of the second kind.

The Riemann-Christoffel tensor plays a fundamental role in differential geometry and in Einstein's relativity theory. It appeared here in the equivalence problem of quadratic forms as an integrability condition of equations (1). The following observations should be made.

The tensor components $R^j{}_{ikl}$ are functions of first and second partial derivatives of the components of the fundamental tensor a_{ij}. It follows that since $a_{ij} = \delta_{ij}$ in the orthogonal cartesian system of coordinates in euclidean three-space, all the components of the R-C tensor are zero. But if a tensor has components zero in one coordinate system, the components are zero in any other coordinate system. Hence, for Euclidean three-space, the R-C tensor is a zero tensor.

Notice that the Christoffel symbols may not be zero for some coordinate system in Euclidean three-space, but that the functions of these symbols in (26) for any coordinate system are all zero in Euclidean space. The tensor $R^i{}_{jkl}$ appears in a measure of curvature of a space, which will be demonstrated for a surface in Section 12-5. Because a euclidean space has zero curvature, it is sometimes called a "flat" space.

Observe that the symbols $\Gamma_{ij}{}^k$ are symmetric in the i and j, but that $R^l{}_{kij}$ are skew-symmetric in i and j. The number of distinct symbols $\Gamma_{ij}{}^k$ in n-space can be calculated easily. Because of symmetry in i and j, the number of combinations of i and j is $(n^2 - n)/2 + n$ or $n(n + 1)/2$. The third index k may be assigned in n ways. Therefore, the total number of $\Gamma_{ij}{}^k$ symbols in n-space is $n^2(n + 1)/2$. For a two-dimensional surface there are six symbols given by $\Gamma_{11}{}^1$, $\Gamma_{11}{}^2$, $\Gamma_{12}{}^1$, $\Gamma_{12}{}^2$, $\Gamma_{22}{}^1$, $\Gamma_{22}{}^2$.

The number of distinct components of the Riemann-Christoffel tensor in n-space can be shown to be $n^2(n^2 - 1)/12$. For an ordinary curved surface ($n = 2$) there is only one non-zero component, which may be written as R_{1212}. It is involved in the formula for the Gaussian curvature of a surface at a point (12-77). Any geometric property of a space which is described in terms of the components $R^i{}_{jkl}$ must be intrinsic because $R^i{}_{jkl}$ involve only the a_{ij} and their derivatives.

EXERCISES

1) Use (26) to show that $R^i{}_{ikl} = 0$.

2) Verify that $R^i{}_{jkl}$ is skew-symmetric in k and l, that is, show that $R^i{}_{jkl} = -R^i{}_{jlk}$.

3) Contraction of $R^i{}_{jkl}$ on i and j yields a zero tensor (Exercise 1). Contraction of $R^i{}_{jkl}$ on i and l yields the Ricci tensor $R_{jk} \equiv R^i{}_{jki}$. Write the component R_{11} in a space for which $n = 2$.

4) Calculate R_{1212} for the sphere given by

$$x^1 = a \sin u^1 \cos u^2, \qquad x^2 = a \sin u^1 \sin u^2, \qquad x^3 = a \cos u^1.$$

(*Hint:* From the formula for ds^2 in three-space, calculate the fundamental metric $g_{\alpha\beta}\, du^\alpha\, du^\beta$ for the surface. Then obtain the Christoffel symbols based on $g_{\alpha\beta}$ for use in the R-C tensor.)

5) Show that the cylinder with Gauss equations
$$x^1 = \cos u^1, \qquad x^2 = \sin u^1, \qquad x^3 = u^2$$
can be mapped, with preservation of arc length, onto a strip of the plane
$$\bar{x}^1 = \bar{u}^1, \qquad \bar{x}^2 = \bar{u}^2, \qquad \bar{x}^3 = 0,$$
by the mapping $\bar{u}^1 = u^1 + c^1$, $\bar{u}^2 = u^2 + c^2$, where c^1 and c^2 are constants. Note that $0 \leq u^1 < 2\pi$ is required.

6) Find a mapping by which the quadratic form ds^2 for the surface
$$x^1 = f(u^1), \qquad x^2 = g(u^1), \qquad x^3 = u^2$$
becomes the form $(d\bar{s})^2 = (d\bar{u}^1)^2 + (d\bar{u}^2)^2$ for the plane of Exercise 5. Use the mapping to show that any curve on the cylindrical surface which makes a constant angle with a generator (that is, a helix) is given by an equation of the type
$$c_1 \int \sqrt{(f')^2 + (g')^2}\, du^1 + c_2 u^2 + c_3 = 0,$$
where c_i are constants.

9–3. Covariant Differentiation; Parallelism of Vectors. Consider a curve $C: x^i = x^i(t)$ in euclidean three-space and a field of parallel vectors with components λ^i in any coordinate system (not necessarily orthogonal cartesian) and components $\bar{\lambda}^i$ in any other coordinate system $\bar{x}^i$. At each point P of C there is a vector of the field. As P moves along C the vector remains parallel to itself. In an orthogonal cartesian system of coordinates the λ^i components are constants at every point of space, but in other coordinates the λ^i are functions of position and therefore functions of t along C. With the transformation $x^i = x^i(\bar{x}^1, \bar{x}^2, \bar{x}^3)$ relating x^i and $\bar{x}^i$, the λ^i and $\bar{\lambda}^i$ are related by

30)
$$\lambda^i = \bar{\lambda}^j \frac{\partial x^i}{\partial \bar{x}^j}.$$

The immediate goal is to find what conditions are imposed upon λ^i and $\bar{\lambda}^i$ if the vector field is a set of parallel vectors. To this end, differentiate (30) with respect to t to find

31)
$$\frac{\partial \lambda^i}{\partial x^p}\frac{dx^p}{dt} = \frac{\partial \bar{\lambda}^j}{\partial \bar{x}^q}\frac{d\bar{x}^q}{dt}\frac{\partial x^i}{\partial \bar{x}^j} + \bar{\lambda}^r \frac{\partial^2 x^i}{\partial \bar{x}^p\, \partial \bar{x}^r}\frac{d\bar{x}^p}{dt}.$$

Equation (15) is used to eliminate the second derivative in (31) to find

32)
$$\frac{\partial \lambda^i}{\partial x^p}\frac{dx^p}{dt} = \frac{\partial \bar{\lambda}^j}{\partial \bar{x}^q}\frac{d\bar{x}^q}{dt}\frac{\partial x^i}{\partial \bar{x}^j} + \bar{\lambda}^r \left(\bar{\Gamma}_{rp}{}^t \frac{\partial x^i}{\partial \bar{x}^t} - \Gamma_{kl}{}^i \frac{\partial x^l}{\partial \bar{x}^p}\frac{\partial x^k}{\partial \bar{x}^r} \right) \frac{d\bar{x}^p}{dt},$$

or

33) $$\frac{\partial \lambda^i}{\partial x^p}\frac{dx^p}{dt} + \bar{\lambda}^r \Gamma_{kl}{}^i \frac{\partial x^l}{\partial \bar{x}^p}\frac{\partial x^k}{\partial \bar{x}^r}\frac{d\bar{x}^p}{dt} = \left(\frac{\partial \bar{\lambda}^j}{\partial \bar{x}^p}\frac{\partial x^i}{\partial \bar{x}^j} + \bar{\lambda}^r \bar{\Gamma}_{rp}{}^j \frac{\partial x^i}{\partial \bar{x}^j}\right)\frac{d\bar{x}^p}{dt}.$$

If $\bar{\lambda}^r$ is replaced by $\lambda^m(\partial \bar{x}^r/\partial x^m)$ in the left-hand member of (33), and suitable adjustments are made in the dummy indices, (33) becomes

34) $$\left(\frac{\partial \lambda^i}{\partial x^p} + \lambda^m \Gamma_{mp}{}^i\right)\frac{dx^p}{dt} = \frac{\partial x^i}{\partial \bar{x}^j}\left(\frac{\partial \bar{\lambda}^j}{\partial \bar{x}^p} + \bar{\lambda}^r \bar{\Gamma}_{rp}{}^j\right)\frac{d\bar{x}^p}{dt}.$$

Now define

35) $$\lambda^i{}_{,p} \equiv \frac{\partial \lambda^i}{\partial x^p} + \lambda^m \Gamma_{mp}{}^i.$$

The expression which $\lambda^i{}_{,p}$ stands for is called the *covariant derivative* of λ^i with respect to x^p. By analogy with the derivative

$$\frac{d\lambda^i}{dt} = \frac{\partial \lambda^i}{\partial x^p}\frac{dx^p}{dt},$$

one may define

36) $$\frac{\delta \lambda^i}{\delta t} \equiv \lambda^i{}_{,p}\frac{dx^p}{dt}.$$

This symbol $\delta\lambda^i/\delta t$ stands for the *intrinsic* derivative of λ^i with respect to t. Note that if the Christoffel symbols in (35) are all zero, the covariant derivative becomes the partial derivative, and the intrinsic derivative in (36) becomes the ordinary derivative. Equation (34) can be written in simpler form now as

37) $$\lambda^i{}_{,p}\frac{dx^p}{dt} = \frac{\partial x^i}{\partial \bar{x}^j}\bar{\lambda}^j{}_{,p}\frac{d\bar{x}^p}{dt},$$

or

38) $$\frac{\delta \lambda^i}{\delta t} = \frac{\delta \bar{\lambda}}{\delta t}\frac{\partial x^i}{\partial \bar{x}^j}.$$

Formula (38) holds for two general coordinate systems. Now suppose the x^i system of coordinates is orthogonal cartesian and the λ^i are constants, that is, the vectors along C are parallel. The intrinsic derivative on the left-hand side in (38) reduces to the ordinary derivative $d\lambda^i/dt$. Because $d\lambda^i/dt = 0$, (38) gives

39) $$\frac{\delta \bar{\lambda}^j}{\delta t} = 0,$$

which, from (34), means

40) $$\left(\frac{\partial \bar{\lambda}^j}{\partial \bar{x}^p} + \bar{\lambda}^r \bar{\Gamma}_{rp}{}^j\right)\frac{d\bar{x}^p}{dt} = 0$$

or

41) $$\frac{d\bar{\lambda}^j}{dt} + \bar{\lambda}^r \bar{\Gamma}_{rp}{}^j \frac{d\bar{x}^p}{dt} = 0.$$

Equations (41) are the differential equations which the $\bar{\lambda}^j$ components of the parallel vector field in the $\bar{x}^i$ system must satisfy as the vector moves along the curve C. The corresponding differential equations in the orthogonal cartesian system are $d\lambda^i/dt = 0$. If a set of values, say c^i, of λ^i at, say, $t = 0$ are given, then the differential equation requires that $\lambda^i = c^i$ all along the curve. Notice that the set of initial values and the differential equation together determine the λ^i components along the curve. The theory of a system of ordinary differential equations like (41) states that if a set of components $\bar{\lambda}^i$ are prescribed at $t = 0$, then the equations determine $\bar{\lambda}^i$ along the curve. Although the λ^i are constants, the $\bar{\lambda}^i$ vary from point to point, and they are solutions of (41) at each point of the curve. Equations (41) may be styled the *conditions of parallelism* of the vector field in the $\bar{x}^i$ coordinates.

It is understood that the parallelism referred to here is the ordinary euclidean parallelism. If one considers tangent vectors to a surface, say a sphere, along a curve on the sphere, he observes that the vectors cannot remain parallel in the euclidean sense. However, a definition of parallelism in curved spaces will be introduced which will reduce to ordinary euclidean parallelism if the V_n is a linear space. And the equations defining such curved-space parallelism will be found to be of the form of the system (41).

Use of the definition of $\lambda^i{}_{,p}$ in (35) and the fact that $d\bar{x}^q = (\partial \bar{x}^q / \partial x^p)\, dx^p$ allows equations (34) to take the form

42) $$\left(\lambda^i{}_{,p} - \bar{\lambda}^j{}_{,q} \frac{\partial x^i}{\partial \bar{x}^j} \frac{\partial \bar{x}^q}{\partial x^p} \right) dx^p = 0.$$

Since (42) must hold for arbitrary dx^p, it follows that

43) $$\lambda^i{}_{,p} = \bar{\lambda}^j{}_{,q} \frac{\partial x^i}{\partial \bar{x}^j} \frac{\partial \bar{x}^q}{\partial x^p},$$

which shows that the covariant derivative $\lambda^i{}_{,p}$ defined in (35) is a mixed tensor of the second order.

9–4. Covariant Derivative of Covariant Tensors. In order to determine the form of the covariant derivative of a covariant vector λ_i with respect to x^p, take the partial derivative of the transformation law

44) $$\bar{\lambda}_j = \lambda_i \frac{\partial x^i}{\partial \bar{x}^j}$$

with respect to $\bar{x}^q$ to obtain

$$45) \quad \frac{\partial \bar{\lambda}_j}{\partial \bar{x}^q} = \frac{\partial \lambda_i}{\partial x^l} \frac{\partial x^l}{\partial \bar{x}^q} \frac{\partial x^i}{\partial \bar{x}^j} + \lambda_i \frac{\partial^2 x^i}{\partial \bar{x}^q \partial \bar{x}^j}.$$

On using equation (15) to eliminate the second derivative in (45), and on replacing λ_i by $\bar{\lambda}_m (\partial \bar{x}^m / \partial x^i)$, equation (45) becomes

$$46) \quad \frac{\partial \bar{\lambda}_j}{\partial \bar{x}^q} - \bar{\lambda}_m \frac{\partial \bar{x}^m}{\partial x^i} \bar{\Gamma}_{jq}{}^t \frac{\partial x^i}{\partial \bar{x}^t} = \left(\frac{\partial \lambda_k}{\partial x^l} - \lambda_i \Gamma_{kl}{}^i \right) \frac{\partial x^l}{\partial \bar{x}^q} \frac{\partial x^k}{\partial \bar{x}^j}$$

which reduces to

$$47) \quad \left(\frac{\partial \bar{\lambda}_j}{\partial \bar{x}^q} - \bar{\lambda}_t \bar{\Gamma}_{jq}{}^t \right) = \left(\frac{\partial \lambda_k}{\partial x^l} - \lambda_i \Gamma_{kl}{}^i \right) \frac{\partial x^l}{\partial \bar{x}^q} \frac{\partial x^k}{\partial \bar{x}^j}.$$

It is evident from (47) that the quantities in the parentheses are components of a covariant tensor of the second order. Use the notation

$$48) \quad \lambda_{k,l} \equiv \frac{\partial \lambda_k}{\partial x^l} - \lambda_i \Gamma_{kl}{}^i$$

to write (47) in the form

$$49) \quad \bar{\lambda}_{j,q} = \lambda_{k,l} \frac{\partial x^k}{\partial \bar{x}^j} \frac{\partial x^l}{\partial \bar{x}^q}.$$

Notice the difference in form between the covariant derivative of a contravariant vector in (35) and the covariant derivative of a covariant vector in (48). (A contravariant derivative of a tensor can be defined, but this will not be necessary.)

The covariant derivative $b_{ij,k}$ of a covariant tensor b_{ij} of the second order will now be considered. Take the partial derivative with respect to $\bar{x}^r$ of the transformation law

$$50) \quad \bar{b}_{pq} = b_{ij} \frac{\partial x^i}{\partial \bar{x}^p} \frac{\partial x^j}{\partial \bar{x}^q}$$

and make use of (15) to eliminate the second derivatives to obtain

$$51) \quad \frac{\partial \bar{b}_{pq}}{\partial x^r} = \frac{\partial b_{ij}}{\partial x^k} \frac{\partial x^i}{\partial \bar{x}^p} \frac{\partial x^j}{\partial \bar{x}^q} \frac{\partial x^k}{\partial \bar{x}^r} + b_{ij} \left[\left(\bar{\Gamma}_{pr}{}^m \frac{\partial x^i}{\partial \bar{x}^m} - \Gamma_{lk}{}^i \frac{\partial x^l}{\partial \bar{x}^p} \frac{\partial x^k}{\partial \bar{x}^r} \right) \frac{\partial x^j}{\partial \bar{x}^q} \right.$$
$$\left. + \left(\bar{\Gamma}_{qr}{}^m \frac{\partial x^j}{\partial \bar{x}^m} - \Gamma_{lk}{}^j \frac{\partial x^l}{\partial \bar{x}^q} \frac{\partial x^k}{\partial \bar{x}^r} \right) \frac{\partial x^i}{\partial \bar{x}^p} \right].$$

Now, by use of (50) and a suitable rearrangement of dummy indices, equation (51) can be written in the form

$$52) \quad \left(\frac{\partial \bar{b}_{pq}}{\partial \bar{x}^r} - \bar{b}_{mq}\bar{\Gamma}_{pr}{}^m - \bar{b}_{pm}\bar{\Gamma}_{qr}{}^m\right)$$
$$= \left(\frac{\partial b_{ij}}{\partial x^k} - b_{lj}\Gamma_{ik}{}^l - b_{il}\Gamma_{jk}{}^l\right)\frac{\partial x^i}{\partial \bar{x}^p}\frac{\partial x^j}{\partial \bar{x}^q}\frac{\partial x^k}{\partial \bar{x}^r}.$$

With the notation

$$53) \quad b_{ij,k} \equiv \frac{\partial b_{ij}}{\partial x^k} - b_{lj}\Gamma_{ik}{}^l - b_{il}\Gamma_{jk}{}^l,$$

equation (52) reads

$$54) \quad \bar{b}_{pq,r} = b_{ij,k}\frac{\partial x^i}{\partial \bar{x}^p}\frac{\partial x^j}{\partial \bar{x}^q}\frac{\partial x^k}{\partial \bar{x}^r},$$

which means that $b_{ij,k}$ and $\bar{b}_{pq,r}$ are components, in the respective coordinate systems x^i and $\bar{x}^i$, of a covariant tensor of the third order. The notation $b_{ij,k}$ indicates the covariant derivative of a second order covariant tensor which results in a covariant tensor of the third order. This shows why the generalized derivative is called a *covariant* derivative.

9–5. Covariant Derivative of a General Tensor. The covariant derivative of a contravariant vector was shown in (35). If the law of transformation of a contravariant tensor of second order, namely,

$$55) \quad \bar{b}^{pq} = b^{ij}\frac{\partial \bar{x}^p}{\partial x^i}\frac{\partial \bar{x}^q}{\partial x^j}$$

is differentiated partially with respect to $\bar{x}^r$, a procedure similar to that used in arriving at (53) can be effected to show that

$$56) \quad b^{ij}{}_{,k} \equiv \frac{\partial b^{ij}}{\partial x^k} + b^{lj}\Gamma_{lk}{}^i + b^{il}\Gamma_{lk}{}^j$$

are components of a mixed tensor of the third order. Observe that, in a covariant derivative of a tensor, a positive term involving a Christoffel Γ is added for each index of contravariance and such a term is subtracted for each index of covariance. The covariant derivative $b_i{}^j{}_{,k}$ of a mixed tensor $b_i{}^j$ of the second order can be shown to be

$$57) \quad b_i{}^j{}_{,k} \equiv \frac{\partial b_i{}^j}{\partial x^k} + b_i{}^l\Gamma_{lk}{}^i - b_l{}^j\Gamma_{ik}{}^l.$$

Now, the formula for the covariant derivative of a more general tensor such as $b_{ijk}{}^{lm}$, indicated by $b_{ijk}{}^{lm}{}_{,p}$, can be constructed readily.

It should be emphasized that the Christoffel symbols $\Gamma_{ij}{}^k$ appearing in the foregoing covariant derivatives have been considered to be evaluated by using the fundamental metric tensor a_{ij} of the space. But note that the Γ's could be evaluated by using any other tensor b_{ij} of the second order. Hence, for clarity in certain investigations one should be careful to state the tensor with respect to which the covariant differentiation is calculated.

It is not difficult to show that the covariant derivative of sums and differences and products of tensors follows the same rule as for ordinary differentiation. For instance, one can show that

58) $$(b_{ij}c_k{}^l)_{,m} = b_{ij,m}c_k{}^l + b_{ij}c_k{}^l{}_{,m}.$$

It can also be shown that the processes of contraction and covariant differentiation are commutative. This means that one may contract first and then take the covariant derivative, or he may take the covariant derivative first and then contract; the results are the same. This may be illustrated by contracting on l and j in (58) and then taking the covariant derivative with respect to x^m to obtain

59) $$(b_{il}c_k{}^l)_{,m} = b_{il,m}c_k{}^l + b_{il}c_k{}^l{}_{,m}.$$

The result is the same if one contracts on l and j in the right-hand member of (58).

9–6. Tensors Which Behave as Constants.

Although the covariant derivative is more cumbersome than the partial derivative, a rewarding feature of the more general covariant differentiation is that

the tensors a_{ij}, a^{ij}, and $\delta_i{}^j$ behave as constants

with respect to covariant differentiation, where a_{ij} is the fundamental metric tensor of the space. In order to show this, recall from formula (9–6) that the Christoffel symbol of first kind is defined by

60) $$2\Gamma_{ki;j} \equiv \frac{\partial a_{ij}}{\partial x^k} + \frac{\partial a_{jk}}{\partial x^i} - \frac{\partial a_{ki}}{\partial x^j}.$$

A cyclic permutation on i, j, k gives

61) $$2\Gamma_{ij;k} \equiv \frac{\partial a_{jk}}{\partial x^i} + \frac{\partial a_{ki}}{\partial x^j} - \frac{\partial a_{ij}}{\partial x^k}.$$

Addition of equations (60) and (61) yields

62) $$\frac{\partial a_{jk}}{\partial x^i} = \Gamma_{ki;j} + \Gamma_{ij;k}.$$

If equation (13) is multiplied by a_{km}, it follows that

63) $$\Gamma_{ij;m} = a_{km}\Gamma_{ij}{}^k,$$

which expresses the Christoffel symbol of first kind in terms of the symbol of second kind. Use of (63) in (62) leads to

$$64)\qquad \frac{\partial a_{jk}}{\partial x^i} = a_{lj}\Gamma_{ki}{}^l + a_{lk}\Gamma_{ij}{}^l.$$

Now the covariant derivative $a_{jk,i}$ can be written from (53), by a permutation of the indices, in the form

$$65)\qquad a_{jk,i} \equiv \frac{\partial a_{jk}}{\partial x^i} - a_{lk}\Gamma_{ij}{}^l - a_{jl}\Gamma_{ki}{}^l.$$

In view of (64), $a_{jk,i}$ in (65) is identically zero.

Next, consider $\delta_i{}^j$. Substitution of $\delta_i{}^j$ for $b_i{}^j$ in equation (57) gives

$$66)\qquad \delta_i{}^j{}_{,k} = \frac{\partial \delta_i{}^j}{\partial x^k} + \delta_i{}^l \Gamma_{lk}{}^j - \delta_l{}^j \Gamma_{ik}{}^l,$$

or

$$67)\qquad \delta_i{}^j{}_{,k} = \Gamma_{ik}{}^j - \Gamma_{ik}{}^j = 0.$$

In order to demonstrate that $a^{ij}{}_{,k} = 0$, take the covariant derivative of the equation

$$68)\qquad a^{ij}a_{ik} = \delta_k{}^j$$

to obtain

$$69)\qquad a^{ij}a_{ik,l} + a^{ij}{}_{,l}a_{ik} = \delta_k{}^j{}_{,l},$$

which, by virtue of $a_{ik,l} = 0$ and $\delta_k{}^j{}_{,l} = 0$, gives

$$70)\qquad a^{ij}{}_{,l}a_{ik} = 0.$$

Multiplication of (70) by a^{km} gives $\delta_i{}^m a^{ij}{}_{,l} = 0$, or finally $a^{mj}{}_{,l} = 0$. The constant behavior of a_{ij}, a^{ij}, and $\delta_i{}^j$ under covariant differentiation has been established.

Because the covariant derivative of a tensor is a tensor, the covariant derivative of a covariant derivative may be calculated, and this is a tensor of order two greater in the covariant indices. For instance, $b_{ij,kl}$ means $(b_{ij,k})_{,l}$. A question naturally arises. Is $b_{ij,kl} = b_{ij,lk}$?

The answer to this question involves the Riemann-Christoffel tensor. To see this, differentiate equation (53) covariantly with respect to x^l to obtain $(b_{ij,k})_{,l}$. Then find $(b_{ij,l})_{,k}$ and form the difference

$$(b_{ij,k})_{,l} - (b_{ij,l})_{,k}.$$

The result can be shown, after some manipulation, to be

$$b_{ij,kl} - b_{ij,lk} = b_{mj}R^m{}_{ikl} + b_{im}R^m{}_{jkl}.$$

More terms of similar type appear in the right-hand side for a tensor of higher order. For a covariant vector, one finds

$$\lambda_{i,kl} - \lambda_{i,lk} = \lambda_m R^m{}_{ikl}.$$

Equations of the type of the last two equations replace for a curved space the usual integrability conditions for a "flat" space, that is, a space in which all components of the Riemann-Christoffel tensor are zero.

Finally, recall that the covariant derivative $\phi_{,i}$ of a tensor of zero order (i.e., a scalar point function) is the same as the partial derivative $\partial \phi / \partial x^i$, and that this covariant vector is the gradient of the scalar field $\phi(x^1, x^2, x^3)$. This is the only case in which the covariant derivative is equivalent to the partial derivative in general coordinates.

EXERCISES

1) Show that $\phi_{,ij} = \phi_{,ji}$, but that $\lambda_{i,j} \neq \lambda_{j,i}$ if λ_i is not a gradient vector field.

2) Calculate the form which equations (41) assume in spherical coordinates.

3) Write out the necessary steps in going from equation (45) to equation (47).

4) Make use of the fact that if $C: x^i = x^i(s)$ is a straight line, dx^i/ds are constants, and use equations (42) to deduce that

$$\frac{d^2 \bar{x}^i}{ds^2} + \bar{\Gamma}_{jk}{}^i \frac{d\bar{x}^j}{ds} \frac{d\bar{x}^k}{ds} = 0$$

are the differential equations of the straight lines in space for any coordinate system.

5) If a_{ij} is the fundamental metric tensor, write the covariant derivative of $a_{ij} b^{kl} c_k$ with respect to x^p.

6) Write out the necessary steps in arriving at the covariant derivative in (56) from the transformation law in (55).

7) Develop the formula in (57).

10
Scalar and Vector Fields

10–1. Fields. A tensor has been defined as an object with a set of components which transform in a specified manner when the coordinates are changed. These components are, in general, functions of position and therefore vary from point to point in space. The totality of these sets of functions at all points of a region of space is called a tensor field. Because the covariant derivative of a tensor is a tensor, it is seen that the covariant derivative of a tensor field is a tensor field of one higher order in the covariant indices.

It has been pointed out that a tensor field of zero order is represented by a scalar point function. As a physical model for this, one may consider the scalar function $\phi(x^1,x^2,x^3)$ as representing the temperature at any point x^i of space. In general, the temperature varies from point to point in space.

A tensor field of the first order is a vector field. With each coordinate system there are associated three functions of position, which may be denoted by λ^i (or by λ_i). As a physical model for this, one may consider the motion of a fluid in three-space. At each point a vector with components λ^i specifies the velocity of a particle of the fluid at the point. A particular kind of vector field is obtained by taking the covariant derivative of a scalar field, which is the same as the partial derivative. For instance, the vector field represented by $\phi_{,i}$ is the gradient vector field of the scalar field ϕ. In the temperature model, $\phi_{,i}$ are the three rates of change of the temperature function with respect to the respective x^i coordinates, and $\phi_{,i}$ are the components of the gradient vector $\nabla\phi$, which will now be shown to point in the direction of the greatest rate of change of the scalar field. If a^i are the components of a unit vector **A**, the scalar projection of $\nabla\phi$ on the direction of **A** is given by $a^i\phi_{,i}$. On the surface $\phi(x^1,x^2,x^3)$ = constant, the function ϕ clearly does not change. The total differential of the equation of the surface gives $\phi_{,i}\,dx^i = 0$, which means that the a^i are proportional to dx^i, the direction of any tangent vector to the surface, so that the scalar projection of $\nabla\phi$ on the surface is zero. This shows that the direction of $\nabla\phi$ is normal to the surface $\phi = $ constant.

One may consider the vectors of a field along a curve $C\colon x^i = x^i(s)$ in space. In particular, consider the vector $\nabla\phi$ along C. The unit tangent vector to C is $x^{i\prime}(s)$. Therefore, the scalar projection of $\nabla\phi$ on the tangent to C at a point is $\phi_{,i}\,(dx^i/ds)$ which is merely $d\phi/ds$, or the rate of change of ϕ with respect to arc length along C. The expression $d\phi/ds$ is called the *directional derivative* of ϕ in the direction of C.

A tensor field of the second order involves a set of nine functions, say b_{ij}, to be evaluated at each point of three-space. As a particular instance, one may consider the product of two vector fields λ^i and μ_j to obtain the dyadic field $\lambda^i \mu_j$. Another particular instance occurs in the second covariant derivative $\phi_{,ij}$ of a scalar field ϕ. A tensor field of the third order involves a set of twenty-seven functions, say b_{ijk}, in three-space. A general tensor field can be defined by use of the general tensor in (5–36).

If the vector $\mathbf{V}$ with components λ_i is differentiated covariantly to obtain $\lambda_{i,j}$ and the tensor product $a^k \lambda_{i,j}$ is contracted on k and j to obtain $a^k \lambda_{i,k}$, a vector results which is called the *derived* vector of vector $\mathbf{V}$ in the direction of A, or the *associate* vector of $\mathbf{V}$ in the direction of A. Hence, the derived vector of $\lambda_i \equiv \phi_{,i}$ in the direction of A is the vector with components $a^k \phi_{,ik}$. This last vector has a scalar projection on any unit vector b^i given by $b^i a^k \phi_{,ik}$. In particular, the projection on A is the invariant (or scalar) $a^i a^k \phi_{,ik}$.

Tensors are important in the differential geometry of higher-dimensional curved spaces and in the general relativity theory. They are also useful in mechanics, hydrodynamics, electricity, electromagnetic theory, and in various other phases of science and engineering. Tensors are a useful tool in the study of dynamical systems with n degrees of freedom. The trajectories in such a system can be shown to correspond to the geodesics (curves of shortest length) in a Riemannian space of $n + 2$ dimensions (see Section 12–10).

In the remainder of this chapter only scalar and vector fields will be considered.

10–2. Divergence of a Vector Field; the Laplacian. Calculate the covariant derivative of a contravariant vector field λ^i to obtain the second-order tensor $\lambda^i{}_{,j}$. Contraction yields the invariant (scalar function) $\lambda^i{}_{,i}$, which is called the *divergence* of the vector λ^i. It is of interest to see what form the divergence takes for a vector in orthogonal cartesian coordinates. Recall that the Christoffel symbols for this coordinate system are all zero so that covariant differentiation reduces to partial differentiation. Hence $\lambda^i{}_{,i}$ yields

$$1) \qquad \frac{\partial \lambda^1}{\partial x^1} + \frac{\partial \lambda^2}{\partial x^2} + \frac{\partial \lambda^3}{\partial x^3}$$

as the divergence of the vector λ^i in orthogonal cartesian coordinates.

In order to obtain the formula for divergence of λ^i (div λ^i) in general coordinates, write the covariant derivative

$$2) \qquad \lambda^i{}_{,j} = \frac{\partial \lambda^i}{\partial x^j} + \lambda^l \Gamma_{lj}{}^i.$$

Contraction in (2) on i and j gives one formula for div λ^i, but an alternative form for the result will be found. The derivative of the determinant $|a_{ij}|$ is given by (3–25) as

$$3) \qquad a_{,k} \equiv \frac{\partial a}{\partial x^k} = A^{ij} \frac{\partial a_{ij}}{\partial x^k} = a^{ij} a \frac{\partial a_{ij}}{\partial x^k},$$

from which it follows, by use of (9–64), that

$$4) \qquad \frac{\partial a}{\partial x^k} = a a^{ij} \frac{\partial a_{ij}}{\partial x^k} = a a^{ij} (\Gamma_{jk;i} + \Gamma_{ki;j}) = a(\Gamma_{jk}{}^j + \Gamma_{ki}{}^i) = 2a \Gamma_{ik}{}^i,$$

and consequently

$$5) \qquad \frac{\partial \log \sqrt{a}}{\partial x^k} = \Gamma_{ik}{}^i,$$

where, as usual, the repeated i in the right-hand member indicates a sum. Contraction of the tensor in (2), and use of the result in (5), produces

$$6) \qquad \lambda^i{}_{,i} = \frac{\partial \lambda^i}{\partial x^i} + \lambda^l \Gamma_{li}{}^i = \frac{\partial \lambda^i}{\partial x^i} + \lambda^i \frac{\partial \log \sqrt{a}}{\partial x^i} = \frac{1}{\sqrt{a}} \frac{\partial}{\partial x^i} (\sqrt{a}\, \lambda^i),$$

which provides a useful formula, in general coordinates, for the divergence of the vector with components λ^i. Observe that a vector field was obtained from a scalar field by differentiating the scalar field, whereas the divergence of a vector field produces a scalar field. The abbreviations div $\mathbf{F}$ and $\nabla \cdot \mathbf{F}$ are commonly employed for the divergence of a vector $\mathbf{F}$.

A physical interpretation of div $\mathbf{F}$ will appear after discussion of the Gauss divergence theorem in Section 11–8. However, in order to give the reader some feeling for the concept of divergence as a measure of change of density in a fluid at a point, the following example is presented.

Let $\mathbf{V} = v^i \mathbf{e}_i$ be the velocity vector of a particle at $P(x^i)$ in a fluid moving through space, where x^i are orthogonal cartesian coordinates. Consider an infinitesimal parallelopiped (Fig. 21) of the volume determined by the increments Δx^i at $P(x^i)$. The net increase in the amount of fluid in this box will be calculated by considering the amount flowing into each face and the outflow through the opposite face in time Δt. The inflow through the face perpendicular to the x^1-direction and through P is given

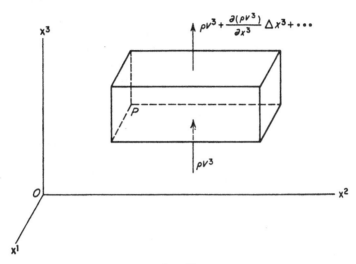

Fig. 21

by the product of the velocity component v^1 times the area $\Delta x^2\,\Delta x^3$ of the face, times the density of the fluid, that is, by $\rho v^1\,\Delta x^2\,\Delta x^3\,\Delta t$ in time Δt. By Taylor's theorem, the value of ρv^1 at the point $x^1 + \Delta x^1$ is given by

$$\rho v^1 + \frac{\partial (\rho v^1)}{\partial x^1}\,\Delta x^1 + \cdots.$$

Hence, the net increase (or decrease) in the mass of volume due to flow in the x^1-direction is given by

$$\frac{\partial (\rho v^1)}{\partial x^1}\,\Delta x^1\,\Delta x^2\,\Delta x^3\,\Delta t + \cdots.$$

A cyclical permutation on the indices gives the change

$$\frac{\partial (\rho v^i)}{\partial x^i}\,\Delta x^1\,\Delta x^2\,\Delta x^3\,\Delta t + \cdots$$

for each direction x^i. The sum of the three amounts of change is now divided by $\Delta x^1\,\Delta x^2\,\Delta x^3\,\Delta t$ to obtain an estimate of the rate of loss of fluid per unit volume per unit time. In the limit as the dimensions of the box diminish, one obtains the sum

$$\frac{\partial (\rho v^1)}{\partial x^1} + \frac{\partial (\rho v^2)}{\partial x^2} + \frac{\partial (\rho v^3)}{\partial x^3},$$

which is the divergence of the vector $\rho \mathbf{V}$ at P. If there is a source of fluid at P, the divergence is positive, and for a sink or point of absorption, the divergence is negative. In case the fluid is incompressible, the density ρ is constant, and the divergence is simply $\partial v^i/\partial x^i$ or $v^i{}_{,i}$. If the divergence of a vector is zero, that is, if $\nabla \cdot \mathbf{V} \equiv \operatorname{div} \mathbf{V} \equiv v^i{}_{,i} = 0$, the vector is said to be *solenoidal*.[1]

Because of the tensor form for the divergence of a vector, it is known to be invariant under a transformation to new coordinates. This is to be expected for an expression having physical significance. The formula (6) is employed for divergence in general coordinates.

The divergence of a vector λ^i was found by differentiating the contravariant components and then contracting. In order to find the divergence in terms of the covariant components, write $\lambda^i = a^{ij}\lambda_j$, and differentiate covariantly with respect to x^k to find $\lambda^i{}_{,k} = a^{ij}\lambda_{j,k}$. (Recall that $a^{ij}{}_{,k}$ is zero.) Contraction now yields $\lambda^i{}_{,i} = \operatorname{div} \lambda^i = a^{ij}\lambda_{j,i} = a^{ij}\lambda_{i,j}$. In particular, let λ_i be the components of $\nabla \phi$, that is, $\lambda_i \equiv \phi_{,i}$. The divergence of $\phi_{,i}$ is then the scalar

$$\text{7)} \qquad a^{ij}\phi_{,ij},$$

which is called the *Laplacian* of ϕ in general coordinates. From (6) it follows, by writing $\lambda^i = a^{ij}\phi_{,j}$, that the expression for the Laplacian may also take the form of the double sum

$$\text{8)} \qquad \frac{1}{\sqrt{a}}\frac{\partial}{\partial x^i}(\sqrt{a}\, a^{ij}\phi_{,j}).$$

The vanishing of the Laplacian gives the important second-order Laplacian partial differential equation of mathematical physics $(\partial/\partial x^i)(\sqrt{a}\, a^{ij}\phi_{,j}) = 0$, or equivalently $a^{ij}\phi_{,ij} = 0$.

10–3. The Curl of a Vector Field. The symbols e_{ijk} and e^{ijk} defined previously are not components of third-order tensors, but it will now be shown that $\epsilon_{ijk} \equiv \sqrt{a}\, e_{ijk}$, and $\epsilon^{ijk} \equiv (1/\sqrt{a})\, e^{ijk}$ are components of third-order covariant and contravariant tensors, respectively, where, as usual, a is the determinant of the fundamental tensor a_{ij}. In order to derive these results, notice first that if one takes the determinant of both sides of

$$\bar{a}_{kl} = a_{ij}\frac{\partial x^i}{\partial \bar{x}^k}\frac{\partial x^j}{\partial \bar{x}^l},$$

it follows that

$$|\bar{a}_{kl}| = |a_{ij}|J^2,$$

---

[1] With vector **V** regarded as the velocity of an incompressible fluid, div $V = 0$ means that there are no sources or sinks, so the stream lines behave much as the lines of force in the magnetic field about a solenoid, hence the term "solenoidal."

where $J \equiv |\partial x^i/\partial \bar{x}^j|$. Therefore, $\sqrt{\bar{a}} = \sqrt{a}\, J$, where J is assumed to be positive. Next verify that

$$e_{ijk}\frac{\partial x^i}{\partial \bar{x}^p}\frac{\partial x^j}{\partial \bar{x}^q}\frac{\partial x^k}{\partial \bar{x}^r} = \bar{e}_{pqr}J.$$

Multiplication of the left-hand side of the last equation by $\sqrt{a}\, J$ and the right-hand side by $\sqrt{\bar{a}}$ gives (after cancelling the non-zero factor J)

$$\sqrt{\bar{a}}\, \bar{e}_{pqr} = \sqrt{a}\, e_{ijk}\frac{\partial x^i}{\partial \bar{x}^p}\frac{\partial x^j}{\partial \bar{x}^q}\frac{\partial x^k}{\partial \bar{x}^r},$$

which means that $\bar{\epsilon}_{pqr} \equiv \sqrt{\bar{a}}\, \bar{e}_{pqr}$ (and $\epsilon_{ijk} \equiv \sqrt{a}\, e_{ijk}$) are components of a third-order covariant tensor.

Similarly, on multiplying

$$\bar{e}^{ijk}\frac{\partial \bar{x}^p}{\partial x^i}\frac{\partial \bar{x}^q}{\partial x^j}\frac{\partial \bar{x}^r}{\partial x^k} = e^{pqr}J$$

on the left by $J/\sqrt{\bar{a}}$ and on the right by $1/\sqrt{a}$, it follows that

$$\frac{\bar{e}^{ijk}}{\sqrt{\bar{a}}}\frac{\partial \bar{x}^p}{\partial x^i}\frac{\partial \bar{x}^q}{\partial x^j}\frac{\partial \bar{x}^r}{\partial x^k} = \frac{e^{pqr}}{\sqrt{a}},$$

which means that $\bar{\epsilon}^{ijk} \equiv \bar{e}^{ijk}/\sqrt{\bar{a}}$ (and $\epsilon^{pqr} \equiv e^{pqr}/\sqrt{a}$) are components of a third-order contravariant tensor.

Calculate the covariant derivative of the vector λ_j, that is, $\lambda_{j,k}$ and form the vector $\mu^i = \epsilon^{ijk}\lambda_{k,j}$. The vector with components μ^i is called the *curl* (or *rotation*) of the vector λ_j. In orthogonal cartesian coordinates $a_{ij} \equiv \delta_{ij}$, so $\sqrt{a} = 1$ and $\epsilon_{ijk} = e_{ijk}$, $\epsilon^{ijk} = e^{ijk}$. Also, because the covariant derivative reduces to the partial derivative, the components of the curl of λ_j are $e^{ijk}(\partial \lambda_k/\partial x^j)$. Let j and k sum, and make use of the definition of e^{ijk} to find the three components μ^i to be

9) $$\frac{\partial \lambda_3}{\partial x^2} - \frac{\partial \lambda_2}{\partial x^3},\quad \frac{\partial \lambda_1}{\partial x^3} - \frac{\partial \lambda_3}{\partial x^1},\quad \frac{\partial \lambda_2}{\partial x^1} - \frac{\partial \lambda_1}{\partial x^2}.$$

Notice that the curl of λ_j can be expressed as

10) $$e^{ijk}\frac{\partial \lambda_k}{\partial x^j}\mathbf{e}_i \equiv \begin{vmatrix} \mathbf{e}_1 & \mathbf{e}_2 & \mathbf{e}_3 \\ \dfrac{\partial}{\partial x^1} & \dfrac{\partial}{\partial x^2} & \dfrac{\partial}{\partial x^3} \\ \lambda_1 & \lambda_2 & \lambda_3 \end{vmatrix},$$

which is a useful form for calculation. Note that the curl of a vector field $\mathbf{F}$ is a vector field. It is commonly denoted by $\nabla \times \mathbf{F}$. The curl in general coordinates has components $1/\sqrt{a}\,(e^{ijk}\lambda_{k,j})$. It should be noted that if $\lambda_i \equiv \phi_{,i}$ then $\lambda_{i,j} = \lambda_{j,i}$ with the consequence that the curl of a gradient vector is identically zero. Any vector for which the curl is a zero vector is called *irrotational*.

The curl will be discussed further after introduction of Stokes' theorem in Chapter 11. A physical interpretation which reveals the reason for the designation "curl" of a vector will now be explained. The coordinates are orthogonal cartesian.

The velocity vector $\mathbf{V} = v^i \mathbf{e}_i$ of a point P in a body rotating with constant angular velocity was shown in Section 7–5 to be $\boldsymbol{\Omega} \times \mathbf{R}$, where $\boldsymbol{\Omega}$ is the constant angular velocity vector of the rotation, and $\mathbf{R}$ is the position vector to P. Write $\mathbf{R} = x^i \mathbf{e}_i$ and $\boldsymbol{\Omega} = \omega^i \mathbf{e}_i$, where ω^i are constants. Then

11) $\quad \mathbf{V} = \boldsymbol{\Omega} \times \mathbf{R} = \omega^i \mathbf{e}_i \times x^j \mathbf{e}_j = \omega^i x^j \mathbf{e}_i \times \mathbf{e}_j = e_{ijk}\omega^i x^j \mathbf{e}^k = v_k \mathbf{e}^k.$

Now the curl of V is given by

12) $\quad \nabla \times \mathbf{V} = e^{ijk} v_{k,j} \mathbf{e}_i = e^{lmk}(v_{k,m}) \mathbf{e}_l = e^{lmk}(e_{ijk}\omega^i x^j)_{,m} \mathbf{e}_l.$

By use of (7–12), further reduction of (12) leads to

13) $\quad \nabla \times \mathbf{V} = e^{lmk} e_{ijk} \omega^i x^j{}_{,m} \mathbf{e}_l = \delta_{ij}{}^{lm} \omega^i \delta_m{}^j \mathbf{e}_l = \delta_{im}{}^{lm}\omega^i \mathbf{e}_l$
$\qquad\qquad = 2\delta_i{}^l \omega^i \mathbf{e}_l = 2\omega^l \mathbf{e}_l = 2\boldsymbol{\Omega}.$

From (13) it follows that $\boldsymbol{\Omega} = \frac{1}{2}$ curl $\mathbf{V}$, which means that the angular velocity vector of a uniformly rotating body is one-half the curl of the velocity vector of any point in the body. If one prefers, he may calculate $\nabla \times \mathbf{V}$ in (13) from the determinant form in (10) with λ_i replaced by v_i. The computation in (13) was chosen in order to give the reader more facility with the notation employed.

10–4. Physical Components. Care must be exercised in applying formula (6) and the formula for the curl, that is, $1/\sqrt{a}\,(e^{ijk}\lambda_{k,j})$. It was seen in Section 6–4 that if a vector $\overrightarrow{PR}$ in any coordinate system has components λ^i, then the scalar projections of the vector on the tangents to the coordinate curves are $\sqrt{a_{ii}}\,\lambda^i$. The three quantities $\sqrt{a_{ii}}\,\lambda^i$ may be defined as *physical* components of vector $\overrightarrow{PR}$ because they are the components of interest in physical applications. Note that the λ^i might even be angles which would not behave as displacements. The components $f^i \equiv \sqrt{a_{ii}}\,\lambda^i$ tangent to the coordinate curves do, however, behave as displacements, which is the reason they are sometimes to be preferred. The following example illustrates the computation in a particular coordinate system.

Example. Find the formula for the divergence of a vector **F** in cylindrical coordinates.

Solution: For convenience the bars will not be written over the cylindrical coordinates x^i. Let λ^i be the contravariant and f^i the physical components of **F**. Then $f^i = \sqrt{a_{ii}}\,\lambda^i$. Calculate $ds^2 = (dx^1)^2 + (x^1)^2(dx^2)^2 + (dx^3)^2$ to find $a_{11} = 1$, $a_{22} = (x^1)^2$, $a_{33} = 1$, $a_{ij} = 0$ ($i \neq j$). Then $f^1 = \lambda^1$, $f^2 = x^1\lambda^2$, $f^3 = \lambda^3$. From formula (6),

$$\text{div } \mathbf{F} = \lambda^i{}_{,i} = \frac{1}{\sqrt{a}}\frac{\partial}{\partial x^i}(\sqrt{a}\,\lambda^i) = \frac{1}{x^1}\frac{\partial}{\partial x^i}(x^1\lambda^i),$$

or, if written out,

$$\text{div } \mathbf{F} = \frac{1}{x^1}\left[\frac{\partial}{\partial x^1}(x^1\lambda^1) + \frac{\partial}{\partial x^2}(x^1\lambda^2) + \frac{\partial}{\partial x^3}(x^1\lambda^3)\right].$$

In terms of f^i one then finds

14) $$\text{div } \mathbf{F} = \frac{1}{x^1}\frac{\partial}{\partial x^1}(x^1 f^1) + \frac{1}{x^1}\frac{\partial f^2}{\partial x^2} + \frac{\partial f^3}{\partial x^3}.$$

If x^1, x^2, x^3 are written as ρ, θ, z, the last expression for div **F** takes the customary form

$$\text{div } \mathbf{F} = \frac{1}{\rho}\frac{\partial}{\partial \rho}(\rho f^1) + \frac{1}{\rho}\frac{\partial f^2}{\partial \theta} + \frac{\partial f^3}{\partial z}.$$

Now let λ_i be covariant components of **F**. In this case div $\mathbf{F} = a^{ij}\lambda_{i,j}$. One finds from the matrix (a_{ij}) that $a^{11} = 1$, $a^{22} = (x^1)^{-2}$, $a^{33} = 1$, $a^{ij} = 0$ ($i \neq j$). The relation between the physical components f^i and the covariant components λ_i is now $f^i = \lambda_i/\sqrt{a_{ii}}$, and the divergence can be found by use of $a^{ij}\lambda_{i,j}$, but it is perhaps more convenient to use contravariant components $\lambda^i = a^{ij}\lambda_j$. Thus,

$$\text{div } \mathbf{F} = \frac{1}{\sqrt{a}}\frac{\partial}{\partial x^i}(\sqrt{a}\,a^{ij}\lambda_j),$$

which yields

$$\text{div } \mathbf{F} = \frac{1}{x^1}\left[\frac{\partial}{\partial x^1}(x^1\lambda_1) + \frac{\partial}{\partial x^2}\left(\frac{\lambda_2}{x^1}\right) + \frac{\partial}{\partial x^3}(x^1\lambda_3)\right],$$

or, on using the same physical components f^i employed previously,

$$\text{div } \mathbf{F} = \frac{1}{x^1}\frac{\partial}{\partial x^1}(x^1 f^1) + \frac{1}{x^1}\frac{\partial f^2}{\partial x^2} + \frac{\partial f^3}{\partial x^3},$$

which is the formula (14) obtained by use of the contravariant components λ^i.

10–5. Some Vector Identities Involving Divergence and Curl. Let the coordinates be orthogonal cartesian. A number of formulas involving the divergence and curl of vectors can be established by means of the notation developed heretofore. Four of these are now shown by way of illustration.

1. div $(\phi \mathbf{V}) = \phi \nabla \cdot \mathbf{V} + \mathbf{V} \cdot \nabla \phi$

Write $(\phi v^i)_{,i} = \phi v^i{}_{,i} + \phi_{,i} v^i = \phi \operatorname{div} \mathbf{V} + \nabla\phi \cdot \mathbf{V} = \phi \nabla \cdot \mathbf{V} + \mathbf{V} \cdot \nabla\phi$.

2. $\operatorname{curl}(\phi \mathbf{V}) = (\nabla \phi) \times \mathbf{V} + \phi(\nabla \times \mathbf{V})$

Write $\nabla \times (\phi V) = e^{ijk}(\phi v_k)_{,j}\mathbf{e}_i = e^{ijk}(\phi_{,j}v_k + \phi v_{k,j})\mathbf{e}_i = e^{ijk}\phi_{,j}v_k\mathbf{e}_i + \phi e^{ijk}v_{k,j}\mathbf{e}_i$. But the last two expressions are respectively $(\nabla \phi) \times \mathbf{V}$ and $\phi(\nabla \times \mathbf{V})$, which shows 2.

3. $\nabla \cdot (\mathbf{U} \times \mathbf{V}) = \mathbf{V} \cdot \nabla \times \mathbf{U} - \mathbf{U} \cdot \nabla \times \mathbf{V}$

Let $\mathbf{U} \times \mathbf{V} = \mathbf{W}$. Then $\nabla \cdot \mathbf{W} = w^i{}_{,i} = (e^{ijk}u_j v_k)_{,i} = e^{ijk}u_j v_{k,i} + e^{ijk}u_{j,i}v_k$. But the last two terms are, respectively, $-\mathbf{U} \cdot \nabla \times \mathbf{V}$ and $\mathbf{V} \cdot \nabla \times \mathbf{U}$, which shows 3.

4. $\nabla \times (\mathbf{U} \times \mathbf{V}) = \mathbf{V} \cdot \nabla \mathbf{U} - \mathbf{U} \cdot \nabla \mathbf{V} + \mathbf{U} \operatorname{div} \mathbf{V} - \mathbf{V} \operatorname{div} \mathbf{U}$.

Again, let $\mathbf{W} = \mathbf{U} \times \mathbf{V}$. Then $\nabla \times \mathbf{W} = e^{ijk}w_{k,j}\mathbf{e}_i = e^{ijk}(e_{klm}u^l v^m)_{,j}\mathbf{e}_i$. By the definition of the symbols involved, the last expression gives

$$\nabla \times \mathbf{W} = (\delta_{lm}{}^{ij}u^l{}_{,j}v^m + \delta_{lm}{}^{ij}u^l v^m{}_{,j})\mathbf{e}_i$$
$$= [(\delta_l{}^i\delta_m{}^j - \delta_m{}^i\delta_l{}^j)u^l{}_{,j}v^m + (\delta_l{}^i\delta_m{}^j - \delta_m{}^i\delta_l{}^j)u^l v^m{}_{,j}]\mathbf{e}_i$$
$$= [\delta_l{}^i u^l{}_{,j}v^j - \delta_l{}^j u^l{}_{,j}v^i + \delta_m{}^j u^i v^m{}_{,j} - \delta_m{}^i u^j v^m{}_{,j}]\mathbf{e}_i$$
$$= [u^i{}_{,j}v^j - u^j{}_{,j}v^i + u^i v^j{}_{,j} - u^j v^i{}_{,j}]\mathbf{e}_i$$
$$= \mathbf{V} \cdot \nabla \mathbf{U} - \mathbf{V} \operatorname{div} \mathbf{U} + \mathbf{U} \operatorname{div} \mathbf{V} - \mathbf{U} \cdot \nabla \mathbf{V},$$

which establishes 4.

10–6. Frenet Formulas in General Coordinates. Let a curve C be given by $x^i = x^i(s)$, and consider the particular vector field dx^i/ds. From the transformation $\bar{x}^i = \bar{x}^i(x^1, x^2, x^3)$, one obtains

$$15) \qquad \frac{d\bar{x}^j}{ds} = \frac{\partial \bar{x}^j}{\partial x^i} \frac{dx^i}{ds}.$$

Let x^i be orthogonal cartesian coordinates. In this case dx^i/ds are the direction cosines α^i of the unit tangent vector at any point P on C. Write $\bar{\alpha}^j$ for the same vector field in general coordinates. From (15) it follows that

$$16) \qquad \alpha^i = \bar{\alpha}^j \frac{\partial x^i}{\partial \bar{x}^j},$$

which shows that α^i are contravariant components of a vector. Similarly, one sees that the direction cosines β^i and γ^i of the principal normal and binormal to C are contravariant vectors, so that if $\bar{\beta}^j$ and $\bar{\gamma}^j$ are the components of these vectors in general coordinates, then

$$17) \qquad \beta^i = \bar{\beta}^j \frac{\partial x^i}{\partial \bar{x}^j}, \qquad \gamma^i = \bar{\gamma}^j \frac{\partial x^i}{\partial \bar{x}^j}.$$

In (8–17) it was seen that the Frenet formulas in cartesian coordinates are given by

18) $$\frac{d\mathbf{T}}{ds} = \kappa \mathbf{N}, \qquad \frac{d\mathbf{N}}{ds} = -\kappa \mathbf{T} + \tau \mathbf{B}, \qquad \frac{d\mathbf{B}}{ds} = -\tau \mathbf{N},$$

where κ and τ are the curvature and torsion of the curve C, and $\mathbf{T}$, $\mathbf{N}$, $\mathbf{B}$ are unit vectors along the tangent, principal normal, and binormal to C at P. If $\mathbf{T}$, $\mathbf{N}$, $\mathbf{B}$ are replaced by their components, equations (18) read as

19) $$\frac{d\alpha^i}{ds} = \kappa \beta^i, \qquad \frac{d\beta^i}{ds} = -\kappa \alpha^i + \tau \gamma^i, \qquad \frac{d\gamma^i}{ds} = -\tau \beta^i.$$

It was shown in (9–34) that if λ^i are contravariant components of a vector field, then

20) $$\left(\frac{\partial \lambda^i}{\partial x^p} + \lambda^m \Gamma_{mp}{}^i\right) \frac{dx^p}{dt}$$

are also contravariant components of a vector field along a curve. With the arbitrary parameter t taken as s, equation (9–34) takes the form

21) $$\frac{d\lambda^i}{ds} + \lambda^m \Gamma_{mp}{}^i \frac{dx^p}{ds} = \frac{\partial x^i}{\partial \bar{x}^j}\left(\frac{\partial \bar{\lambda}^j}{\partial \bar{x}^p} + \bar{\lambda}^r \bar{\Gamma}_{rp}{}^j\right)\frac{d\bar{x}^p}{ds}.$$

Because the Γ's on the left-hand side of (21) are all zero in the x^i coordinates, the last equation becomes

22) $$\frac{d\lambda^i}{ds} = \frac{\partial x^i}{\partial \bar{x}^j}\left(\frac{\partial \bar{\lambda}^j}{\partial \bar{x}^p} + \bar{\lambda}^r \bar{\Gamma}_{rp}{}^j\right)\frac{d\bar{x}^p}{ds} = \frac{\partial x^i}{\partial \bar{x}^j}\left(\frac{d\bar{\lambda}^j}{ds} + \bar{\lambda}^r \bar{\Gamma}_{rp}{}^j \frac{d\bar{x}^p}{ds}\right).$$

If λ^i are now taken to be $\alpha^i \equiv dx^i/ds$, (22) shows that

23) $$\frac{d\alpha^i}{ds} = \frac{d^2 x^i}{ds^2} = \frac{\partial x^i}{\partial \bar{x}^j}\left(\frac{d\bar{\alpha}^j}{ds} + \bar{\alpha}^r \bar{\Gamma}_{rp}{}^j \frac{d\bar{x}^p}{ds}\right),$$

which means that the vector with components $d\alpha^i/ds \equiv \kappa \beta^i$ in orthogonal cartesian coordinates has components

24) $$\kappa \bar{\beta}^j = \frac{d\bar{\alpha}^j}{ds} + \bar{\alpha}^r \bar{\Gamma}_{rp}{}^j \frac{d\bar{x}^p}{ds}$$

in any other coordinate system.

By use of the notation introduced in (9–35, 9–36) equation (24) may be written, on interchanging members, as

25) $$\bar{\alpha}^i{}_{,l}\frac{d\bar{x}^l}{ds} \equiv \frac{\delta \bar{\alpha}^i}{\delta s} \equiv \frac{d\bar{\alpha}^i}{ds} + \bar{\Gamma}_{jk}{}^i \bar{\alpha}^j \frac{d\bar{x}^k}{ds} = \kappa \bar{\beta}^i.$$

On replacing λ^i in (22) by β^i and γ^i, in turn, a similar development yields the two remaining Frenet formulas. The three resulting Frenet formulas in general coordinates are

$$
\bar{\alpha}^i{}_{,l}\frac{d\bar{x}^l}{ds} \equiv \frac{\delta\bar{\alpha}^i}{\delta s} = \kappa\bar{\beta}^i,
$$

26)
$$
\bar{\beta}^i{}_{,l}\frac{d\bar{x}^l}{ds} \equiv \frac{\delta\bar{\beta}^i}{\delta s} = -\kappa\bar{\alpha}^i + \tau\bar{\gamma}^i,
$$

$$
\bar{\gamma}^i{}_{,l}\frac{d\bar{x}^l}{ds} \equiv \frac{\delta\bar{\gamma}^i}{\delta s} = -\tau\bar{\beta}^i,
$$

in which the meaning of the intrinsic derivative $\delta\bar{\alpha}^i/\delta s$ is clear from (25). Observe the similarity between equations (26) and (19). If the Γ's are all zero in (26), the intrinsic derivatives reduce to ordinary derivatives, and, in this case, equations (26) are identically those in (19). Note that the $\bar{\alpha}^i$, $\bar{\beta}^i$, $\bar{\gamma}^i$ are contravariant components of unit vectors along the directions of the tangent, principal normal, and binormal to C.

10–7. The Acceleration Vector. In Section 8–5 an example was shown to illustrate the fact that if the velocity vector components are known in one coordinate system, and if a transformation to another coordinate system is given, then the components of the velocity vector can be computed for the new system by use of the tensor law of transformation. Ordinary differentiation sufficed for the case of velocity components, because if dx^i/dt are the components in the x^i-system, and $d\bar{x}^i/dt$ are the components of the same vector in the $\bar{x}^i$-system,

27)
$$
\frac{d\bar{x}^i}{dt} = \frac{\partial\bar{x}^i}{\partial x^j}\frac{dx^j}{dt}
$$

is the tensor law of transformation which relates the velocity components in the two systems. On differentiating (27) with respect to t, one finds

28)
$$
\frac{d^2\bar{x}^i}{dt^2} = \frac{\partial\bar{x}^i}{\partial x^j}\frac{d^2x^j}{dt^2} + \frac{\partial^2\bar{x}^i}{\partial x^k\,\partial x^j}\frac{dx^k}{dt}\frac{dx^j}{dt}
$$

which is not a tensor law of transformation unless the coordinate change $x^i = x^i(\bar{x}^1,\bar{x}^2,\bar{x}^3)$ is linear. If one knows the components d^2x^i/dt^2 in the x^i-system, equation (28) does not suffice to determine the acceleration components directly in another system. The intrinsic derivative will be found useful here.

Let the velocity vector **V** have components $v^i \equiv dx^i/dt$. Intrinsic differentiation of v^i gives the acceleration vector **A** with components a^i, that is,

29)
$$
a^i \equiv \frac{\delta v^i}{\delta t} \equiv \frac{d}{dt}v^i + \Gamma_{jk}{}^i v^j\frac{dx^k}{dt} = \frac{d^2x^i}{dt^2} + \Gamma_{jk}{}^i\frac{dx^j}{dt}\frac{dx^k}{dt},
$$

where the Christoffel symbols are calculated with respect to the fundamental metric tensor of the coordinate system employed. If x^i are cartesian, the Γ's are all zero and the components a^i are the ordinary derivatives d^2x^i/dt^2.

Example. Calculate the components of the acceleration vector in plane polar coordinates.

Solution: This solution was promised in Section 8-4. Equations (29) furnish the desired result. One must, of course, think of the x^i in (29) as polar coordinates. That is, $x^1 = r, x^2 = \theta$. The first task is to calculate the Christoffel symbols. The arc element ds given by $ds^2 = (dx^1)^2 + (x^1)^2(dx^2)^2$ furnishes the metric tensor a_{ij} for use in the Christoffel symbols. The only non-zero symbols are found to be

$$\Gamma_{22}^1 = -x^1, \qquad \Gamma_{12}^2 = \Gamma_{21}^2 = \frac{1}{x^1}.$$

From (29), with the indices ranging over 1, 2, ($x^3 = 0$ here), one finds

30) $$a^1 = \frac{d}{dt}\left(\frac{dx^1}{dt}\right) + \Gamma_{22}^1 \frac{dx^2}{dt}\frac{dx^2}{dt} = \frac{d^2x^1}{dt^2} - x^1\left(\frac{dx^2}{dt}\right)^2 = \frac{d^2r}{dt^2} - r\left(\frac{d\theta}{dt}\right)^2,$$

$$a^2 = \frac{d}{dt}\left(\frac{dx^2}{dt}\right) + 2\,\Gamma_{12}^2 \frac{dx^1}{dt}\frac{dx^2}{dt} = \frac{d^2x^2}{dt^2} + \frac{2}{x^1}\frac{dx^1}{dt}\frac{dx^2}{dt} = \frac{d^2\theta}{dt^2} + \frac{2}{r}\frac{dr}{dt}\frac{d\theta}{dt},$$

for the contravariant components of the acceleration vector in polar coordinates. It remains to find the physical components, which are given by $a^i/\sqrt{a^{ii}}$, that is, by $a^1/1$, $a^2/(r)^{-1}$. Hence, the desired acceleration components are

31) $$a_r = \frac{d^2r}{dt^2} - r\omega^2,$$

$$a_\theta = r\frac{d\omega}{dt} + 2\omega\frac{dr}{dt},$$

where $\omega \equiv d\theta/dt$. Equations (31) are precisely those obtained in (8-27). The reader will observe that equations (29) which were used to obtain (31) serve for any coordinate system, while the formulas in (8-27) were derived in a rather special manner.

10-8. Equations of Motion. Newton's second law of motion was introduced in Section 8-3. It states that for the motion of a free particle in space, the mass m of the particle times the acceleration is equal (in appropriate units) to the magnitude of the force acting. Let the force field be represented by the vector **F** with components f^i, and the acceleration vector by **A** with components a^i. In orthogonal cartesian coordinates, Newton's law, $ma^i = f^i$ takes the form

32) $$m\frac{d^2x^i}{dt^2} = f^i,$$

which can be written equivalently as

33) $$ma_i = f_i$$

if covariant components are used. If both sides of (33) are multiplied by the displacement vector dx^i, the sum $f_i\, dx^i$ is an invariant which represents the element of work dW required to move the particle from the point x^i to the point $x^i + dx^i$. Thus,

34) $$dW = m\, \frac{d^2 x^i}{dt^2}\, \frac{dx^i}{dt}\, dt.$$

Integration of (34) from t_1 to t_2 gives

35) $$W = m \int_{t_1}^{t_2} \frac{dx^i}{dt}\, \frac{d^2 x^i}{dt^2}\, dt = \frac{m}{2} \int_{t_1}^{t_2} \frac{d}{dt}\left[\sum_{i=1}^{3} \left(\frac{dx^i}{dt}\right)^2 \right] dt = \frac{m}{2} \sum_{i=1}^{3} \left(\frac{dx^i}{dt}\right)^2 \Big|_{t_1}^{t_2}.$$

Equation (35) states that the work done in moving the particle from the point where $t = t_1$ to the point where $t = t_2$ is equal to the change in the kinetic energy of the particle in going from the first to the second position.

It is seen in (34) that $dW \equiv f_i\, dx^i$ is an exact differential, so that a function W exists for which the total differential is $f_i\, dx^i$. A vector field of force for which this is true is called a *conservative* field. From (35), the work is independent of the curve traversed by the particle in going from the point t_1 to the point t_2. Equation (35) does not hold for a non-conservative force field. It is customary to introduce a function, say U, called the potential function, so that instead of $f_i \equiv \partial W/\partial x^i$, one uses $f_i \equiv -\partial U/\partial x^i$.

In order to return to contravariant components, write $f^i = a^{ij} f_j = -a^{ij}\,(\partial U/\partial x^j)$. Now Newton's law in (32) becomes

36) $$m\, \frac{d^2 x^i}{dt^2} = -a^{ij}\, \frac{\partial U}{\partial x^j}$$

in tensor form. Observe that from (36) the tensor components

37) $$m\, \frac{d^2 x^i}{dt^2} + a^{ij}\, \frac{\partial U}{\partial x^j}$$

are all zero. Recall that if the components of a tensor are all zero in one coordinate system, the tensor law of transformation determines the components as zero in any other coordinate system. Hence, by use of the form of the acceleration vector in general coordinates as given by (29), one can write Newton's law in general coordinates as

38) $$m\left(\frac{d^2 x^i}{dt^2} + \Gamma_{jk}^{\ i}\, \frac{dx^j}{dt}\, \frac{dx^k}{dt} \right) = -a^{ij}\, \frac{\partial U}{\partial x^j},$$

which, of course, reduces to (36) for cartesian coordinates. As soon as the potential U is known, the equations of motion (38) can be written for any coordinate system. These equations must be integrated to find the trajectory $x^i = x^i(t)$ of the particle.

10–9. The Lagrange Form of the Equations of Motion. The kinetic energy of a particle, usually denoted by T, is $\frac{1}{2}mv^2$ where v is the magnitude of the velocity vector. Now

$$39) \qquad v^2 = \left(\frac{ds}{dt}\right)^2 = a_{ij}\frac{dx^i}{dt}\frac{dx^j}{dt} = a_{ij}\dot{x}^i\dot{x}^j,$$

where a dot is used to indicate differentiation with respect to the time t. Hence,

$$40) \qquad T = \tfrac{1}{2}ma_{ij}\dot{x}^i\dot{x}^j.$$

Partial differentiation of T with respect to $\dot{x}^j$ yields

$$41) \qquad \frac{\partial T}{\partial \dot{x}^j} = ma_{ij}\dot{x}^i,$$

and the time derivative of (41) is

$$42) \qquad \frac{d}{dt}\left(\frac{\partial T}{\partial \dot{x}^j}\right) = m\left(a_{ij}\ddot{x}^i + \frac{\partial a_{ij}}{\partial x^k}\dot{x}^i\dot{x}^k\right).$$

The derivative of T with respect to x^j is given by

$$43) \qquad \frac{\partial T}{\partial x^j} = \frac{m}{2}\frac{\partial a_{ik}}{\partial x^j}\dot{x}^i\dot{x}^k.$$

Now subtract equation (43) from (42), and make use of the fact that

$$44) \qquad \frac{\partial a_{ij}}{\partial x^k}\dot{x}^i\dot{x}^k = \frac{1}{2}\frac{\partial a_{ij}}{\partial x^k}\dot{x}^i\dot{x}^k + \frac{1}{2}\frac{\partial a_{kj}}{\partial x^i}\dot{x}^k\dot{x}^i$$

to obtain

$$45) \qquad \frac{d}{dt}\left(\frac{\partial T}{\partial \dot{x}^j}\right) - \frac{\partial T}{\partial x^j} = m\left[a_{ij}\ddot{x}^i + \frac{1}{2}\left(\frac{\partial a_{ij}}{\partial x^k} + \frac{\partial a_{kj}}{\partial x^i} - \frac{\partial a_{ik}}{\partial x^j}\right)\dot{x}^i\dot{x}^k\right].$$

With the definition of the Christoffel symbol of the first kind which was introduced in (9–6), equation (45) becomes

$$46) \qquad \frac{d}{dt}\left(\frac{\partial T}{\partial \dot{x}^j}\right) - \frac{\partial T}{\partial x^j} = m[a_{ij}\ddot{x}^i + \Gamma_{ik;j}\dot{x}^i\dot{x}^k].$$

Equation (9–13) may be used to express the symbol $\Gamma_{ik;j}$ in terms of the symbol of second kind. Multiply

$$\Gamma_{ij}{}^k = a^{kl}\Gamma_{ij;l}$$

on both sides by a_{km} and sum on k to obtain

47) $$a_{km}\Gamma_{ij}{}^{k} = a_{km}a^{kl}\Gamma_{ij;l} = \delta_m{}^l\Gamma_{ij;l} = \Gamma_{ij;m}.$$

Hence, on changing dummy indices, it follows from (47) that

48) $$\Gamma_{ik;j} = a_{lj}\Gamma_{ik}{}^{l},$$

which can be placed in (46) to obtain

49) $$\frac{d}{dt}\left(\frac{\partial T}{\partial \dot{x}^j}\right) - \frac{\partial T}{\partial x^j} = m[a_{ij}\ddot{x}^i + a_{lj}\Gamma_{ik}{}^{l}\dot{x}^i\dot{x}^k],$$

which, with an obvious change of dummy indices, gives

50) $$\frac{d}{dt}\left(\frac{\partial T}{\partial \dot{x}^j}\right) - \frac{\partial T}{\partial x^j} = ma_{ij}[\ddot{x}^i + \Gamma_{lk}{}^{i}\dot{x}^l\dot{x}^k].$$

Equations (50) are clearly in covariant form. In order to change equations (38) to covariant form, multiply by a_{im} to have

51) $$ma_{im}[\ddot{x}^i + \Gamma_{jk}{}^{i}\dot{x}^j\dot{x}^k] = -a_{im}a^{ij}\frac{\partial U}{\partial x^j} = -\frac{\partial U}{\partial x^m}.$$

Equations (50) may now be written in the Lagrange form for a conservative force field, that is,

52) $$\frac{d}{dt}\left(\frac{\partial T}{\partial \dot{x}^j}\right) - \frac{\partial T}{\partial x^j} = -\frac{\partial U}{\partial x^j},$$

where U is the potential function. The Lagrange function $L \equiv T - U$ may be used to allow equations (52) to be written in the form

53) $$\frac{d}{dt}\left(\frac{\partial L}{\partial \dot{x}^j}\right) - \frac{\partial L}{\partial x^j} = 0,$$

where it is assumed that U is not a function of $\dot{x}^j$. (*Note:* The potential function is customarily denoted by V, but the letter U has been used here to avoid confusion with the velocity vector **V**.)

It is important to notice that since the covariant components of the acceleration vector appear in the right-hand member of (50), the expression on the left-hand side (for unit mass) may be very useful in obtaining the acceleration vector components in a particular coordinate system.

On the other hand, equations (50) are sometimes useful in evaluating the Christoffel symbols for a particular coordinate system. For, as soon as the equations of motion (50) are known, the coefficients $\Gamma_{lk}{}^{i}$ of the quadratic form in $\dot{x}^i$ are identified as the Christoffel symbols.

Example 1. Find the covariant components of the acceleration vector in spherical coordinates.

Solution: It has been shown that in spherical coordinates
$$ds^2 = (dx^1)^2 + (x^1)^2 (dx^2)^2 + (x^1 \sin x^2)^2 (dx^3)^2.$$

The kinetic energy is therefore
$$T = \frac{1}{2} m(\dot{s})^2 = \frac{m}{2} [(\dot{x}^1)^2 + (x^1)^2 (\dot{x}^2)^2 + (x^1 \sin x^2)^2 (\dot{x}^3)^2].$$

One calculates readily
$$\frac{d}{dt}\left(\frac{\partial T}{\partial \dot{x}^1}\right) = m \frac{d}{dt}(\dot{x}^1) = m\ddot{x}^1, \qquad \frac{\partial T}{\partial x^1} = m[x^1(\dot{x}^2)^2 + x^1(\sin x^2)^2(\dot{x}^3)^2],$$

and the further derivatives involving $\dot{x}^2$, $\dot{x}^3$ and x^2, x^3. Use of these derivatives in (52) yields

$$ma_1 = \frac{d}{dt}\left(\frac{\partial T}{\partial \dot{x}^1}\right) - \frac{\partial T}{\partial x^1} = m[\ddot{x}^1 - x^1(\dot{x}^2)^2 - x^1(\sin x^2)^2(\dot{x}^3)^2],$$

$$ma_2 = \frac{d}{dt}\left(\frac{\partial T}{\partial \dot{x}^2}\right) - \frac{\partial T}{\partial x^2} = m[(x^1)^2\ddot{x}^2 + 2x^1\dot{x}^1\dot{x}^2 - (x^1)^2 \sin x^2 \cos x^2 (\dot{x}^3)^2],$$

$$ma_3 = \frac{d}{dt}\left(\frac{\partial T}{\partial \dot{x}^3}\right) - \frac{\partial T}{\partial x^3} = m[(x^1 \sin x^2)^2 \ddot{x}^3 + 2x^1 (\sin x^2)^2 \dot{x}^1 \dot{x}^3 + (x^1)^2 \sin 2x^2 \dot{x}^2 \dot{x}^3].$$

Hence, the covariant components a_i of the acceleration vector are identified as the expressions in brackets in the last three equations. The equations of motion $ma_i = f_i$ can be written on assigning the vector f_i of the force field.

It was mentioned that the Christoffel symbols can be deduced from the acceleration components. The symbols of first kind $\Gamma_{lk;j} \equiv a_{ij}\Gamma_{lk}{}^i$ can be read from (50). For the instance of spherical coordinates, note that from the acceleration component a_1 just derived, it is evident that

$$\Gamma_{22;1} = -x^1, \qquad \Gamma_{33;1} = -x^1(\sin x^2)^2.$$

From the form of a_2, it follows that

$$\Gamma_{12;2} = x^1, \qquad \Gamma_{33;2} = -(x^1)^2 \sin x^2 \cos x^2,$$

and from a_3 that

$$\Gamma_{13;3} = x^1 (\sin x^2)^2, \qquad \Gamma_{23;3} = \tfrac{1}{2}(x^1)^2 \sin 2x^2.$$

All other symbols have zero values in spherical coordinates. (*Caution:* Recall that, for instance, the coefficient of $x^1 x^2$ in the quadratic form $(x^1)^2 + 5x^1 x^2 + \cdots$ is $\tfrac{5}{2}$, and the coefficient of $x^2 x^1$ is also $\tfrac{5}{2}$. This explains the appearance of the factor $\tfrac{1}{2}$ in the symbols $\Gamma_{ij;k}$ for which $i \neq j$.)

In order to obtain the Christoffel symbols of the first kind, use

$$\Gamma_{ij}{}^h = a^{hk} \Gamma_{ij;k}.$$

The a^{hk} have been found previously to be given by $a^{11} = 1$, $a^{22} = (x^1)^{-2}$, $a^{33} = (x^1 \sin x^2)^{-2}$, $a^{hk} = 0$ for $h \neq k$. The non-zero Christoffels of the second kind are therefore

$$\Gamma_{22}{}^1 = -x^1, \qquad \Gamma_{33}{}^1 = -x^1 (\sin x^2)^2,$$
$$\Gamma_{12}{}^2 = 1/x^1, \qquad \Gamma_{33}{}^2 = -\sin x^2 \cos x^2,$$
$$\Gamma_{13}{}^2 = 1/x^1, \qquad \Gamma_{23}{}^3 = \cot x^2.$$

The reader may wish to check these results by calculating the values of the symbols directly from their definition.

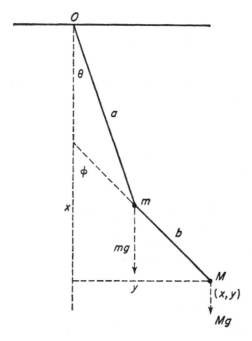

Fig. 22

This section will close with two examples to illustrate the use of the Lagrange form of the equations of motion. It should be observed that it is a simple matter to write the Lagrange equations of motion of a dynamical system once the expressions for the kinetic and potential energies are determined.

Example 2. Find the differential equations which describe the motion of the double pendulum shown in Fig. 22. The strings of length a and b are considered as weightless. Weights of magnitude mg and Mg are the two pendulum bobs which move in a vertical plane.

Solution: Relative to the fixed horizontal line through O, the potential energy of the system is $-mga \cos \theta - Mg(a \cos \theta + b \cos \phi)$. The kinetic energy for the weight mg is $(m/2)a^2\dot\theta^2$. For the second weight Mg it is $(M/2)(\dot x^2 + \dot y^2)$, where, from the figure,

$$x = a \cos \theta + b \cos \phi, \qquad y = a \sin \theta + b \sin \phi.$$

Application of Lagrange's equations for θ and ϕ yields

$$\frac{d}{dt}[ma^2\dot\theta + Ma^2\dot\theta + Mab\dot\phi \cos(\phi - \theta)] - Mab\dot\theta\dot\phi \sin(\phi - \theta) = -(m+M)ag \sin \theta,$$

$$\frac{d}{dt}[Mb^2\dot\phi + Mab\dot\theta \cos(\phi - \theta)] + Mab\dot\theta\dot\phi \sin(\phi - \theta) = -Mgb \sin \phi.$$

Note: This problem can be made more complicated by allowing the bobs to oscillate in space to yield a system with four degrees of freedom. *Query:* Why are there four degrees of freedom?

One integral of Lagrange's equations is the equation of energy which states that the potential energy plus the kinetic energy is constant. That is P.E. + K.E. = constant. This equation can be written without the use of Lagrange's equations, and it is to be preferred in solving problems concerning the motions of the two masses. For discussion of oscillations involving small displacements θ and ϕ, the Lagrange equations are preferred.

Suppose $\theta, \phi, \dot\theta, \dot\phi$ are all small enough to allow neglect of second-order terms. Then the first of the Lagrange equations in Example 2 reduces to

$$(M + m)a^2\ddot\theta + Mab\ddot\phi = -(M + m)ag\theta,$$

and the second to

$$Mb^2\ddot\phi + Mab\ddot\theta = -Mbg\,\phi.$$

If both bobs execute small oscillations of period $2\pi/p$, then θ and ϕ can be expressed by $A \cos pt + B \sin pt$, so that $\ddot\theta = -p^2\theta$, $\ddot\phi = -p^2\phi$. Use of these expressions for $\ddot\theta$ and $\ddot\phi$ in the last two differential equations gives

$$(M + m)ap^2\theta + Mbp^2\phi = (M + m)g\theta,$$

$$bp^2\phi + ap^2\theta = g\phi.$$

On equating the constant value of the ratio θ/ϕ from each of the last two equations, there results

$$(M + m)(g - ap^2)(g - bp^2) = Mabp^4,$$

a quadratic in p^2, which reveals that the system can have two periods. The two periods are actually both real because the roots p^2 can be shown to be positive for all m and M.

Some special cases are interesting to consider. First, let m go to zero to realize a simple pendulum of length $(a + b)$. The quadratic equation in p^2 gives $p = [g/(a + b)]^{1/2}$, so the period in this case is $2\pi/p = 2\pi[a + b/g]^{1/2}$, which gives the known result $2\pi[l/g]^{1/2}$, where $l = a + b$. Next, let $m \to \infty$ with the result that $p^2 = g/a$ or $p^2 = g/b$, which gives the periods for two distinct pendulums. Finally, if $a \to 0$, $p^2 = g/b$; and if $b \to 0$, $p^2 = g/a$, as is to be expected.

Example 3. A smooth circular wire with a smooth bead of weight $w = mg$ on it rotates freely about a vertical diameter of the wire with constant angular speed ω. Investigate the motion of the bead.

Solution: Let θ measure the angular displacement of the bead at any point P in the motion (Fig. 23). Suppose that $\theta = 0$ and $a\dot\theta = v$ at time $t = 0$. Denote

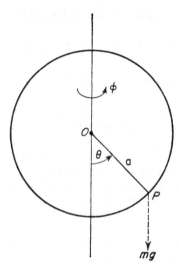

Fig. 23

time derivatives by dots. Let ϕ measure the rotation of the plane of the wire from a fixed position. Then $\dot\phi = \omega$. This is a problem with two degrees of freedom, so two Lagrange equations can be written. Equations (53) will be used. In order to find $L = T - U$, observe that the kinetic energy of the system is the K.E. of the bead plus the K.E. of the wire, that is, the total

$$\text{K.E.} \equiv T = \frac{m}{2}(a^2 \sin^2\theta\,\dot\phi^2 + a^2\dot\theta^2) + \frac{1}{2}I\dot\phi^2,$$

where $I = (M/2)a^2 \equiv Mk^2$ is the moment of inertia of the wire. The potential energy U of the bead is $-Mga\cos\theta$. Hence,

$$L = T - U = \frac{m}{2}(a^2\sin^2\theta\,\dot\phi^2 + a^2\dot\theta^2) + \frac{M}{2}k^2\dot\phi^2 + mga\cos\theta.$$

In equations (53), $x^1 = \theta$, and $x^2 = \phi$. One calculates

$$\frac{d}{dt}\left(\frac{\partial L}{\partial \dot\theta}\right) - \frac{\partial L}{\partial \theta} = m(a^2\ddot\theta - a^2\sin\theta\cos\theta\,\dot\phi^2 + ga\sin\theta) = 0,$$

and

$$\frac{d}{dt}\left(\frac{\partial L}{\partial \dot\phi}\right) - \frac{\partial L}{\partial \phi} = (ma^2\sin^2\theta + Mk^2)\ddot\phi + 2ma^2\sin\theta\cos\theta\,\dot\theta\dot\phi = 0.$$

The second of the two Lagrange equations can be written as

$$\frac{\ddot{\phi}}{\dot{\phi}} dt + \frac{2ma^2 \sin\theta \cos\theta}{ma^2 \sin^2\theta + Mk^2} d\theta = 0,$$

and integrated to obtain

$$\dot{\phi} = \frac{C}{ma^2 \sin^2\theta + Mk^2}.$$

Since at $\theta = 0$, $\dot{\phi} = \omega$, so that $C = Mk^2\omega$, and

$$\dot{\phi} = \frac{Mk^2\omega}{ma^2 \sin^2\theta + Mk^2}.$$

This value of $\dot{\phi}$ may be substituted into the first of the Lagrange equations, which can then be multiplied by $\dot{\theta}$ and integrated to find

$$ma^2\dot{\theta}^2 - 2mag \cos\theta + \frac{(Mk^2\omega)^2}{ma^2 \sin^2\theta + Mk^2} = C'.$$

By the fact that $\dot{\theta} = v/a$ when $\theta = 0$, the constant C' is determined as

$$C' = mv^2 - 2mag + Mk^2\omega^2.$$

Hence,

$$ma^2\dot{\theta}^2 - 2mag \cos\theta + \frac{(Mk^2\omega)^2}{ma^2 \sin^2\theta + Mk^2} = mv^2 - 2mag + Mk^2\omega^2.$$

If the bead just reaches the highest point of the wire (where $\theta = \pi$), then $\dot{\theta} = 0$. Substitution of these values in the last equation gives $v = 2\sqrt{ag}$ as the minimum magnitude of the initial velocity of the bead which will carry it around the wire. Note that this result is independent of ω, but that the time of transit does depend upon ω, as is shown by the last differential equation.

EXERCISES

1) Calculate the divergence of the vector with components $[x^2x^3 + (x^1)^2, x^3x^1 + (x^2)^2, x^1x^3 + (x^3)^2]$ in orthogonal cartesian coordinates at the point (1,1,1).

2) Show that the formula for div **F** in physical components f^i for spherical coordinates is given by

$$\text{div } \mathbf{F} = \frac{1}{(x^1)^2} \frac{\partial}{\partial x^1}[(x^1)^2 f^1] + \frac{1}{x^1 \sin x^2} \frac{\partial}{\partial x^2}(\sin x^2 f^2) + \frac{1}{x^1 \sin x^2} \frac{\partial f^3}{\partial x^3}.$$

(*Note:* The spherical coordinates x^i are related to orthogonal cartesian coordinates $\bar{x}^i$ by $\bar{x}^1 = x^1 \sin x^2 \cos x^3$, $\bar{x}^2 = x^1 \sin x^2 \sin x^3$, $\bar{x}^3 = x^1 \cos x^2$. This notation is chosen to avoid writing the bars in the formula for div **F**.)

3) If ∇ stands for the differential operator $\mathbf{e}^i(\partial/\partial x^i)$ and if $\mathbf{V} = \lambda^j \mathbf{e}_j$ in orthogonal cartesian coordinates, show that div $\mathbf{V} = \nabla \cdot \mathbf{V}$.

4) Use the ∇ operator $\mathbf{e}^i (\partial/\partial x^i)$ in orthogonal cartesian coordinates and the vector $\mathbf{V} = \lambda^j \mathbf{e}_j$ to show that $\mathbf{W} \equiv \text{curl } \mathbf{V} = \nabla \times \mathbf{V}$.

5) Show that the Laplacian $a^{ij}\phi_{,ij}$ gives

$$\sum_{i=1}^{3} \frac{\partial^2 \phi}{\partial x^i \partial x^i}$$

in orthogonal cartesian coordinates, and

$$\frac{\partial^2 \phi}{\partial x^1 \partial x^1} + \frac{1}{(x^1)^2} \frac{\partial^2 \phi}{\partial x^2 \partial x^2} + \frac{1}{(x^1 \sin x^2)^2} \frac{\partial^2 \phi}{\partial x^3 \partial x^3} + \frac{2}{x^1} \frac{\partial \phi}{\partial x^1} + \frac{1}{(x^1)^2} \cot x^2 \frac{\partial \phi}{\partial x^2}$$

in spherical coordinates.

6) Calculate curl $\mathbf{V}$, if $\mathbf{V}$ has orthogonal cartesian components $\lambda_i = (x^3, x^1, x^2)$.

7) If $\lambda_i = \sqrt{a_{ii}} f^i$ are the covariant components of a vector $\mathbf{V}$, show that the physical components of curl $\mathbf{V}$ are given by

$$\sqrt{a_{ii}} \, \mu^i = -\sqrt{a_{ii}} \, \frac{1}{\sqrt{a}} \, e^{ijk} \lambda_{j,k}.$$

Show that in spherical coordinates these components are

$$\frac{1}{x^1 \sin x^2} \left[\frac{\partial f^2}{\partial x^3} - \frac{\partial}{\partial x^2} (\sin x^2 f^3) \right], \quad \frac{1}{x^1 \sin x^2} \left[\sin x^2 \frac{\partial}{\partial x^1} (x^1 f^3) - \frac{\partial f^1}{\partial x^3} \right],$$

$$\frac{1}{x^1} \left[\frac{\partial f^1}{\partial x^2} - \frac{\partial}{\partial x^1} (x^1 f^2) \right].$$

8) If $\mathbf{R}$ is the position vector $x^i \mathbf{e}_i$ and $r \equiv |\mathbf{R}|$, show that $\nabla r^n = n r^{n-2} \mathbf{R}$.

9) Find the value of n for which $\nabla \cdot (r^n \mathbf{R})$ is identically zero.

10) Show that $\kappa = \bar{\beta}_i \dfrac{\delta \bar{\alpha}^i}{\delta s}$ and $\tau = -\bar{\beta}_i \dfrac{\delta \bar{\gamma}^i}{\delta s}$.

11) Show that $\dfrac{\delta^2 \bar{\alpha}^i}{\delta s^2} = \dfrac{d\kappa}{ds} \bar{\beta}^i + \kappa(-\kappa \bar{\alpha}^i + \tau \bar{\gamma}^i)$.

12) Form the tensor

$$\epsilon_{pqr} \bar{\alpha}^p \frac{\delta \bar{\alpha}^q}{\delta s} \frac{\delta^2 \bar{\alpha}^r}{\delta s^2}$$

and reduce it to $\kappa^2 \tau \epsilon_{pqr} \bar{\alpha}^p \bar{\beta}^q \bar{\gamma}^r = -\kappa^2 \tau$ to show that

$$\tau = -\frac{1}{\kappa^2} \epsilon_{pqr} \bar{\alpha}^p \frac{\delta \bar{\alpha}^q}{\delta s} \frac{\delta^2 \bar{\alpha}^r}{\delta s^2}$$

in general coordinates.

13) Write the Frenet-Serret formulas in cylindrical coordinates. The only non-zero Christoffels are $\Gamma_{22}^1 = -x^1$, $\Gamma_{12}^2 = \Gamma_{21}^2 = 1/x^1$.

14) Draw a circle of radius a on a plane rectangle. Roll the rectangle into a circular cylinder of radius b. Show that the parametric equations of the de-

formed circle may be written as

$$x = b\cos\left[\frac{a}{b}\cos\frac{s}{a}\right], \qquad y = b\sin\left[\frac{a}{b}\cos\frac{s}{a}\right], \qquad z = a\sin\frac{s}{a},$$

where s is arc length on the deformed circle. The formula

$$\kappa^2 = \left(\frac{d^2x}{ds^2}\right)^2 + \left(\frac{d^2y}{ds^2}\right)^2 + \left(\frac{d^2z}{ds^2}\right)^2$$

can be used to show that the curvature κ is given by $\kappa^2 = a^{-2} + b^{-2}\sin^4 s/a$.

15) Use cylindrical coordinates to solve for κ in Exercise 14. Write $x^1 = b$, $x^2 = \theta$, $x^3 = z$ so that the equations of the defined circle are $x^1 = b$, $x^2 = a/b \cos s/a$, $x^3 = a \sin s/a$. The Frenet formulas give $\kappa\beta^i = \delta\alpha^i/\delta s$, i.e.,

$$\kappa\beta^i = \frac{d\alpha^i}{ds} + \alpha^m \Gamma_{ml}{}^i \frac{dx^l}{ds}.$$

Note that α^i are

$$\left(\frac{dx^1}{ds}, \frac{dx^2}{ds}, \frac{dx^3}{ds}\right) = \left(0, -\frac{1}{b}\sin\frac{s}{a}, \cos\frac{s}{a}\right).$$

On using the Christoffels in Exercise 13, it follows that $\kappa\beta^1 = -1/b \sin^2 s/a$, $\kappa\beta^2 = -1/ab \cos s/a$, $\kappa\beta^3 = -1/a \sin s/a$. Now use the fact that β^i is a unit vector, i.e., $a_{ij}\beta^i\beta^j = 1$, where the a_{ij} are obtained from $ds^2 = (dx^1)^2 + (x^1)^2(dx^2)^2 + (dx^3)^2$, to find $(\beta^1)^2 + (b)^2(\beta^2)^2 + (\beta^3)^2 = 1$. Therefore, deduce that κ is given by the formula in Exercise 14.

16) Use cylindrical coordinates to show that κ and τ for the helix $x = a\cos\theta$, $y = a\sin\theta$, $z = b\theta$ are given by $(a^2 + b^2)\kappa = a$, $(a^2 + b^2)\tau = b$. (Compare with Exercise 5 following Section 8–10.)

17) A bar of weight mg and length $2l$ slides down the helicoidal surface represented by

$$x^1 = u^2 \cos u^1, \qquad x^2 = u^2 \sin u^1, \qquad x^3 = au^1.$$

One end of the bar is constrained to move smoothly along the axis of the helicoid, and the bar always remains horizontal. Write Lagrange's equation and deduce that

$$\ddot{u}^1 = -\frac{3ag}{4l^2 + 3a^2}.$$

18) Work out the details of Example 2 on the double pendulum.

19) A simple pendulum is realized by a weight mg on the end of a weightless string of length a. Let θ measure the angular displacement of the pendulum bob from the vertical. Use the Lagrange method to find the differential equation of the motion. Ans. $a\ddot{\theta} + g\sin\theta = 0$.

20) Consider a homogeneous rod of weight mg and length $2b$ with one end attached to a weightless string of length a which is suspended from a fixed point O.

The rod moves in a vertical plane. Find the Lagrange equations for the system. Then assume small displacements and find the periods.

21) If x^i are rectangular cartesian coordinates and

$$x^1 = u^1 u^2 \cos u^3, \qquad x^2 = u^1 u^2 \sin u^3, \qquad 2x^3 = (u^1)^2 - (u^2)^2,$$

the u^i are called paraboloidal coordinates. Note that $x^1 = (\tan u^3) x^2$ gives planes through the x^3 axis as u^3 varies, and that the curve $u^2 =$ constant, $u^3 =$ constant is a parabola. Write Laplace's partial differential equation in paraboloidal coordinates.

Ans. $u^1 u^2 \left[\dfrac{\partial}{\partial u^1} \left(u^1 u^2 \dfrac{\partial \phi}{\partial u^1} \right) + \dfrac{\partial}{\partial u^2} \left(u^1 u^2 \dfrac{\partial \phi}{\partial u^2} \right) \right] + (u^{1^2} + u^{2^2}) \dfrac{\partial^2 \phi}{\partial u^3 \partial u^3} = 0.$

11
Integration of Vectors

11–1. Line Integrals. One may consider the integration of a tensor (scalar function, vector, dyadic, etc.) over a space (curve, surface, three-space, etc.), but attention will be restricted in this chapter to the integration of scalar functions and vectors along segments of curves, portions of surfaces, and domains of three-space. Of immediate concern is the integration of a function along a curve. This will be called a *line* integral.

In order to make the concept of a line integral meaningful, it is well to begin by pointing out that the ordinary definite integral

$$\int_a^b f(x)\, dx$$

over the segment from $x = a$ to $x = b$ of the x-axis is a line integral. The meaning attached to this integral was shown in the elementary calculus to be the following: Partition the interval from a to b (Fig. 24) into n

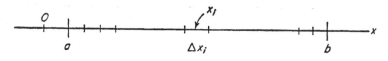

Fig. 24

subintervals, choose a point x_i in the ith subinterval, the length of which is Δx_i, and form the product $f(x_i)\, \Delta x_i$. Take the sum of such products for $i = 1, 2, \cdots, n$ to obtain

$$\sum_{i=1}^{n} f(x_i)\, \Delta x_i.$$

The limit (if it exists) of this sum as $n \to \infty$ and as each $\Delta x_i \to 0$ in length, is evaluated by the definite integral

$$\int_a^b f(x)\,dx.$$

Of course, the resulting integral may be interpreted as an area, but it need not be so interpreted. It is preferable at the moment to think of the result of the integration as a number. The result of taking the limit of the sum of the products of numbers $f(x_i)$ and Δx_i is a number, if the limit exists. Notice that if $b = x$, a variable, then

$$\int_a^x f(x)\,dx$$

is a function of the upper limit. Each of the line, surface, and volume integrals to appear below may be obtained as a limit of a sum, just as

$$\int_a^b f(x)\,dx$$

was obtained in the calculus; but for the sake of brevity, the integral notation will be used.

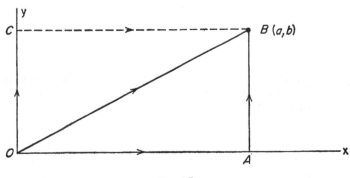

Fig. 25

As a first attempt at generalizing the line integral of a function along a segment of the x-axis, proceed as follows. Integrate a function $f(x,y)$ of two variables over a segment, say OA, on the x-axis, and then over the line AB parallel to the y-axis (Fig. 25). Note that $y = 0$ along OA, and $x = a = $ constant along AB, $ds = dx$ on OA, and $ds = dy$ on AB, so that the integral desired is expressed in terms of ordinary integrals as

1) $$\int_0^B f(x,y)\,ds = \int_0^a f(x,0)\,dx + \int_0^b f(a,y)\,dy.$$

The number obtained in (1) is the result of integrating the function $f(x,y)$ along a "curve" between two points O and B in the plane, the "curve" or path of integration being in this case the elbow path from O to A and then from A to B. It is assumed that the function $f(x,y)$ is sufficiently well behaved to allow the integrals to exist. It is important to realize that if the direction of integration along OA is reversed, the sign of the integral is changed. The same is true with respect to the integral along AB. However, the sign of the integral on the left-hand side in (1) is unchanged upon reversing the direction of integration along the path OAB, for the sign of ds is considered as positive in both directions.

Example 1. Integrate the function $x^2 + xy + y^2$ over the elbow path $OC + CB$ in Fig. 25.

Solution: Along OC, $x = 0$, and along CB, $y = b$, so the integral is

$$\int_0^b y^2\, dy + \int_0^a (x^2 + bx + b^2)\, dx = \tfrac{1}{6}(2a^3 + 3a^2 b + 6ab^2 + 2b^3).$$

It is natural to ask for the result of integrating $f(x,y)$ over an arbitrary path from O to B. Probably the first path to occur to one is the straight line from O to B with equation $ay = bx$. Whereas the element ds of arc length was dx along OA and dy along AB, the expression for ds at any point of OB is given by

2) $$ds^2 = dx^2 + dy^2 = dx^2 + \left[d\left(\frac{bx}{a}\right)\right]^2 = \frac{a^2 + b^2}{a^2}\, dx^2.$$

Three distinct line integrals of $f(x,y)$ along the straight line from O to B may be defined. First, on using the equation of the line and the expression for ds from (2) in terms of dx, write

3) $$\int_O^B f(x,y)\, ds = \int_{x=0}^{x=a} f\left(x, \frac{b}{a}x\right) \frac{\sqrt{a^2 + b^2}}{a}\, dx.$$

Observe that the integration in (3) may be in terms of y as the variable of integration. That is,

4) $$\int_O^B f(x,y)\, ds = \int_{y=0}^{y=b} f\left(\frac{a}{b}y, y\right) \frac{\sqrt{a^2 + b^2}}{b}\, dy$$

gives the same result as that found in (3).

A second type of integration of $f(x,y)$ over the line OB is

5) $$\int_{x=0}^{x=a} f(x,y)\, dx,$$

where y is to be replaced by its function of x on the line before integrating with respect to x. A third type is

6) $$\int_{y=0}^{y=b} f(x,y)\, dy,$$

where x may be replaced as a function of y before integration takes place. Actually, in some instances it may be more convenient to perform the integration in (6) with respect to x. Instead of the straight line OB, any curve joining O and B may be used for each of the integrals defined in (4), (5), (6). In general, the result varies with the choice of path.

Example 2. Integrate the function $x^2 + xy + y^2$ with respect to x over the segment of the parabola $a^2 y = bx^2$ joining $O(0,0)$ to $B(a,b)$.

Solution: The value of

$$\int_O^B (x^2 + xy + y^2)\, dx$$

is desired. On using $y = bx^2/a^2$ from the equation of the curve to express the integral as a function of x only, one finds

$$\int_{x=0}^{x=a} \left(x^2 + \frac{bx^3}{a^2} + \frac{b^2 x^4}{a^4} \right) dx = \frac{a}{60}(20a^2 + 15ab + 12b^2).$$

(*Note:* If integration with respect to s were required in this example, one would have a very unwelcomed task of integration to perform. The reader is not advised to pursue this.)

In order to generalize further the concept of integrating a function along a curve, consider integrating $f(x,y,z)$ along a curve in space.

In particular, consider first the elbow path in Fig. 26 from O to A, A to B, and B to C, where C has coordinates (a,b,c). Along OA, $y = z = 0$, along AB, $x = a$, $z = 0$, and along BC, $x = a$, $y = b$. Therefore, the integral is

7) $$\int_O^C f(x,y,z)\, ds = \int_{x=0}^{x=a} f(x,0,0)\, dx$$
$$+ \int_{y=0}^{y=b} f(a,y,0)\, dy + \int_{z=0}^{z=c} f(a,b,z)\, dz.$$

Example 3. Integrate the function $x + y + z$ along the elbow path O-A-B-C shown in Fig. 26.

Solution: With $f \equiv x + y + z$, (7) yields

$$\int_0^a x\, dx + \int_0^b (a + y)\, dy + \int_0^c (a + b + z)\, dz = \frac{(a+b+c)^2}{2}.$$

Consider next the integration of $f(x,y,z)$ along the straight line joining O to C in Fig. 26. The symmetric equations of line OC are $x/a = y/b = z/c\,(=t)$, so, in terms of the general parameter t, $x = at$, $y = bt$, $z = ct$, and $ds^2 = (a^2 + b^2 + c^2)\,dt^2$. Hence,

8) $$\int_O^C f(x,y,z)\,ds = \int_{t=0}^{t=1} f(at,bt,ct)\sqrt{a^2+b^2+c^2}\,dt.$$

Observe that when the equations of the path are known the line integral reduces to the integral of a function of a single variable.

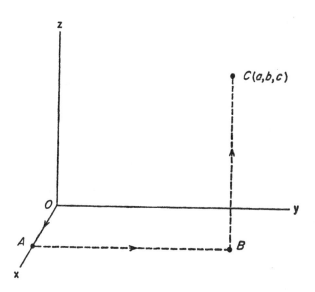

Fig. 26

Example 4. Integrate $x + y + z$ along the straight line from $A(1,2,3)$ to $B(4,-1,6)$.

Solution: The line AB has equations

$$\frac{x-1}{1} = \frac{y-2}{-1} = \frac{z-3}{1}\;(=t),$$

so $x = 1+t$, $y = 2-t$, $z = 3+t$, and $ds^2 = 3dt^2$. Hence,

$$\int_A^B (x+y+z)\,ds = \int_{t=0}^{t=3} (6+t)\sqrt{3}\,dt = 45\sqrt{3}/2.$$

The integral in (8) may be taken along a curve joining two points. Of course, in this case the coordinates x, y, z are not linear functions of a parameter.

Analogous to the types (5) and (6) in the plane, one may consider the integrals

9) $\quad \int_A^B U_1(x,y,z)\, dx, \qquad \int_A^B U_2(x,y,z)\, dy, \qquad \int_A^B U_3(x,y,z)\, dz$

along a curve joining two points A and B in space. But more often one encounters a line integral which appears as the sum of the integrals in (9), that is a line integral of the form

10) $\quad \int_A^B U_1(x,y,z)\, dx + U_2(x,y,z)\, dy + U_3(x,y,z)\, dz.$

Apparently the value of this integral depends upon the choice of functions U_1, U_2, U_3, and upon the points A and B. In general, the result of the integration depends also upon the choice of path connecting A and B.

Example 5. Find the value of the integral

$$\int_{(0,0,0)}^{(1,2,3)} x\, dx + 2z\, dy + y\, dz$$

(a) along the straight line path, (b) along an elbow path as in Fig. 26.

Solution: (a) Along the straight line connecting $(0,0,0)$ and $(1,2,3)$, $x = t$, $y = 2t$, $z = 3t$, so the integral is

$$\int_0^1 t\, dt + 2(3t)(2dt) + (2t)(3dt) = \int_0^1 19t\, dt = \tfrac{19}{2}.$$

(b) Along OA, $y = z = 0$, on AB, $x = 1$, $z = 0$, and on BC, $x = 1$, $y = 2$, so the three contributions sum to

$$\int_0^1 x\, dx + 0 + \int_0^3 2dz = \tfrac{13}{2}.$$

The line integral in Example 5 is evidently not independent of the path. Compare this with the following example in which the integral is independent of the path.

Example 6. Find the value of the line integral

$$\int_{(0,0,0)}^{(1,2,3)} x\, dx + 2y\, dy + z\, dz.$$

Solution: Observe that $x\, dx + 2y\, dy + z\, dz \equiv d[x^2/2 + y^2 + z^2/2]$, so that the integral may be written as

$$\int_{(0,0,0)}^{(1,2,3)} d\left(\frac{x^2}{2} + y^2 + \frac{z^2}{2}\right) = \frac{x^2}{2} + y^2 + \frac{z^2}{2}\bigg|_{(0,0,0)}^{(1,2,3)} = \frac{1}{2} + 4 + \frac{9}{2} = 9.$$

In this case it is not necessary to specify the path joining (0,0,0) and (1,2,3), for the answer is the same for any path. One might be led to make the conjecture that if the integrand $U_1\,dx + U_2\,dy + U_3\,dz$ in (10) is the total differential of some function $f(x,y,z)$, then the result of the integration is the same for all paths joining points A and B. This is not always true. Necessary and sufficient conditions for independence of path will be discussed later (see Section 11–7).

11–2. Vector Form of Line Integrals. Let the integral in (10) be written in the form

11) $$\int_A^B U_1(x^1,x^2,x^3)\,dx^1 + U_2(x^1,x^2,x^3)\,dx^2 + U_3(x^1,x^2,x^3)\,dx^3,$$

and consider the functions U_i as the components of a vector $\mathbf{F}$, so that $\mathbf{F} = U_i \mathbf{e}^i$. If $\mathbf{R} = x^j \mathbf{e}_j$ is the position vector to any point $P(x^j)$, then $d\mathbf{R} = dx^j \mathbf{e}_j$, and $\mathbf{F}\cdot d\mathbf{R} = U_i\,dx^j \mathbf{e}^i\cdot\mathbf{e}_j = U_i\,dx^j\,\delta_i{}^j = U_i\,dx^i$. Thus, the integral in (11) becomes

12) $$\int_A^B \mathbf{F}\cdot d\mathbf{R} \equiv \int_A^B U_i\,dx^i.$$

It is understood that the path of integration C joining A and B is made up of a set of arcs on each of which the vector function $\mathbf{F}$ has a continuous derivative. That is, conditions necessary for the existence of the integral in (12) are assumed.

If the vector function $\mathbf{R}(s) = x^i(s)\mathbf{e}_i$ giving the position vector of points on the curve C takes on the same value for two or more different values of the parameter s, the point on C where this occurs is called a *multiple* point. The curve intersects itself at this point. It will be assumed that C is *simple*, that is, it has no multiple point. As the parameter s varies continuously over the interval $a < s < b$ the points of the arc C are in continuous one-to-one correspondence with the values of s. In case $\mathbf{R}(a) = \mathbf{R}(b)$, the curve C is *closed*, and in the absence of multiple points C will be called a *simple closed* curve.

In the sequel mention will be made frequently of a curve or a surface enclosing a *region*. It will be useful, in the interest of clarity, to give a precise definition of a region. One begins with the undefined notion of a collection or set of points. Consider a point set in ordinary three-space. A point is *interior* to the set if there is a sphere with the point as center which contains only points of the set.[1] A point is *exterior* to the set if there is a sphere with center at the point which contains no points of the set. A point is a *boundary* point of the set if every sphere with center at the point contains both interior and exterior points. The set is *closed* or *open* according as it contains all or none of its boundary points. The set

---

[1] It is to be understood here that the sphere is the inside of a spherical surface.

is *connected* if for every two points of the set there exists a continuous curve joining the two points such that all points of the curve are in the set. A *region* (or *open* region) is a point set which is connected and open. A *closed* region is a region augmented by the set of its boundary points. Finally, a *simply connected* region is a region with the property that every simple closed curve in it can be deformed continuously to a point of the region without any point of the curve becoming a boundary point of the region. One may speak of a region of three-space with a simple closed surface as its boundary. Similarly, one may define a region on a plane with a simple closed curve as boundary. A *multiply connected* region is one which is not simply connected (see Section 11–5).

If the integral in (12) is to be calculated around a simple closed curve C in a plane, then the curve C is the boundary of a region $\mathcal{K}$, and the positive direction of advance along the boundary is taken such that the points of the region $\mathcal{K}$ are on the left. If the integral in (12) is along a simple closed space curve, one can imagine a curved surface spanning the curve so that a region of the surface is bounded by the curve, and the positive direction along C is again that for which the points of the surface region are on the left. The concept of a *region on a surface* is made more precise as follows. The Gauss equations of a surface S were introduced in Section 2–4 and the vector representation appeared in Section 8–6 in the form $\mathbf{R} = x^i(u^1, u^2)\mathbf{e}_i$. A region on S may be defined as the point set in space with coordinates x^i where the functions $x^i(u^1, u^2)$ are defined in a simply connected region $\overline{\mathcal{K}}$ of the $u^1 u^2$ plane. To each point (u^1, u^2) of $\overline{\mathcal{K}}$ there corresponds an associated point x^i given by the mapping $x^i = x^i(u^1, u^2)$. Of course, the usual conditions on the functions $x^i(u^1, u^2)$ are assumed, as in Section 2–4, to ensure that the surface does not degenerate to a curve and that certain operations of the calculus may be performed.

An interpretation of a line integral from A to B along a curve C as representing work is realized if the vector field F is a force field. Recall that the unit vector tangent to C at any point P is $\mathbf{T} \equiv d\mathbf{R}/ds$ if $\mathbf{R}$ is the position vector of P. The component of the force field in the direction of C is the scalar projection $\mathbf{F} \cdot \mathbf{T}$, and the element of work dw of the force on a particle in moving it a distance ds along C is $\mathbf{F} \cdot \mathbf{T}\, ds$. Hence, the work done by the force in moving the particle from A to B on C is given by the line integral in (12). If the integral (12) has the same value along all paths joining A and B, the field $\mathbf{F}$ is *conservative* (see Section 10–8).

It is important that the reader solve some exercises to gain technique in evaluating line integrals. In general, the line integral (12) is evaluated for a curve $C: x^i = x^i(t)$ by substituting the $x^i(t)$ functions for x^i and then integrating with respect to t. Thus,

$$13) \quad \int_A^B \mathbf{F} \cdot d\mathbf{R} = \int_A^B \mathbf{F} \cdot \frac{d\mathbf{R}}{dt}\, dt = \int_{t_1}^{t_2} U_i[x^1(t), x^2(t), x^3(t)] \frac{dx^i}{dt}\, dt,$$

where t ranges from t_1 at A to t_2 at B. If $U_i\,dx^i$ is the exact differential of a single-valued function ϕ, one should save labor by writing the integral as follows

$$\int_A^B \mathbf{F}\cdot d\mathbf{R} = \int_A^B d\phi = \phi(x^1,x^2,x^3)\Big|_A^B = \phi(B) - \phi(A),$$

where $\phi(B)$, for instance, means that the coordinates b^i of B are substituted for x^i in ϕ.

EXERCISES

1) Integrate the function $x + y + z$ with respect to x from the origin to the point where $t = 1$ on the curve $x = t$, $y = t^2$, $z = t^3$. Ans. $1\frac{3}{5}$.

2) Evaluate

$$\int_A^B xyz\,ds$$

along the curve $x = \cos t$, $y = \cos t$, $z = \sqrt{2}\sin t$ if $t = 0$ at A and $t = \pi/2$ at B. Ans. $\frac{2}{3}$.

3) Evaluate

$$\int_A^B z\,dx + x\,dy + dz$$

along the straight line joining the points $A(1,1,2)$ and $B(2,-1,3)$. Ans. $\frac{1}{2}$.

4) Evaluate

$$\int_A^B x\,dx + y\,dy + z\,dz$$

along any path you choose between $A(1,1,2)$ and $B(2,-1,3)$. Ans. 4.

5) Find the value of

$$\int_A^B \mathbf{F}\cdot d\mathbf{R}$$

if $\mathbf{F}$ is the vector field $2xyz\mathbf{i} + x^2z\mathbf{j} + x^2y\mathbf{k}$, and A and B are the same as in Exercise 4. (You may choose the path arbitrarily, or find the function ϕ such that $\mathbf{F} \equiv \nabla\phi$.)

6) Calculate the total work done in moving a particle in a force field given by $\mathbf{F} = (2xy - z)\mathbf{i} + (z - 2x^3)\mathbf{j} + x^2\mathbf{k}$ along the curve $x = t$, $y = t^2 - 1$, $z = 2t^3$ from $t = 0$ to $t = 1$. Ans. $\frac{1}{5}$.

7) A region $\mathcal{R}$ is enclosed in the plane $z = 0$ by lines in this plane with equations $y = 0$, $x = 4$, $4y = 3x$. Find the value of the integral $\int y^2\,dx + x^2\,dy + z^2\,dz$ around the boundary of $\mathcal{R}$ in the positive sense. (Note that if the fingers of the right hand point in the direction of advance along the boundary, the thumb

points in the direction of the positive z-axis. What would be the value of the integral around the boundary in the opposite direction?) *Ans.* 20.

8) Calculate $\oint_C y^2\,dx + x^2\,dy$, where C is the circle $x^2 + y^2 = a^2$. You may use a parametric representation of the circle such as $x = a\cos\theta$, $y = a\sin\theta$, although this is not necessary in this exercise.

9) Evaluate $\oint_C (x^2 - y^2)\,ds$ around the circle C: $x^2 + y^2 = a^2$. A parametric representation of the circle is useful here. *Ans.* 0.

10) Evaluate $\int_\Gamma \mathbf{F}\cdot d\mathbf{R}$, where $\mathbf{F} = (x^2 + y^2)\mathbf{i} + 2xy\mathbf{j}$, and (a) Γ is the circle with radius a and center at the origin, (b) Γ is the square with vertices at $(1,1)$, $(-1,1)$, $(-1,-1)$, $(1,-1)$. *Ans.* (a) 0, (b) 0.

11–3. Surface and Volume Integrals. Consider first a simply connected region $\mathcal{K}$ in the plane with a simple closed curve Γ as boundary. A *unit outward* normal vector $\mathbf{N}$ is defined at each point P of Γ (Fig. 27). At P, the tangential component of a vector field $\mathbf{F} = f_1(x,y)\mathbf{i} + f_2(x,y)\mathbf{j}$ is $\mathbf{F}\cdot\mathbf{T}$. The integral of this scalar function $\mathbf{F}\cdot\mathbf{T}$ around the contour Γ is given by

$$\int_\Gamma \mathbf{F}\cdot\mathbf{T}\,ds \equiv \int_\Gamma \mathbf{F}\cdot d\mathbf{R}$$

and is called the *circulation* of the vector $\mathbf{F}$ along the curve Γ. Instances of this appeared in the foregoing exercises.

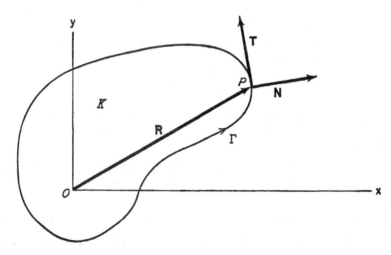

Fig. 27

Instead of the tangential component of **F**, consider the normal component **F·N** of the vector field and integrate this scalar function around Γ to obtain

$$\int \mathbf{F} \cdot \mathbf{N} \, ds,$$

which is called the *flux* of **F** along Γ. Observe from Fig. 27 that the unit normal vector **N** is 90° behind the unit tangent vector **T**. If the tangent to Γ at any point makes an angle α with the x-axis, then

$$\mathbf{T} = (\cos \alpha)\mathbf{i} + (\sin \alpha)\mathbf{j} = \frac{dx}{ds}\mathbf{i} + \frac{dy}{ds}\mathbf{j}$$

and

$$\mathbf{N} = \left[\cos\left(\alpha - \frac{\pi}{2}\right)\right]\mathbf{i} + \left[\sin\left(\alpha - \frac{\pi}{2}\right)\right]\mathbf{j}$$

$$= (\sin \alpha)\mathbf{i} - (\cos \alpha)\mathbf{j} = \frac{dy}{ds}\mathbf{i} - \frac{dx}{ds}\mathbf{j}.$$

As an illustration, consider the following.

Example 1. Find (a) the circulation of the vector field $\mathbf{F} = (x+y)\mathbf{i} + (x-y)\mathbf{j}$ in the first quadrant along the circle $x^2 + y^2 = a^2$, (b) the integral of the normal component of the vector field **F** along the same curve.

Solution: (a) For the circulation

$$\int_C \mathbf{F} \cdot d\mathbf{R} = \int_C \mathbf{F} \cdot \mathbf{T} \, ds = \int_C \left[(x+y)\frac{dx}{ds} + (x-y)\frac{dy}{ds}\right] ds$$

$$= \int_C (x+y)\, dx + (x-y)\, dy$$

$$= a^2 \int_0^{\pi/2} [(\cos\theta + \sin\theta)(-\sin\theta) + (\cos\theta - \sin\theta)\cos\theta]\, d\theta = 0.$$

(b) For the integral of the normal component of the field

$$\int_C \mathbf{F} \cdot \mathbf{N} \, ds = \int_C \left[(x+y)\frac{dy}{ds} - (x-y)\frac{dx}{ds}\right] ds$$

$$= a^2 \int_0^{\pi/2} [(\sin\theta - \cos\theta)(-\sin\theta) + (\cos\theta + \sin\theta)\cos\theta]\, d\theta = 0.$$

The circulation of a vector field along a curve in three-space is readily obtained. The scalar function **F·T** is integrated along the curve in the usual way.

Now the integral of the *normal* component of a vector field along a curve may be extended to apply to a curve in three-space, but here any one of an infinitude of normals to the curve at any point could be chosen.

For instance, the component **F·B** of **F** along the binormal could be integrated along the curve to obtain $\int_C \mathbf{F \cdot B}\, ds$, which is again a case of integration of a function along a curve.

Another way of extending to three-space the integral of the normal component of a vector field is to generalize the curve to a surface. One has to consider then the integral of the normal component of a vector field over a region $\mathcal{K}$ of a surface (Fig. 28). The question of orientation

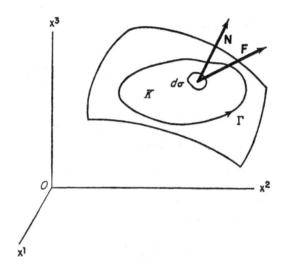

Fig. 28

of the surface must be settled. The surface is assumed to be closed, and the unit normal vector **N** is taken along the *outward* normal to the surface. The right-hand rule can be used to orient the boundary Γ of the region $\mathcal{K}$. If the fingers of the right hand point in the positive direction along the boundary Γ of the region $\mathcal{K}$ (that is, with the region on the left), the thumb will point in the direction of the *outward* normal to the surface. The component **F·N** of the space vector field at a point P in the region $\mathcal{K}$ on the surface is multiplied by an element of surface area $d\sigma$ surrounding P. The result of integrating the scalar function **F·N** over the surface region $\mathcal{K}$ is given by

14)
$$\iint_\mathcal{K} \mathbf{F \cdot N}\, d\sigma,$$

which is called the *flux* of the field **F** over $\mathcal{K}$. The region $\mathcal{K}$ may encompass the entire area of the surface. The integral (14) can be arrived at as a limit of a sum analogous to the manner in which the line integral

$$\int_a^b f(x)\, dx$$

was introduced at the beginning of this chapter.

A surface integral need not represent only the flux of a vector field. Among others, the following forms of surface integrals may be considered

15) $$\iint_{\mathcal{K}} \mathbf{F} \times \mathbf{N}\, d\sigma, \qquad \iint_{\mathcal{K}} f(x^1, x^2, x^3)\, d\sigma.$$

Before showing an example of a surface integral, it will be useful to indicate the form which the surface element $d\sigma$ and the unit normal assume if the surface is represented in the Gaussian form $x^i = x^i(u^1, u^2)$, or by $\mathbf{R} = x^i(u^1, u^2)\mathbf{e}_i$ as in Section 8–6. It was seen there that the unit tangents to the coordinate curves $u^\alpha = $ constant ($\alpha = 1,2$) on the surface S are given by

16) $$\mathbf{T}_\alpha = \frac{1}{\sqrt{g_{\alpha\alpha}}} \frac{\partial \mathbf{R}}{\partial u^\alpha} = \frac{1}{\sqrt{g_{\alpha\alpha}}} \frac{\partial x^i}{\partial u^\alpha} \mathbf{e}_i \qquad (\alpha = 1,2),$$

where $ds^2 = g_{\alpha\beta}\, du^\alpha\, du^\beta$ gives the metric on the surface S. The direction of the outward normal to S at P is given by

17) $$\mathbf{T}_1 \times \mathbf{T}_2 = \frac{1}{\sqrt{g_{11}g_{22}}} \frac{\partial x^i}{\partial u^1} \frac{\partial x^j}{\partial u^2} \mathbf{e}_i \times \mathbf{e}_j = \frac{e_{ijk}}{\sqrt{g_{11}g_{22}}} \frac{\partial x^i}{\partial u^1} \frac{\partial x^j}{\partial u^2} \mathbf{e}^k.$$

The magnitude of $\mathbf{T}_1 \times \mathbf{T}_2$ is given by

18) $$\frac{1}{\sqrt{g_{11}g_{22}}} \left[\begin{vmatrix} \dfrac{\partial x^2}{\partial u^1} & \dfrac{\partial x^3}{\partial u^1} \\ \dfrac{\partial x^2}{\partial u^2} & \dfrac{\partial x^3}{\partial u^2} \end{vmatrix}^2 + \begin{vmatrix} \dfrac{\partial x^3}{\partial u^1} & \dfrac{\partial x^1}{\partial u^1} \\ \dfrac{\partial x^3}{\partial u^2} & \dfrac{\partial x^1}{\partial u^2} \end{vmatrix}^2 + \begin{vmatrix} \dfrac{\partial x^1}{\partial u^1} & \dfrac{\partial x^2}{\partial u^1} \\ \dfrac{\partial x^1}{\partial u^2} & \dfrac{\partial x^2}{\partial u^2} \end{vmatrix}^2 \right]^{1/2}.$$

The unit normal $\mathbf{N}$ is expressed by

19) $$\mathbf{N} = \frac{\mathbf{T}_1 \times \mathbf{T}_2}{|\mathbf{T}_1 \times \mathbf{T}_2|},$$

and is calculated by use of (17) and (18).

The surface element $d\sigma$ is determined as follows. Take a parallelogram determined by segments of lengths ds_1 and ds_2 along the tangents to the u^1 and u^2 curves. In Section 8–7 these were shown to be

20) $$ds_1 = \sqrt{g_{11}}\, du^1, \qquad ds_2 = \sqrt{g_{22}}\, du^2.$$

The area $d\sigma$ is then $ds_1\,ds_2 \sin \omega$, where ω is the angle between the coordinate curves at any point P. From formula (8–45), one finds

21) $$\sin^2 \omega = 1 - \cos^2 \omega = \frac{g_{11}g_{22} - (g_{12})^2}{g_{11}g_{22}}.$$

Hence, the area element $d\sigma$ takes the form

22) $$d\sigma = \sqrt{g_{11}g_{22} - g_{12}^2}\,du^1\,du^2.$$

Example 2. Find the total area of the surface

$$x^1 = a \sin u^1 \cos u^2, \qquad x^2 = a \sin u^1 \sin u^2, \qquad x^3 = a \cos u^1.$$

Solution: This is the sphere with radius a and center at the origin. The parameters u^1 and u^2 represent colatitude and longitude, respectively. The required area is found by integrating the function 1 over the surface, that is

23) $$\iint_S (1)\,d\sigma = \int_0^{2\pi}\!\!\int_0^{\pi} \sqrt{g_{11}g_{22} - g_{12}^2}\,du^1\,du^2.$$

Because the parametric curves are orthogonal, $g_{12} = 0$, so it is necessary to calculate only g_{11} and g_{22}. From the matrix of partial derivatives

24) $$\left(\frac{\partial x^i}{\partial u^a}\right) \equiv \begin{pmatrix} a \cos u^1 \cos u^2 & a \cos u^1 \sin u^2 & -a \sin u^1 \\ -a \sin u^1 \sin u^2 & a \sin u^1 \cos u^2 & 0 \end{pmatrix}$$

and the definitions (8–43), one obtains

25) $$g_{11} = a^2, \qquad g_{22} = a^2 \sin^2 u^1.$$

Therefore, the area in (23) is found to be

26) $$a^2 \int_0^{2\pi}\!\!\int_0^{\pi} \sin u^1\,du^1\,du^2 = 4\pi a^2.$$

Example 3. Find the total flux of the vector field $\mathbf{F} = x^2\mathbf{i} + x^3\mathbf{j} + x^1\mathbf{k}$ over the first-octant portion of the area of the sphere of Example 2. (*Note:* The superscripts in the vector are not exponents.)

Solution: The integral to be evaluated is

$$\iint_S \mathbf{F}\cdot\mathbf{N}\,d\sigma.$$

In order to find $\mathbf{N}$ write the three second-order determinants in (24), which are proportional to the components of $\mathbf{T}_1 \times \mathbf{T}_2$. These are $a^2 \sin^2 u^1 \cos u^2$, $a^2 \sin^2 u^1 \sin u^2$, $a^2 \sin u^1 \cos u^1$. On changing these to the components of a unit vector one obtains

$$\mathbf{N} = (\sin u^1 \cos u^2)\mathbf{i} + (\sin u^1 \sin u^2)\mathbf{j} + (\cos u^1)\mathbf{k}.$$

(Of course, for the sphere the expression for $\mathbf{N}$ could be written without calculation, but the method shown here indicates the general procedure for an arbitrary surface.) On placing the values of x^1, x^2, x^3 as functions of u^1, u^2 in $\mathbf{F}$, the integral becomes

$$a^3 \int_0^{\pi/2} \int_0^{\pi/2} (\sin^3 u^1 \sin u^2 \cos u^2 + \sin^2 u^1 \cos u^1 \sin u^2$$
$$+ \sin^2 u^1 \cos u^1 \cos u^2)\, du^1\, du^2 = a^3.$$

A volume integral is conceived as follows: Consider a region of space enclosed by a surface, and a function f defined at each point of the region. Note that the function may be a scalar or a vector. The value of the function at a point is multiplied by an element of volume surrounding the point. The limit of the sum of such products can be shown to be given by the triple integral

$$\iiint_V f\, dV,$$

where dV is the differential of volume in the coordinate system employed, and V is the region of space over which the integral is evaluated.

Example 4. Integrate the function xy over the volume enclosed by the planes $x = 0, y = 0, z = 0, x + y + z = 1$.

Solution: The integral $\iiint_V xy\, dV$ may be evaluated by the iterated integral

$$\int_0^1 \int_0^{1-x} \int_0^{1-x-y} xy\, dz\, dy\, dx = \frac{1}{120}.$$

The result here can be interpreted as the mass of the region if the variable density ρ is expressed by xy.

Example 5. Integrate $\mathbf{F} = y\mathbf{i} + z\mathbf{j} + x\mathbf{k}$ over the volume enclosed by the planes $x = 0, x = 2, y = 0, y = 3, z = 0$, and the surface $z = x^2$.

Solution: The required integral is expressed by

$$\iiint \mathbf{F}\, dV = \mathbf{i} \int_0^2 \int_0^3 \int_0^{x^2} y\, dz\, dy\, dx + \mathbf{j} \int_0^2 \int_0^3 \int_0^{x^2} z\, dz\, dy\, dx$$
$$+ \mathbf{k} \int_0^2 \int_0^3 \int_0^{x^2} x\, dz\, dy\, dx = 12\mathbf{i} + \tfrac{48}{5}\mathbf{j} + 12\mathbf{k}.$$

EXERCISES

1) Calculate the total flux of the vector field $\mathbf{F} = x^1\mathbf{i} + x^2\mathbf{j} + x^3\mathbf{k}$ over the surface of the sphere with radius a and center at the origin. *Ans.* $4\pi a^3$.

2) Find the flux of the vector field $\mathbf{F} = x\mathbf{i} + y\mathbf{j} + z\mathbf{k}$ over the plane $\pi: x + 2y + 3z = 6$ in the region of space where x, y, z are all positive. (Note that the Gauss equations of the plane π may be taken as $x = x, y = y, z = \frac{1}{3}(6 - x - 2y)$, where the parameters u^1, u^2 are x, y. Calculate g_{11}, g_{12}, g_{22} and show that the element of area $d\sigma$ on the plane π as given by formula (22) is $(\sqrt{14}/3) \, dy \, dx$. This could be attained otherwise as follows: Multiply the area element $d\sigma$ on π by the cosine of the angle γ between π and the xy-plane to obtain the area element $dy \, dx$ as the projection of $d\sigma$ on the xy-plane. Since $\cos \gamma = 3/\sqrt{14}$, $d\sigma = (\sqrt{14}/3) \, dy \, dx$.)

3) Integrate the function xz over the entire surface of the cylinder which is bounded by $z = 0, z = h, x^2 + y^2 = a^2$.

4) Integrate the function xz over the volume of the cylinder in Exercise 3.

5) Show that the flux of a vector field $\mathbf{F}$ over a region S of a surface in the Monge form $z = f(x,y)$ is given by

$$\iint_S \mathbf{F} \cdot \mathbf{N} \, d\sigma = \iint_{S_{xy}} \mathbf{F} \cdot (p\mathbf{i} + q\mathbf{j} - \mathbf{k}) \, dx \, dy,$$

where $p \equiv \partial z/\partial x$, $q \equiv \partial z/\partial y$, and S_{xy} is the projection of S onto the xy-plane.

6) Show that the flux of a vector field $\mathbf{F}$ over a region S of a surface with equation $H(x,y,z) = 0$ is given by

$$\iint_S \mathbf{F} \cdot \mathbf{N} \, d\sigma = \iint_{S_{xy}} \mathbf{F} \cdot \frac{\nabla H}{|\nabla H|} \, d\sigma = \iint_{S_{xy}} (\mathbf{F} \cdot \nabla H) \frac{1}{\left|\frac{\partial H}{\partial z}\right|} \, dx \, dy,$$

where it is assumed that ∇H does not vanish at any point on S.

11–4. Green's Theorem in the Plane. Two important theorems involving vectors in three-space will be introduced later. They are the theorem of Stokes and the divergence theorem of Gauss. It will be useful to study first a specialization of these theorems to the case of a region in the plane $z = 0$. This specialization, called Green's theorem in the plane, expresses a relation between a line integral around the boundary Γ' of a region $\mathcal{K}'$ in the plane and an integral over $\mathcal{K}'$. The boundary Γ' is understood to be a simple closed curve such as that shown in Fig. 29. For simplicity, the theorem will be proved for a region $\mathcal{K}$ which is bounded by a curve Γ with the property that a line parallel to the x-axis or to the y-axis will intersect Γ in at most two points (Fig. 30). The theorem can then be extended to a more general contour, such as the one shown in Fig. 29.

Theorem: If $\mathcal{K}$ is a region of the xy-plane bounded by a simple closed curve Γ, and if $U_1(x,y), U_2(x,y)$ are scalar functions which together with

$\partial U_2/\partial x$, $\partial U_1/\partial y$ are continuous in $\mathcal{K}$ and on Γ, then

27) $$\oint_\Gamma U_1\,dx + U_2\,dy = \iint_\mathcal{K} \left(\frac{\partial U_2}{\partial x} - \frac{\partial U_1}{\partial y}\right) dx\,dy.$$

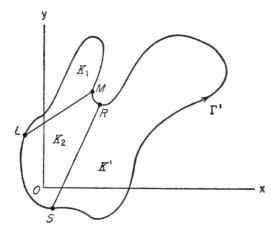

Fig. 29

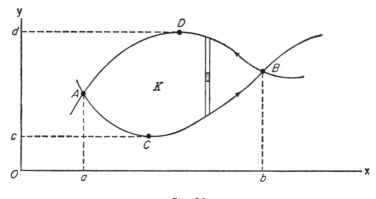

Fig. 30

(Note that the direction of integration on Γ is counterclockwise, as indicated by the directed circle on the integral sign.)

Proof: It will be shown first that

28) $$\iint_\mathcal{K} \frac{\partial U_1}{\partial y}\,dy\,dx = -\int_\Gamma U_1\,dx.$$

Consider Γ as made up of the upper arc ADB (Fig. 30) with equation $y = f_2(x)$, and of the lower arc ACB with equation $y = f_1(x)$. The double integral in (28) can be expressed as the iterated integral

$$29) \quad \int_a^b \int_{f_1(x)}^{f_2(x)} \frac{\partial U_1}{\partial y} \, dy \, dx = \int_a^b U_1(x,y) \Big|_{f_1(x)}^{f_2(x)} dx$$

$$= \int_a^b U_1[x, f_2(x)] \, dx - \int_a^b U_1[x, f_1(x)] \, dx,$$

where the abscissas of A and B are a and b, respectively. Now, by using the fact that

$$\int_a^b U_1[x, f_2(x)] \, dx = -\int_b^a U_1[x, f_2(x)] \, dx = -\int_{BDA} U_1(x,y) \, dx$$

along the curve BDA with equation $y = f_2(x)$, and that

$$\int_a^b U_1[x, f_1(x)] \, dx = \int_{ACB} U_1(x,y) \, dx$$

along the curve ACB with equation $y = f_1(x)$, equation (29) gives

$$30) \quad -\iint_{\mathcal{K}} \frac{\partial U_1}{\partial y} \, dy \, dx = \int_{ACB} U_1(x,y) \, dx + \int_{BDA} U_1(x,y) \, dx = \oint_\Gamma U_1(x,y) \, dx.$$

In a similar manner, if C and D are the points with least and greatest ordinates c and d on the contour Γ, let the part of Γ along DAC be represented by $x = g_1(y)$ and the part CBD by $x = g_2(y)$. The remaining term in the right-hand member of (27) is evaluated as

$$31) \quad \iint_{\mathcal{K}} \frac{\partial U_2}{\partial x} \, dx \, dy = \int_c^d \int_{g_1(y)}^{g_2(y)} \frac{\partial U_2}{\partial x} \, dx \, dy$$

$$= \int_c^d U_2[g_2(y), y] \, dy - \int_c^d U_2[g_1(y), y] \, dy$$

$$= \int_{CBD} U_2(x,y) \, dy + \int_{DAC} U_2(x,y) \, dy = \oint_\Gamma U_2(x,y) \, dy.$$

Addition of equations (30) and (31) yields

$$\iint_{\mathcal{K}} \left(\frac{\partial U_2}{\partial x} - \frac{\partial U_1}{\partial y} \right) dx \, dy = \oint_\Gamma U_1 \, dx + U_2 \, dy,$$

which completes the proof.

Remark: The graph in Fig. 30 suggests that the part DAC of the boundary of $\mathcal{K}$ may be described by two analytical representations—one for DA and another for AC.

Now that Green's theorem in the plane is proved for a restricted type of bounding curve, the theorem can be extended to a more general contour Γ' such as that shown in Fig. 29. Actually, consideration is restricted here to types of regions which can be subdivided by a finite number of arcs into regions of the type used in the proof of the theorem. Such regions suffice for most practical applications. Let arcs such as LM and RS be inserted into the region $\mathcal{K}'$ in Fig. 29 to form subregions $\mathcal{K}_1, \mathcal{K}_2, \cdots$ each of which is of the type used in the proof of the theorem. Note that when the theorem is applied to contiguous regions such as $\mathcal{K}_1$ and $\mathcal{K}_2$ in Fig. 29, the line integral along ML as part of the boundary of $\mathcal{K}_2$ cancels the line integral along LM as part of the boundary of $\mathcal{K}_1$. If this argument is applied to the entire dissected region $\mathcal{K}'$, all line integrals across the cross-cuts will contribute zero and the result is the line integral around the boundary Γ' of $\mathcal{K}'$. Hence, Green's theorem in the plane holds for a simple closed curve Γ' bounding a region $\mathcal{K}'$ of the type considered here.

Caution: One must be sure that the hypotheses of Green's theorem are satisfied before applying the theorem. If U_1 or U_2 is not defined at a point of a region, or if $\partial U_2/\partial x$ or $\partial U_1/\partial y$ does not exist at a point, the theorem does not necessarily hold. Further discussion of this appears in the next section.

It should be observed that if $\mathbf{F} = U_1\mathbf{i} + U_2\mathbf{j} + U_3\mathbf{k}$, with U_1 and U_2 functions of x and y only, and with $U_3 \equiv 0$, then

$$\text{curl } \mathbf{F} \equiv \nabla \times \mathbf{F} = \mathbf{k}\left(\frac{\partial U_2}{\partial x} - \frac{\partial U_1}{\partial y}\right).$$

Further, the unit normal $\mathbf{N}$ to the xy-plane is $(0,0,1)$ so that $\mathbf{N} \cdot \text{curl } \mathbf{F} = (\partial U_2/\partial x - \partial U_1/\partial y)$. Hence, equation (27) can be written in vector form as

$$32) \qquad \iint_{\mathcal{K}} (\mathbf{N} \cdot \text{curl } \mathbf{F})\, d\sigma = \oint_\Gamma \mathbf{F} \cdot d\mathbf{R} = \oint_\Gamma \mathbf{F} \cdot \mathbf{T}\, ds,$$

where $d\mathbf{R}$ is $\mathbf{i}\, dx + \mathbf{j}\, dy + \mathbf{k}\, dz$ and $d\sigma$ is the element of area $dx\, dy$ in the xy-plane. Equation (32) states the theorem of Stokes (see Section 11–10) if $\mathcal{K}$ is a region on a surface enclosed by a contour Γ on the surface. Equation (32) was developed by using the plane $z = 0$, but because of the vector form of the equation the result holds for any system of coordinates. Therefore, one may state the result in (32) without symbols as follows. The integral, over any plane region, of the normal component of the curl of a vector field is equal to the circulation of the vector field (taken in the appropriate direction) around the boundary of the region. In the generalization to Stokes' theorem, the plane region becomes a region on a curved surface.

Equation (27) can also be interpreted as a special case of the divergence theorem of Gauss which will be proved in Section 11-8 for three-space. To do this, let the vector field **F** be given by $U_2\mathbf{i} - U_1\mathbf{j} + U_3\mathbf{k}$ with $U_3 \equiv 0$. The divergence of **F** is given by

$$\text{div } \mathbf{F} \equiv \nabla \cdot \mathbf{F} = \frac{\partial U_2}{\partial x} - \frac{\partial U_1}{\partial y} + \frac{\partial U_3}{\partial z} = \frac{\partial U_2}{\partial x} - \frac{\partial U_1}{\partial y},$$

which is the integrand of the right-hand member in (27). Consider the component of **F** along the normal to the curve Γ. The unit vector along the normal is $(dy/ds)\mathbf{i} - (dx/ds)\mathbf{j}$. Hence, $\mathbf{F} \cdot \mathbf{N}\, ds = U_1\, dx + U_2\, dy$, which appears in the line integral of (27). Therefore, Green's theorem in the plane can be written in the form

33) $$\iint_{\mathcal{K}} \text{div } \mathbf{F}\, d\sigma = \oint_{\Gamma} \mathbf{N} \cdot \mathbf{F}\, ds.$$

If the region $\mathcal{K}$ is generalized to a region D of three-space and the bounding curve Γ is generalized to the bounding *surface* S of the region D, one might hazard the conjecture that

34) $$\iiint_{D} \text{div } \mathbf{F}\, dV = \iint_{S} \mathbf{N} \cdot \mathbf{F}\, d\sigma,$$

where $d\sigma$ in (33) becomes the differential of volume dV in (34) and ds in (33) becomes the differential of area $d\sigma$ on S in (34). Equation (34) expresses the Gauss divergence theorem in space which will appear in Section 11-8.

EXERCISES

1) Verify Green's theorem in the plane for the finite region bounded by $2y = x$ and $4y = x^2$, where the line integral is $\int_C (x - y)\, dx + x\, dy$. *Ans.* The common value of the two integrals is $\tfrac{2}{3}$.

2) Verify Green's theorem in the plane for $\int_C (x - 2y)\, dx + (x + y)\, dy$ taken around the circle $x^2 + y^2 = a^2$. *Ans.* The common value is $3\pi a^2$.

3) Evaluate $\int (x^2 \sin x - y)\, dx + (y^2 \cos y + 3x)\, dy$ around the ellipse $b^2 x^2 + a^2 y^2 = a^2 b^2$. (*Hints:* Use Green's theorem, and save further time by using πab for the area of the ellipse.) *Ans.* $4\pi ab$.

4) Show that $\int x\, dy$ around any simple closed plane curve gives the area

enclosed by the curve. Do the same for $\int (-y)\, dx$. Then show that $\frac{1}{2}\int x\, dy - y\, dx$ gives the same area.

5) (a) If **R** is the position vector to any point on a simple closed curve C, evaluate $\int_C \mathbf{R} \cdot \mathbf{T}\, ds$ around the curve. (b) Show that $\frac{1}{2}\int \mathbf{R} \cdot \mathbf{N}\, ds$ is the area enclosed by C, where **N** is the outward normal to C at any point.

6) Evaluate the circulation $\int_C \mathbf{F} \cdot d\mathbf{R}$ of the vector field $\mathbf{F} = (x - 2y)\mathbf{i} + (y + 2x)\mathbf{j}$ around the circle $(x - a)^2 + (y - b)^2 = r^2$, $z = 0$.

7) Evaluate $\int \mathbf{F} \cdot d\mathbf{R}$ around the curve $(x - a)^2 + (y - b)^2 = r^2$, $z = 0$, if **F** is the vector field $(2xy + z^2)\mathbf{i} + (x^2 + 2yz)\mathbf{j} + (y^2 + 2xz)\mathbf{k}$.

8) Use Green's theorem to evaluate $\int_C (x^2 + xy)\, dx + (y^2 + x^2)\, dy$, where C is the boundary of the square with sides $x = \pm 1$, $y = \pm 1$.

9) Given that P and Q are continuous functions of x and y in a region $\mathcal{K}$ and that $\partial P/\partial y$ and $\partial Q/\partial x$ are single-valued and continuous in $\mathcal{K}$, show that $\partial P/\partial y = \partial Q/\partial x$ at all points of $\mathcal{K}$ is a sufficient condition for $\int_C P\, dx + Q\, dy$ to be zero around any closed curve in $\mathcal{K}$.

10) Show that the condition $\partial P/\partial y = \partial Q/\partial x$ at all points of $\mathcal{K}$ is a necessary condition for the integral $\int_C P\, dx + Q\, dy$ to be zero around an arbitrarily chosen closed curve in $\mathcal{K}$, provided the conditions of continuity in Exercise 9 are satisfied.

11) Find the area of the four-cusped hypocycloid $x = a \cos^3 t$, $y = a \sin^3 t$.

12) In (11-27) put $U_1 = -\partial u/\partial y$, $U_2 = \partial u/\partial x$ to show that

$$\iint_\mathcal{K} \left(\frac{\partial^2 u}{\partial x^2} + \frac{\partial^2 u}{\partial y^2}\right) dy\, dx = \int_\Gamma \frac{\partial u}{\partial n}\, ds,$$

where n is the outward normal to Γ.

13) In (11-27) put $U_1 = -v\,(\partial u/\partial y)$, $U_2 = v\,(\partial u/\partial x)$ to arrive at

$$\iint_\mathcal{K} \left(\frac{\partial u}{\partial x}\frac{\partial v}{\partial x} + \frac{\partial u}{\partial y}\frac{\partial v}{\partial y}\right) dx\, dy + \iint_\mathcal{K} \left(\frac{\partial^2 u}{\partial x^2} + \frac{\partial^2 u}{\partial y^2}\right) v\, dx\, dy = \int_\Gamma v\,\frac{\partial u}{\partial n}\, ds.$$

11-5. Simply and Multiply Connected Regions. A more complex region than the type considered heretofore will now be introduced. Consider a region with two simple closed curves C_1 and C_2 as boundaries (Fig. 31). One may think of starting with a simply connected region with boundary C_1 and then of removing the point set inside the boundary C_2. The positive direction along the boundary C_2 is clockwise while along the boundary C_1 it is counterclockwise. It is intuitively evident that if a simply connected region is cut along an arbitrary path joining two arbitrarily chosen boundary points, the region will be severed into two regions. (*Note:* It is understood that the cut does not intersect itself.) On the other hand, if the region $\mathcal{K}$ in Fig. 31 (which is not simply connected) is cut

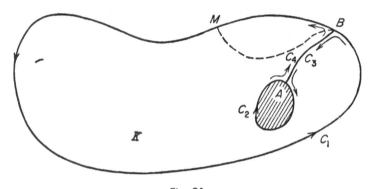

Fig. 31

along a path from boundary point B to boundary point A, such a cut does not sever the region into two pieces. Although a cut joining boundary points B and M (along the dotted line in Fig. 31) severs $\mathcal{K}$, a cut from B to A does not sever the region $\mathcal{K}$. After the cut from B to A is effected, the two sides (or banks) of the cut must be considered as additional bounding segments C_3 and C_4 for the region $\mathcal{K}$. The arrows in Fig. 31 indicate the positive traverse of the boundary of $\mathcal{K}$. From B to A the left bank C_3 is followed and from A to B the left bank C_4 is the path. The complete boundary Γ of $\mathcal{K}$ is now C_1, C_3, C_2, C_4. If a cut is made between two arbitrarily chosen points of this complete boundary, the region $\mathcal{K}$ will now fall into two pieces. Such a region which is severed by a cut joining any two boundary points is called *simply* connected. The cut from B to A in Fig. 31 was necessary to render the region simply connected. If two shaded areas are removed from the interior of C_1 as in Fig. 32, two cuts are necessary to render $\mathcal{K}$ simply connected. These cuts may be made in a variety of ways. One possibility is to cut from A to B and from L to M, as in Fig. 32. Another procedure is to cut from A to B and then from L to M, as in Fig. 33. Again, the arrows indicate the total boundary of $\mathcal{K}$.

If n cuts are necessary to sever a region, it is called *n-tuply* connected. Thus, for a region with one hole or shaded area, as in Fig. 31 before the cut BA is made, the region is doubly connected. With two holes as in Fig. 32 before cuts AB and LM are made, the region is triply connected.

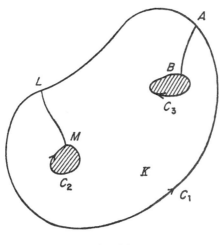

Fig. 32

The region inside any simple closed curve such as an ellipse or a rectangle is simply connected. The definition of simple connectedness may be extended so that the whole plane, or a sector of the whole plane between

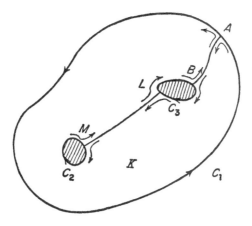

Fig. 33

two half rays issuing from a point, or the whole plane minus the points on the positive x-axis, furnish instances of simply connected regions.

The connectivity of a region may be characterized in a different manner from that of the cuts used in the foregoing discussion. Consider first two simple closed curves C_1 and C_2 lying in a simply connected region $\mathcal{K}$ with boundary C_3 as in Fig. 34. One may think of C_1 and C_2 as elastic threads which can be deformed at will. It is possible to deform C_1 into C_2, for instance, without any point of C_1 leaving the region $\mathcal{K}$. The curves C_1 and C_2 are said to be *homologous* to each other. Actually, either C_1 or C_2 can be shrunk to a point in $\mathcal{K}$ without points of C_1 or C_2 leaving the

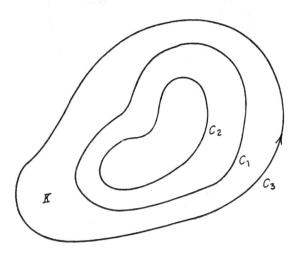

Fig. 34

region $\mathcal{K}$. A curve which can be shrunk to a point while always in the region is *homologous to zero*. It is evident that all simple closed curves in $\mathcal{K}$ are homologous to each other and to a point (i.e., to zero) if and only if $\mathcal{K}$ is simply connected. Observe that it is possible, for instance, that $\mathcal{K}$ is the entire plane.

Consider now Fig. 35 which shows a doubly connected region $\mathcal{K}$. The shaded area H has been removed. All simple closed curves which can be drawn in $\mathcal{K}$ are divided into two classes. The first consists of all those which are homologous to any simple closed curve C_1 which does not surround the hole H, and the second class consists of all curves homologous to any simple closed curve C_2 which does encircle H. Note that C_2 cannot be shrunk to a point without leaving the region $\mathcal{K}$, and that C_2 is homologous to the boundary of H.

If two holes H_1 and H_2 are punched out of a simply connected region, the resulting region is triply connected and four classes of homologous curves exist. Representatives of these four classes are homologous respectively to zero, to the boundary of H_1, to the boundary of H_2, and finally to any curve which surrounds both H_1 and H_2.

Instead of continuing this discussion to plane regions of higher connectivity, it will be of more interest to see how the notion of connectivity carries over to three-space. Actually, the ideas can be extended in two ways. To illustrate, think of the region of space between two concentric spheres. It is clear that a simple closed space curve in this region can be shrunk to a point without any point of the curve leaving the region. Thus, all simple closed curves are homologous to zero and the region is simply connected *relative to one-dimensional subspaces*, i.e., to curves. But

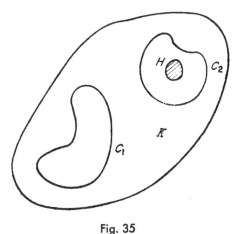

Fig. 35

think of generalizing the closed curve to a closed surface and consider a simple closed surface lying in the region and surrounding the inner sphere. It is not possible to shrink the surface to a point in the region. Hence, the region in this example is not simply connected *relative to two-dimensional subspaces*, i.e., surfaces. Consider the torus (anchor ring or doughnut) with $\mathcal{K}$ as the region inside it. This region is multiply connected relative to both one- and two-dimensional subspaces.

In order to see why the foregoing discussion on connectivity was introduced, consider the line integral

$$33') \qquad \int_C \frac{y\, dx - x\, dy}{x^2 + y^2},$$

where C is a simple closed curve not yet specified. For a given closed curve C the line integral might be difficult to evaluate, so it would be advisable to appeal to Green's theorem in the plane, that is

$$\int_C U_1\, dx + U_2\, dy = \iint_{\mathcal{K}} \left(\frac{\partial U_2}{\partial x} - \frac{\partial U_1}{\partial y} \right) dy\, dx.$$

One finds from (33') that

35) $$\frac{\partial U_1}{\partial y} = \frac{x^2 - y^2}{(x^2 + y^2)^2} = \frac{\partial U_2}{\partial x},$$

so the line integral in (33') apparently has the value zero. This is true for any simple closed curve C (such as C_1 in Fig. 36) which bounds a region $\mathcal{K}$ of the plane that does not contain the origin, because the hypotheses of

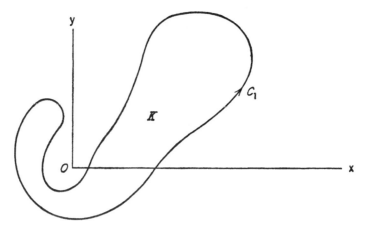

Fig. 36

Green's theorem are satisfied in such a region. But the partial derivatives in (35) do not exist at the origin, so this is a singular point which must be deleted from the plane before applying Green's theorem to the integral (33'). This means that the plane with a point removed is not simply connected. A curve encircling the origin is not homologous to a closed curve which does not encircle the origin. Note that the value of the integral in (33') is zero for the circle $C: (x-2)^2 + (y-2)^2 = 1$. The value is also zero for all curves homologous to this circle.

Now consider the value of the integral in (33') around a curve which does surround the origin. Try the circle $x^2 + y^2 = a^2$. With $x = a\cos\theta$, $y = a\sin\theta$ one has

36) $$\int_C \frac{y\,dx - x\,dy}{x^2 + y^2} = -\int_0^{2\pi} d\theta = -2\pi.$$

Another way to evaluate the integral in (36) along the chosen circle is the following. On the circle the function $x^2 + y^2$ may be replaced by a^2, so the integral is

$$\tfrac{1}{2} \int_C y\,dx - x\,dy.$$

Green's theorem now applies, so that

$$\frac{1}{a^2}\int_C y\,dx - x\,dy = \frac{1}{a^2}\iint_{\mathcal{K}}(-1-1)\,dy\,dx = -\frac{2}{a^2}\text{(area of circle)} = -2\pi.$$

Note that since the result is independent of the radius a, the value of a can be taken as arbitrarily small.

It will be shown next that the value of the integral (33') along any curve Γ homologous to the circle $C: x^2 + y^2 = a^2$ is the same as that of the integral along C, namely, -2π. Choose some curve Γ as in Fig. 37, and then render the doubly connected region $\mathcal{K}$ between Γ and the circle C simply connected by making a cut from any point L on Γ to any point M on C. The circuit shown by the arrows in Fig. 37 is now homologous

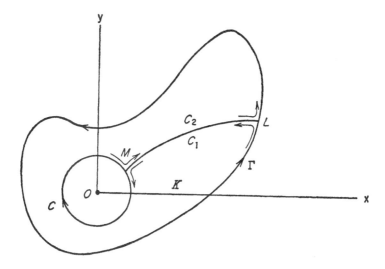

Fig. 37

to zero or to any curve not encircling the origin. Green's theorem applies to yield

37) $\quad\oint_\Gamma [\ ] + \int_{C_1}[\ ] + \oint_C[\ ] + \int_{C_2}[\ ] = \iint_{\mathcal{K}}(0)\,dy\,dx = 0,$

where use was made of (35) in the double integral, and, for brevity, the brackets [ ] all indicate the integrand in (33'). Now, because the line integrals along C_1 and C_2 have opposite signs and therefore cancel in (37), there results

38) $\quad\oint_\Gamma [\ ] = -\oint_C[\ ] = +\oint_C[\ ] = -2\pi$

which shows that the line integral (33′) is the same along all curves homologous to $C: x^2 + y^2 = a^2$, where a is arbitrarily small.

If, instead of one singular point (as the origin in the foregoing instance), there are k singular points $Q_1, \cdots, Q_k$ for a given integral (which render the plane multiply connected), these points can be enclosed separately by small circles $C_1, \cdots, C_k$. If the values of the given line integral along these circles (in the counterclockwise sense) are respectively $v_1, \cdots, v_k$, it is not difficult to show by an argument similar to that leading to (38) that if $\partial U_2/\partial x = \partial U_1/\partial y$, the integral $\int_\Gamma U_1\, dx + U_2\, dy$ along any simple closed curve Γ which encloses all of the singularities $Q_1, \cdots, Q_k$ is given by

39) $$\oint_\Gamma U_1\, dx + U_2\, dy = v_1 + \cdots + v_k.$$

It is assumed, of course, that the hypotheses of Green's theorem are satisfied everywhere within and on Γ except at $Q_1, \cdots, Q_k$. A formal proof of (39) is left to the exercises.

11–6. Independence of the Path of Integration. If the line integral $\int_A^B \mathbf{F} \cdot d\mathbf{R}$ has the same value along all curves joining points A and B in some region $\mathcal{K}$ of three-space, the integral is described as being *independent of the path* in $\mathcal{K}$. If $\int_A^B \mathbf{F} \cdot d\mathbf{R}$ is independent of the path in $\mathcal{K}$, it follows that $\int_C \mathbf{F} \cdot d\mathbf{R} = 0$, where C is an arbitrary closed curve in $\mathcal{K}$. To show this, consider two points A and B on an arbitrary closed curve C in $\mathcal{K}$ (Fig. 38). Because the integral is independent of the path

40) $$\int_{ALB} \mathbf{F} \cdot d\mathbf{R} = \int_{AMB} \mathbf{F} \cdot d\mathbf{R} = -\int_{BMA} \mathbf{F} \cdot d\mathbf{R}$$

from which

41) $$\int_{ALB} \mathbf{F} \cdot d\mathbf{R} + \int_{BMA} \mathbf{F} \cdot d\mathbf{R} = \int_C \mathbf{F} \cdot d\mathbf{R} = 0.$$

Conversely, if $\int_C \mathbf{F} \cdot d\mathbf{R} = 0$, where C is an arbitrary closed curve,

42) $$\int_{ALB} \mathbf{F} \cdot d\mathbf{R} + \int_{BMA} \mathbf{F} \cdot d\mathbf{R} = 0,$$

from which

43) $$\int_{ALB} \mathbf{F} \cdot d\mathbf{R} = -\int_{BMA} \mathbf{F} \cdot d\mathbf{R} = \int_{AMB} \mathbf{F} \cdot d\mathbf{R},$$

Sec. 11-6]   INTEGRATION OF VECTORS   175

which means that the integral is the same along all paths in $\mathcal{K}$ joining A to B.

As an example to illustrate the foregoing result for the case of a plane region, consider again the integral in (33′) of the preceding section. Remember that condition (35) is a special feature of this example. If one starts from a point A in the plane and follows a path which encircles the origin n times in the counterclockwise direction before returning to A, the form

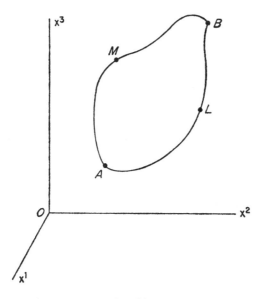

Fig. 38

of the integrand in (36) shows that the result of the integration around the closed curve is $-2n\pi$. Integration in the clockwise direction along the same curve would yield $2n\pi$. It is evident that the result depends upon the number of times the singularity at the origin is encircled. The angle $\theta = \arctan(-y/x)$ is not single-valued for a curve which encloses the origin, so the integral is not independent of the path for a region which contains the origin. However, if the plane is rendered simply connected by a cut (say, along the positive x-axis) so a closed curve cannot encircle the origin, then the integral in this example becomes independent of the path in the entire plane, that is, the integral is zero around all closed curves in the resulting simply connected plane.

The foregoing example shows that a necessary condition for independence of path in a region is simple connectedness of the region. It can be shown that if $\mathbf{F} \cdot d\mathbf{R}$ is an exact differential, say $d\phi(x^1,x^2,x^3)$, and if the region $\mathcal{K}$ is simply connected, then ϕ is a single-valued function in $\mathcal{K}$.

11-7. Test for Independence of Path. For a given line integral $\int \mathbf{F} \cdot d\mathbf{R}$ it is of interest to know if the integral is independent of the path. In this connection, a proof is offered for the following

Theorem 1: A necessary and sufficient condition for the line integral

$$\int_A^B \mathbf{F} \cdot d\mathbf{R}$$

of a continuous vector function $\mathbf{F}$ to be independent of the path joining points A and B in a simply connected space region $\mathcal{K}$ is that the vector $\mathbf{F}$ be the gradient of a scalar function ϕ, that is $\mathbf{F} = \nabla \phi$.

Proof of the sufficiency of the condition will be given first. Let $\mathbf{F} \equiv U_i \mathbf{e}^i = \phi_{,i} \mathbf{e}^i \equiv \nabla \phi$, and $d\mathbf{R} = \mathbf{e}_j \, dx^j$, so that $\mathbf{F} \cdot d\mathbf{R} = \phi_{,i} \, dx^i = d\phi$. Then

44) $$\int_A^B \mathbf{F} \cdot d\mathbf{R} = \int_A^B d\phi = \phi(x^1, x^2, x^3) \Big]_A^B = \phi(B) - \phi(A).$$

The result depends only upon the end points A and B and not upon the choice of the curve Γ connecting A and B. Note that if B coincides with A, the integral around an arbitrary closed path is zero.

To prove the necessity of the condition, suppose that the integral $\int \mathbf{F} \cdot d\mathbf{R}$ is independent of the path in $\mathcal{K}$, and integrate $\mathbf{F} \cdot d\mathbf{R}$ from an arbitrary fixed point a^i to a variable point x^i in $\mathcal{K}$. Because of independence of path the integral $\int_{(a^i)}^{(x^i)} \mathbf{F} \cdot d\mathbf{R}$ is a function of the upper limit only. Call this function $\phi(x^1, x^2, x^3)$. One has

45) $$\phi(x^1, x^2, x^3) = \int_{(a^1, a^2, a^3)}^{(x^1, x^2, x^3)} \mathbf{F} \cdot d\mathbf{R}.$$

Now let x^1 take the increment Δx^1 while x^2 and x^3 are held fixed. From (45), the value of ϕ at the neighboring point $(x^1 + \Delta x^1, x^2, x^3)$ is given by

46) $$\phi(x^1 + \Delta x^1, x^2, x^3) = \int_{(a^1, a^2, a^3)}^{(x^1 + \Delta x^1, x^2, x^3)} \mathbf{F} \cdot d\mathbf{R}.$$

From (45) and (46), the increment in ϕ is

47) $$\phi(x^1 + \Delta x^1, x^2, x^3) - \phi(x^1, x^2, x^3) = \int_{(a^1, a^2, a^3)}^{(x^1 + \Delta x^1, x^2, x^3)} \mathbf{F} \cdot d\mathbf{R}$$
$$- \int_{(a^1, a^2, a^3)}^{(x^1, x^2, x^3)} \mathbf{F} \cdot d\mathbf{R} = \int_{(x^1, x^2, x^3)}^{(x^1 + \Delta x^1, x^2, x^3)} \mathbf{F} \cdot d\mathbf{R} = \int_{x^1}^{x^1 + \Delta x^1} U_1(x^1, x^2, x^3) \, dx^1,$$

and on dividing (47) by Δx^1 there results

$$48) \quad \frac{\phi(x^1+\Delta x^1,x^2,x^3) - \phi(x^1,x^2,x^3)}{\Delta x^1} = \frac{1}{\Delta x^1}\int_{x^1}^{x^1+\Delta x^1} U_1(x^1,x^2,x^3)\,dx^1$$

$$= \frac{U_1(\bar{x}^1,x^2,x^3)}{\Delta x^1}\int_{x^1}^{x^1+\Delta x^1} dx^1,$$

where $\bar{x}^1$ is a properly chosen value between x^1 and $x^1 + \Delta x^1$. The last expression in (48) follows by use of the mean value theorem for integrals (see 11–65). Passing to the limit as $\Delta x^1 \to 0$ in (48) yields

$$49) \quad \frac{\partial \phi}{\partial x^1} = U_1(x^1,x^2,x^3).$$

Similarly, by considering increments on x^2 and x^3 in turn, one shows that

$$\frac{\partial \phi}{\partial x^2} = U_2(x^1,x^2,x^3), \qquad \frac{\partial \phi}{\partial x^3} = U_3(x^1,x^2,x^3).$$

Hence, it has been established that $\mathbf{F} = \nabla \phi$. One must realize here that the function ϕ must be defined and continuous in $\mathcal{K}$ and that the path of integration is in $\mathcal{K}$.

In essence, the theorem just proved says that $\int \mathbf{F}\cdot d\mathbf{R}$ is independent of the path if and only if $\mathbf{F} = \nabla\phi$. This is equivalent to saying that $\mathbf{F}\cdot d\mathbf{R} = d\phi$, the total differential of a scalar function. For, $\mathbf{F}\cdot d\mathbf{R} = U_i\,dx^i = \phi_{,i}\,dx^i = d\phi$.

In order to use the preceding theorem relative to $\int \mathbf{F}\cdot d\mathbf{R}$ it is necessary to be able to determine that $\mathbf{F}\cdot d\mathbf{R} = U_i\,dx^i$ is or is not a perfect differential $d\phi$. For this purpose attention is directed to the following

Theorem 2: If, in a region $\mathcal{K}$ of space, the functions U_i, together with their partial derivatives which appear in the curl of the vector $U_i\mathbf{e}^i$, are continuous, and $U_i\,dx^i$ is the total differential of a scalar function ϕ, then the curl of $U_i\mathbf{e}^i$ is zero.

To prove this, note that if

$$U_i\,dx^i \equiv \frac{\partial \phi}{\partial x^i}\,dx^i,$$

then

$$50) \quad U_i = \frac{\partial \phi}{\partial x^i} \quad (i = 1,2,3).$$

Differentiation of (50) with respect to x^j gives

51) $$\frac{\partial U_i}{\partial x^j} = \frac{\partial^2 \phi}{\partial x^j \, \partial x^i} = \frac{\partial^2 \phi}{\partial x^i \, \partial x^j} = \frac{\partial U_j}{\partial x^i}.$$

(Note that the continuity of the partial derivatives of U_i insures the validity of the change of order in the differentiation of ϕ.) From (51), the conclusion

52) $$\frac{\partial U_i}{\partial x^j} - \frac{\partial U_j}{\partial x^i} = 0$$

means that the components of curl $(U_i \mathbf{e}^i)$ are all zero.

The converse of Theorem 2 is not valid unless the region $\mathcal{K}$ is simply connected. A proof of the converse is exhibited in Section 11–11 following the curl theorem of Stokes. Meanwhile, the condition curl $(U_i \mathbf{e}^i) = \mathbf{0}$ may be used as a sufficient condition for $U_i \, dx^i$ to be an exact differential in a simply connected region.

Recall that a vector field $\mathbf{F}$ is conservative (see Section 10–8) in a region of space if $\int \mathbf{F} \cdot d\mathbf{R}$ is independent of the path in the region. For an integral to be independent of the path in a region, the region must be simply connected relative to curves. Note that the condition curl $\mathbf{F} = \mathbf{0}$ is not sufficient for $\mathbf{F}$ to be a conservative field.

As a mnemonic device the conditions (52) may be written in the form

$$\left| \begin{array}{ccc} \dfrac{\partial}{\partial x^1} & \dfrac{\partial}{\partial x^2} & \dfrac{\partial}{\partial x^3} \\ U_1 & U_2 & U_3 \end{array} \right| = 0,$$

which is to be interpreted as

$$\frac{\partial U_3}{\partial x^2} - \frac{\partial U_2}{\partial x^3} = 0, \quad -\frac{\partial U_1}{\partial x^3} - \frac{\partial U_3}{\partial x^1}, \quad \frac{\partial U_2}{\partial x^1} - \frac{\partial U_1}{\partial x^2} = 0.$$

Example 1. Show that the vector field $\mathbf{F} = (yz^2)\mathbf{i} + (xz^2)\mathbf{j} + (2xyz)\mathbf{k}$ is conservative and determine ϕ such that $\mathbf{F} = \nabla \phi$.

Solution: A simple calculation shows that curl $\mathbf{F} = \mathbf{0}$, so there exists a function $\phi(x,y,z)$ for which

53) $$\frac{\partial \phi}{\partial x} = yz^2, \quad \frac{\partial \phi}{\partial y} = xz^2, \quad \frac{\partial \phi}{\partial z} = 2xyz,$$

or therefore for which $d\phi = yz^2 \, dx + xz^2 \, dy + 2xyz \, dz$. In this case the function ϕ can be determined by inspection to be xyz^2, but the following procedure may be used to find ϕ. Integrate equations (53) with respect to x, y, z, respectively, to obtain

54)
$$\phi = xyz^2 + f(y,z),$$
$$\phi = xyz^2 + g(z,x),$$
$$\phi = xyz^2 + h(x,y),$$

where f, g, h are arbitrary functions of the variables appearing in them. Since the three functional forms for ϕ in (54) must be identical, it follows that f, g, h must all be taken as the same constant. Hence, $\phi(x,y,z) \equiv xyz^2 + C$, where C is an arbitrary constant. There is no singularity for the components of $\mathbf{F}$ so the integral $\int \mathbf{F} \cdot d\mathbf{R}$ is everywhere independent of the path. To find $\int_A^B \mathbf{F} \cdot d\mathbf{R}$ from $A(1,0,1)$ to $B(1,2,3)$, for instance, write

$$\int_A^B \mathbf{F} \cdot d\mathbf{R} = \int_A^B d\phi = xyz^2 + C \Big|_A^B = 18.$$

Example 2. Show that $\mathbf{F} = (e^y - 2xz)\mathbf{i} + (xe^y + z)\mathbf{j} + (y - x^2)\mathbf{k}$ is a conservative field, and determine ϕ such that $\mathbf{F} = \nabla\phi$.

Solution: It is readily verified that curl $\mathbf{F} = 0$. Integration of

$$\frac{\partial \phi}{\partial x} = e^y - 2xz, \qquad \frac{\partial \phi}{\partial y} = xe^y + z, \qquad \frac{\partial \phi}{\partial z} = y - x^2,$$

respectively, with respect to x, y, z, gives

$$\phi = xe^y - x^2z + f(y,z),$$
$$\phi = xe^y + yz + g(z,x),$$
$$\phi = yz - x^2z + h(x,y).$$

Inspection of the three forms of ϕ shows that $f(y,z) = yz$, $g(z,x) = -x^2z$, $h(x,y) = xe^y$. Therefore,

$$\phi(x,y,z) = xe^y + yz - x^2z + \text{constant}.$$

The field $\mathbf{F}$ is therefore conservative, for there are no singularities of the components.

The next example is offered to show that although curl $\mathbf{F}$ is the zero vector, the integral $\int \mathbf{F} \cdot d\mathbf{R}$ is not everywhere independent of the path.

Example 3. Consider the region of space inside the torus obtained by revolving the circle $y = 0$, $(x - 4)^2 + z^2 = 4$ about the z-axis. Let the vector field be given by

$$\mathbf{F} = \frac{(-yz)\mathbf{i} + (xz)\mathbf{j} + (xy)\mathbf{k}}{x^2 + y^2z^2}.$$

Show that $\int_{C_1} \mathbf{F} \cdot d\mathbf{R} = 0$, where C_1 is the circle $z = 0$, $x^2 + (y-4)^2 = 1$, and that $\int_{C_2} \mathbf{F} \cdot d\mathbf{R} = -2\pi$, where C_2 is the circle $z = 1$, $x^2 + y^2 = 9$.

Solution: All three components of curl $\mathbf{F}$ are zero, so $\mathbf{F} = \nabla \phi$. The scalar ϕ is found to be arc tan yz/x + constant, which is single-valued on C_1 so the integral is zero. Direct calculation of $\int_{C_1} \mathbf{F} \cdot d\mathbf{R}$ on C_1, where $z = 0$, confirms the zero result. On C_2, where $z = 1$, the integral $\int_{C_2} \mathbf{F} \cdot d\mathbf{R}$ reduces to

$$\int_{C_2} \frac{-y\, dx + x\, dy}{x^2 + y^2} = -2\pi$$

if C_2 is traversed once in the counterclockwise direction. The path C_2 is not homologous to zero in the region inside the torus, so the integral is not independent of the path in that region.

EXERCISES

1) Evaluate

$$\oint_C \frac{-y^3\, dx + xy^2\, dy}{(x^2 + y^2)^2},$$

where C is the ellipse $x^2 + 2y^2 = 1$. (*Hint:* Show that the path may be changed to the circle $x^2 + y^2 = 1$, and then use Green's theorem.) *Ans.* π.

2) Evaluate

$$\oint \frac{(y^2 - 3x^2 + 4xy)\, dx - (x^2 - 3y^2 + 4xy)\, dy}{(x^2 + y^2)^3}$$

around the square with vertices $(\pm 2, \pm 2)$. (*Hint:* Change the path to $x^2 + y^2 = 1$ and use Green's theorem.) *Ans.* 0.

3) Evaluate

$$\oint_C \frac{-y\, dx + (x+2)\, dy}{(x+2)^2 + y^2}$$

if C is the circle $x^2 + y^2 = 16$. (*Hint:* Use the circle $x + 2 = \cos\theta$, $y = \sin\theta$.)

4) Show that the integral $\int (x^2 + xy)\, dx + (x^2 + y^2)\, dy$ around the square with sides $x = \pm 1$, $y = \pm 1$ is zero, but that the integrand is not an exact differential. Why does this not contradict the theorem on independence of the path?

5) In illustrative Example 2 of this section, $\int \mathbf{F} \cdot d\mathbf{R}$ is independent of the path. The function $\phi(x,y,z)$ can be defined by

$$\int_{(a,b,c)}^{(x,y,z)} \mathbf{F} \cdot d\mathbf{R} = \int_{(a,b,c)}^{(x,y,z)} (e^y - 2xz)\,dx + (xe^y + z)\,dy + (y - x^2)\,dz,$$

where (a,b,c) is an arbitrary fixed point. Use an elbow path from $(0,0,0)$ to (x,y,z) to obtain

$$\phi(x,y,z) = \int_0^x 1\,dx + \int_0^y xe^y\,dy + \int_0^z (y - x^2)\,dz$$

and thus find $\phi = xe^y + yz - x^2z + \text{constant}$, as in the example. The constant changes with the choice of point (a,b,c) for the lower limit.

6) Consider the integral

$$I = \int \frac{x\,dx + y\,dy}{x^2 + y^2}.$$

Show that the origin must be deleted from the plane and therefore that any region containing the origin is doubly connected. However, show that I is independent of the path in the doubly connected region, provided the path does not pass through the origin. In particular, if $r \equiv |\mathbf{R}|$, show that

$$I = \int_A^B d(\log r) = \log r \Big|_A^B.$$

What is the result if B coincides with A?

7) Find the value of

$$\int_{(1,2)}^{(2,1)} (9x^2y + 8xy^2 + 5y^3)\,dx + (3x^3 + 8x^2y + 15xy^2)\,dy$$

along the curve $x^4 + y^4 = 17$. (*Hint:* The integral is independent of the path. Choose therefore a simpler path of integration.) *Ans.* -12.

8) Show that if a region of space is multiply connected relative to curves, it must be multiply connected relative to surfaces.

11-8. Green's Theorem in Three-Space (The Divergence Theorem).

The theorem to be presented in this section is variously referred to as the divergence theorem, the theorem of Green in three-space, and the theorem of Gauss. Because of the important application to the divergence of a vector field, the theorem will be referred to here as the divergence theorem. Essentially it states that the integral (over a volume of space) of the divergence of a vector field is equal to the integral (over the surface enclosing the volume) of the normal component of the vector field. The importance of the theorem lies not so much in the computational aspect but in the fact that it relates a surface integral to a volume integral. This fact is useful, for instance, in deriving partial differential equations in mathematical physics.

The theorem will be proved for a restricted type of region, but extension can be effected for a more general region with the property that it can be subdivided into regions of the type used in the proof. The region of space

$\mathcal{K}$ to be considered is enclosed by (1) a cylindrical surface with elements parallel to the z-axis of an orthogonal cartesian system, (2) a surface convex upward forming the upper cap, and (3) a surface convex downward forming the lower cap. A line inside the cylindrical surface and parallel to the z-axis cuts the boundary of the space region in two points, one on the lower cap and one on the upper.

Theorem: If $\mathbf{F}$ is a continuous vector field with continuous first partial derivatives in a region $\mathcal{K}$ of space, then

$$55) \qquad \iiint_{\mathcal{K}} \nabla \cdot \mathbf{F} \, dV = \iint_{S} \mathbf{N} \cdot \mathbf{F} \, d\sigma,$$

where $\mathbf{N}$ is the unit vector normal to the surface S which encloses $\mathcal{K}$, dV is the volume element, and $d\sigma$ is the element of surface area.

Proof: If $\mathbf{F} = U_i \mathbf{e}^i$, the left-hand side of (55) is

$$56) \qquad \iiint_{\mathcal{K}} \left(\frac{\partial U_1}{\partial x} + \frac{\partial U_2}{\partial y} + \frac{\partial U_3}{\partial z} \right) dz \, dy \, dx.$$

Consider first the integral

$$57) \qquad \iiint_{\mathcal{K}} \frac{\partial U_3}{\partial z} \, dz \, dy \, dx.$$

The total bounding surface S of $\mathcal{K}$ (Fig. 39) is composed of the lower cap S_1 with equation $z = z_1(x,y)$, the upper cap S_2 with equation $z = z_2(x,y)$, and the lateral (cylindrical) surface S_3. Note that both S_1 and S_2 map by vertical projection into the region S_{xy} in the xy-plane. Let a vertical line through P in the xy-plane intersect the surfaces S_1 and S_2 in P_1 and P_2, respectively. The integral in (57) may be evaluated by the iterated integral

$$58) \qquad \iint_{S_{xy}} \int_{z_1}^{z_2} \frac{\partial U_3}{\partial z} \, dz \, dy \, dx = \iint_{S_{xy}} [U_3(x,y,z_2(x,y)) - U_3(x,y,z_1(x,y))] \, dy \, dx.$$

Observe next that the element of area $dx \, dy$ in the xy-plane projects upward into an element of area $d\sigma_1$ on S_1 and $d\sigma_2$ on S_2. If the normal $\mathbf{N}$ to S has direction cosines $\cos \alpha$, $\cos \beta$, $\cos \gamma$, it is seen that the angle between the tangent plane to S_2 at P_2 and the xy-plane is γ. Hence, the projection of the element of area $d\sigma_2$ at P_2 onto the xy-plane is given by $d\sigma_2 \cos \gamma = dx \, dy$. Because of the convexity of the surface S_2 the angle γ is always acute. However, at P_1 on S_1 the normal makes an obtuse angle with the positive z-axis. The angle between the surface element $d\sigma_1$ at P_1 and the xy-plane is $\gamma' = \pi - \gamma$. As at P_2 one has at P_1 the projection

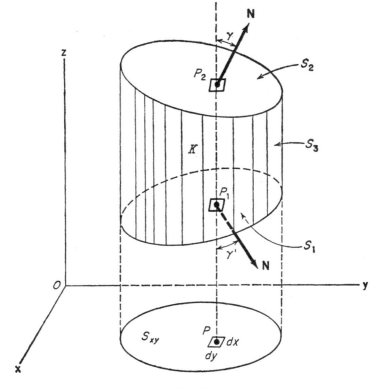

Fig. 39

$d\sigma_1 \cos \gamma' = dx\, dy$, or $d\sigma_1 \cos \gamma = -dx\, dy$. Now the right-hand member of (58) may be written as

$$\text{59)} \qquad \iint_{S_2} U_3(x,y,z) \cos \gamma \, d\sigma_2 + \iint_{S_1} U_3(x,y,z) \cos \gamma \, d\sigma_1.$$

Because $\cos \gamma = 0$ at all points on S_3, so that $\iint_{S_3} U_3(x,y,z) \cos \gamma \, d\sigma = 0$ on S_3, one sees from the result in (59) that (57) can be expressed as

$$\text{60)} \qquad \iiint_{\mathcal{K}} \frac{\partial U_3}{\partial z} dz\, dy\, dx = \iint_{S} U_3(x,y,z) \cos \gamma \, d\sigma,$$

where S is the total surface of the region $\mathcal{K}$. Similarly, by using projections on the yz- and zx-planes, the remaining terms in (56) are expressible in the form

61) $$\iiint_{\mathcal{K}} \frac{\partial U_1}{\partial x} dx\, dz\, dy = \iint_S U_1(x,y,z) \cos \alpha\, d\sigma,$$

62) $$\iiint_{\mathcal{K}} \frac{\partial U_2}{\partial y} dy\, dx\, dz = \iint_S U_2(x,y,z) \cos \beta\, d\sigma.$$

Addition of equations (60), (61), (62) yields

63) $$\iiint_{\mathcal{K}} \left(\frac{\partial U_1}{\partial x} + \frac{\partial U_2}{\partial y} + \frac{\partial U_3}{\partial z} \right) dx\, dy\, dz$$

$$= \iint_S (U_1 \cos \alpha + U_2 \cos \beta + U_3 \cos \gamma)\, d\sigma.$$

The integrand in the left-hand member of (63) is div $\mathbf{F}$ and that in the right-hand member is $\mathbf{F} \cdot \mathbf{N}$, where $\mathbf{N}$ is the unit normal vector $\mathbf{i} \cos \alpha + \mathbf{j} \cos \beta + \mathbf{k} \cos \gamma$ at any point of S. Hence, equation (63) is precisely (55), and the theorem is proved.

The divergence theorem was proved under rather restrictive conditions. Actually, the integrals exist and the theorem holds even if a finite number of points exist on the bounding surface where the unit normal $\mathbf{N}$ is indeterminate. Also a finite number of finite jump discontinuities may occur in the partial derivatives without invalidating the theorem.

The divergence theorem was proved for a region S made up of an upper cap, a lower cap, and a cylindrical surface. If the region is not of this type, but can be subdivided by cylindrical surfaces into a finite number of regions each of the type used in the theorem, then the theorem can be applied to each subregion. The surface integral over a common bounding surface of two adjacent subregions is zero because the area is covered once with the outward normal on one side and once with the outward normal pointing in the opposite direction. The surface integral over the totality of the surfaces of the subregions is then merely the surface integral over the complete boundary of the space region. Most regions which occur in practical applications can be treated by the device suggested. However, in case a given region cannot be subdivided as suggested into a finite number of regions of the type used in the theorem, it may be necessary to resort to a limit process to conclude the proof for such a region. The reader is referred to the treatment by O. D. Kellogg, Foundations of Potential Theory (Berlin: Springer, 1929).

A doubly connected region relative to surfaces can be rendered simply connected relative to surfaces by an appropriate cutting surface. For instance, the doubly connected region inside the torus obtained by revolving the circle $y = 0$, $(x - 4)^2 + z^2 = 4$ about the z-axis can be made simply connected for surfaces by the cutting half plane described by $x = 0, y > 0$. The set of points common to the cutting plane and the inside of the torus is covered twice by the surface integral of the theorem, and these two contributions annul each other. It is proper then to apply the divergence theorem to multiply connected regions, but the surface integral must apply to the complete boundary of the space region.

As was shown (11–33) the divergence theorem for a plane region can be interpreted as Green's theorem in the plane. In this case, a single integral (line integral) around a simple closed curve is related to a double integral (surface integral) over the region enclosed by the curve. In three-space the divergence theorem relates a double integral over a simple closed surface to a triple integral throughout the volume enclosed. Actually, the theorem states that the surface integral of an invariant scalar function is equal to the volume integral of a certain differential invariant function.

In order to show that the divergence theorem is invariant with respect to a change of coordinates, the theorem will be expressed in tensor form. Let λ^i be a vector field $\mathbf{F}$ in general coordinates, and denote by μ_i the covariant components of the *unit* outward normal at any point of a surface S which bounds a region $\mathcal{K}$. Now recall that div $\mathbf{F} = \lambda^i{}_{,i}$ and $\mathbf{F} \cdot \mathbf{N} = \lambda^i \mu_i$. The divergence theorem in (55) becomes

$$64) \qquad \iiint_{\mathcal{K}} \lambda^i{}_{,i} \, dV = \iint_{S} \lambda^i \mu_i \, d\sigma.$$

Because of the tensor character of equation (64), it can be stated that the divergence theorem holds for every coordinate system if it holds for any one system. But it was proved for the orthogonal cartesian system, so it is true for all systems.

Formula (64) can be extended to n-space where the left-hand member becomes an n-tuple integral and the right-hand member an $(n - 1)$-tuple integral.

When the divergence of a field was first defined (Section 10–2) a physical application of the concept was given. A more meaningful definition can now be offered by use of the divergence theorem. For this, an understanding of the mean value theorem is necessary. The reader will recall from the calculus that if $f(x)$ is a continuous function on the interval $a \leq x \leq b$ there exists a point $\bar{x}$ between a and b for which $f(\bar{x})(b - a) = \int_a^b f(x) \, dx$, from which $f(\bar{x})$, given by

$$f(\bar{x}) = \frac{1}{b - a} \int_a^b f(x) \, dx$$

is called the mean value of $f(x)$ over the interval $[a,b]$. The mean value concept can be extended to a function $f(x,y,z)$ which is continuous in a closed region $\mathcal{K}$ with bounding surface S. There exists a point $(\bar{x},\bar{y},\bar{z})$ in $\mathcal{K}$ for which

$$65) \qquad f(\bar{x},\bar{y},\bar{z}) = \frac{1}{V} \iiint_{\mathcal{K}} f(x,y,z)\, dx\, dy\, dz.$$

In particular, if $f(x,y,z)$ is taken as the divergence of a vector field $\mathbf{F}$, and if the divergence theorem is applied, (65) becomes

$$66) \qquad \operatorname{div} \mathbf{F}\big|_{(\bar{x},\bar{y},\bar{z})} = \frac{1}{V} \iiint_{\mathcal{K}} \operatorname{div} \mathbf{F}\, dx\, dy\, dz = \frac{1}{V} \iint_{S} \mathbf{F}\cdot\mathbf{N}\, d\sigma.$$

If the surface S is allowed to shrink to a point (x,y,z) in such a manner that the distance between any two points of $\mathcal{K}$ tends to zero, the point $(\bar{x},\bar{y},\bar{z})$ must tend to (x,y,z) and in the limit one has

$$67) \qquad \operatorname{div} \mathbf{F}\big|_{(x,y,z)} = \lim_{V \to 0} \frac{1}{V} \iint_{S} \mathbf{F}\cdot\mathbf{N}\, d\sigma.$$

The volume V may be taken as that of a small sphere of radius r about the point (x,y,z). The integral in the right-hand member of (67) measures the total flux of the field through the surface of the sphere, and this quantity divided by V gives the flux per unit volume. Equation (67) may, therefore, be interpreted as follows.

> The divergence of a vector field at a point is the limit of the total flux per unit volume for a sphere about the point as the radius of the sphere tends to zero at the point.

It is possible to deduce the formula for div $\mathbf{F}$ as given in (10–1) from the foregoing definition of divergence.

11–9. An Application of the Divergence Theorem. There are various physical applications of the divergence theorem. Only one, however, that of fluid flow, will be discussed here. Consider a fluid of variable density ρ flowing through a region $\mathcal{K}$ bounded by a surface S. Let $\mathbf{U}$ represent the velocity vector field. The mass of fluid passing through an element of area $\Delta\sigma$ of the surface S in time Δt is given by $(\rho \mathbf{U}\cdot\mathbf{N})\,\Delta\sigma\,\Delta t$, where $\mathbf{N}$ is the outer unit normal to S. Summing over the surface S gives

$$68) \qquad \iint_{S} (\rho \mathbf{U}\cdot\mathbf{N})\, d\sigma$$

as the time rate of mass of fluid passing *outward* through S. Let $\rho\Delta V$ be the mass of an element of volume of $\mathcal{K}$. Then $(\partial \rho/\partial t)\, \Delta V\, \Delta t$ is the increase of the mass of the fluid in time Δt, and

$$\text{69)} \qquad \iiint_{\mathcal{K}} \frac{\partial \rho}{\partial t}\, dV$$

is the total rate of increase in the fluid mass within $\mathcal{K}$. If there are no sources or sinks in $\mathcal{K}$, the rate of mass flow *inward* through S is the negative of the rate in (68). Therefore, for continuous flow,

$$\text{70)} \qquad \iint_{S} (\rho \mathbf{U} \cdot \mathbf{N})\, d\sigma + \iiint_{\mathcal{K}} \frac{\partial \rho}{\partial t}\, dV = 0.$$

Now change the surface integral in (70) to a volume integral by use of the divergence theorem to obtain

$$\iiint_{\mathcal{K}} \text{div}\, (\rho \mathbf{U})\, dV + \iiint_{\mathcal{K}} \frac{\partial \rho}{\partial t}\, dV = 0,$$

or

$$\text{71)} \qquad \iiint_{\mathcal{K}} \left[\nabla \cdot (\rho \mathbf{U}) + \frac{\partial \rho}{\partial t} \right] dV = 0.$$

If one assumes that the integral in (71) vanishes over every closed subregion of $\mathcal{K}$ in the fluid, it follows that the integrand must be zero at every point of $\mathcal{K}$. Hence,

$$\text{72)} \qquad \nabla \cdot (\rho \mathbf{U}) + \frac{\partial \rho}{\partial t} = 0.$$

This may be seen as follows. Denote the integrand in (71) by I. Suppose I is not zero at some point P. For definiteness, assume it to be positive at P. Then, because I is assumed to be a continuous function, there must exist a region M of space about P in which I is everywhere positive. Integration of the positive function I over the region M gives a positive result, but this contradicts the assumption in (71) that the integral vanishes over M. Therefore, I cannot be positive at P. One can show similarly that if I is negative at a point then (71) is contradicted. Hence, I must be zero at P.

Equation (72) is the continuity equation of hydrodynamics. The density ρ is constant for an incompressible fluid, and in this case (72) reduces to $\nabla \cdot \mathbf{U} = 0$. In case $\mathbf{U} = \text{grad}\, \phi = \nabla \phi$, $\nabla \cdot \mathbf{U} = 0$ becomes $\nabla \cdot \nabla \phi = 0$, or $\nabla^2 \phi = 0$, which is Laplace's equation.

Finally, note that the tensor form of equation (72) is

$$(\rho u^i)_{,i} + \frac{\partial \rho}{\partial t} = 0,$$

which is invariant under a change of space coordinates.

EXERCISES

1) If S is a closed surface, show that $\iint_S \mathbf{R} \cdot \mathbf{N}\, d\sigma$ is three times the volume enclosed by S.

2) Verify the divergence theorem for $\mathbf{F} = x^2\mathbf{i} + y^2\mathbf{j} + z^2\mathbf{k}$ over the cube with faces $x = 0, x = 2, y = 0, y = 2, z = 0, z = 2$.

3) Verify the divergence theorem for the field $\mathbf{F} = x\mathbf{i} + z^2\mathbf{j} + y^2\mathbf{k}$ in the region of space bounded by the coordinate planes and the plane $x + y + z = 1$. Ans. $\frac{1}{6}$.

4) Verify the divergence theorem for the field $\mathbf{F} = y\mathbf{i} + z\mathbf{j} + x\mathbf{k}$ over the region in the first octant of the sphere $x^2 + y^2 + z^2 = a^2$.

5) Find the value of the flux of the field $\mathbf{F} = x\mathbf{i} + y\mathbf{j} + z\mathbf{k}$ over the entire surface of the sphere with center at the origin and radius a.

6) Write the divergence theorem for orthogonal cartesian coordinates.

7) Write the equation of continuity in (a) cylindrical coordinates x^i, (b) in spherical coordinates x^i.

8) Use the divergence theorem to evaluate the total flux of the field $\mathbf{F} = f^i \mathbf{e}_i$, where f^i are physical components $(r, r\theta, z)$ in cylindrical coordinates, over the entire surface of the cylinder bounded by $r = a$, $z = 0$, $z = h$. Ans. $4\pi a^2 h$.

9) Show that if u and v are scalar functions possessing second partial derivatives

$$\iiint_\mathcal{K} (\nabla u \cdot \nabla v + u \nabla^2 v)\, dV = \iint_S \mathbf{N} \cdot u \nabla v\, d\sigma$$

and

$$\iiint_\mathcal{K} (\nabla v \cdot \nabla u + v \nabla^2 u)\, dV = \iint_S \mathbf{N} \cdot v \nabla u\, d\sigma,$$

where $\nabla^2 \phi$ is the Laplacian $a^{ij}\phi_{,ij} = \delta^{ij}\phi_{,ij} = \phi_{,ii}$, and $\mathbf{N}$ is the unit normal to the surface. [*Aid:* In the divergence theorem (64), put $\lambda_i = uv_{,i}$ to obtain

$$\iiint_\mathcal{K} (u_{,i}v_{,i} + uv_{,ii})\, dV = \iint_S uv_{,i}\mu^i\, d\sigma.$$

Interchange u and v for the second integral.]

10) If $\nabla^2\phi = 0$ in a region of space, ϕ is said to be *harmonic* in the region. Show by Exercise 9 that, in general,

$$\iiint_{\mathcal{K}} (v\nabla^2 v - u\nabla^2 u)\, dV = \iint_S \mathbf{N} \cdot (u\nabla v - v\nabla u)\, d\sigma,$$

but that if u and v are harmonic in $\mathcal{K}$,

$$\iint_S \mathbf{N} \cdot (u\nabla v - v\nabla u)\, d\sigma = 0.$$

11) Verify that $\nabla \cdot (r^{-3}\mathbf{R}) = 0$, where $r \equiv |\mathbf{R}|$, and use the divergence theorem to show that

$$\iint_S (r^{-3}\mathbf{R}) \cdot \mathbf{N}\, d\sigma = 0$$

for any simple closed surface S if the origin of the vector $\mathbf{R}$ is outside S. (See Exercise 9 in Section 10-9.)

12) Calculate the value of the integral in Exercise 11 if the origin of vector $\mathbf{R}$ is inside S. (*Suggestion:* Delete the origin from the region $\mathcal{K}$ inside S by a sphere Σ with radius ϵ and center at the origin, with the radius ϵ small enough to allow Σ to be inside S. The total boundary of the region $\mathcal{K}'$ between Σ and S is now the surface S plus the surface of Σ. The integral in Exercise 11 can now be applied to $\Sigma + S$, but notice that the exterior unit normal to Σ at any point is $-\mathbf{R}/\epsilon$, and that $\mathbf{N} \cdot \mathbf{R} = -\mathbf{R}/\epsilon \cdot \mathbf{R} = -\epsilon^2/\epsilon = -\epsilon$ on Σ. Complete the calculation to obtain 4π as the value of the integral. The result holds for arbitrarily small ϵ.)

11-10. The Theorem of Stokes (The Curl Theorem). It has been seen (11-32) that Green's theorem in the plane can be interpreted in the form

$$\int_\Gamma \mathbf{F} \cdot d\mathbf{R} = \iint_{\mathcal{K}} (\text{curl } \mathbf{F}) \cdot \mathbf{k}\, d\sigma$$

(where $\mathbf{k} \equiv \mathbf{e}^3$ is the unit vector normal to the x^1x^2-plane) which states that the circulation of vector field $\mathbf{F}$ around a simple closed plane curve Γ is equal to the flux of the vector field curl $\mathbf{F}$ over the plane surface $\mathcal{K}$ bounded by Γ. The following question arises. Does the statement hold if the plane surface is replaced by a curved surface and the constant unit normal vector $\mathbf{k}$ is replaced by the variable unit normal vector $\mathbf{N}$ to the curved surface? An affirmative answer is afforded by the following

Theorem of Stokes: If on a curved surface a region $\mathcal{K}$ is bounded by a curve Γ, and if a vector field $\mathbf{F}$ has continuous first partial derivatives on $\mathcal{K}$, then the circulation of the field $\mathbf{F}$ around Γ is equal to the flux of the curl of **F** over the surface region $\mathcal{K}$, that is

73) $$\int_\Gamma \mathbf{F} \cdot d\mathbf{R} = \iint_{\mathcal{K}} (\text{curl } \mathbf{F}) \cdot \mathbf{N} \, d\sigma.$$

Proof: It is assumed that a line parallel to the x^3-axis intersects $\mathcal{K}$ in only one point. Let the surface be represented in the Gauss form by single-valued, continuous, differentiable functions $x^i = x^i(u^1, u^2)$, and consider the region $\overline{\mathcal{K}}$ bounded by $\overline{\Gamma}$ in the $u^1 u^2$-plane which maps in a one-to-one continuous manner into the region $\mathcal{K}$ on the surface, that is, to each point $\overline{P}$ of $\overline{\mathcal{K}}$ there corresponds a unique point P of $\mathcal{K}$, and to each point of $\mathcal{K}$ there is a unique image point in $\overline{\mathcal{K}}$ (see Fig. 40). By means of

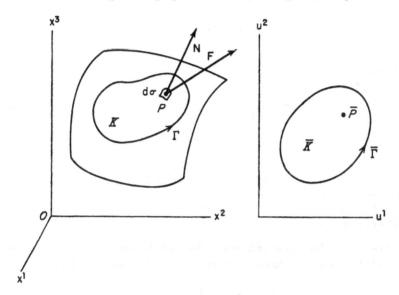

Fig. 40

formulas (19) and (22) in Section 11–3, the expression $\mathbf{N} \, d\sigma$ in (73) is given by

74) $$\mathbf{N} \, d\sigma = \frac{\partial \mathbf{R}}{\partial u^1} \times \frac{\partial \mathbf{R}}{\partial u^2} du^1 \, du^2 = e_{ijk} \mathbf{e}^i x^j_{,1} x^k_{,2} \, du^1 \, du^2,$$

where $\mathbf{R} = x^i \mathbf{e}_i$ is the position vector of a point on the surface, and the commas indicate partial derivatives with respect to u^1 and u^2. It will be recalled that if the field **F** is $F_k \mathbf{e}^k$ then curl $\mathbf{F} = e^{ijk} F_{k,j} \mathbf{e}_i = e^{pqr} F_{r,q} \mathbf{e}_p$. Hence, the right-hand member of (73) becomes

75) $$\iint_{\overline{\mathcal{K}}} (e^{pqr} F_{r,q} \mathbf{e}_p) \cdot (e_{ijk} \mathbf{e}^i x^j_{,1} x^k_{,2}) \, du^1 \, du^2$$

which, by use of $\mathbf{e}_p \cdot \mathbf{e}^i = \delta_p{}^i$, and the identity (7-12), that is

$$e^{iqr}e_{ijk} \equiv \delta_{jk}{}^{qr} \equiv \delta_j{}^q \delta_k{}^r - \delta_k{}^q \delta_j{}^r,$$

can be reduced to

77) $$\iint_{\overline{\mathcal{K}}} (F_{k,j} - F_{j,k}) x^j{}_{,1} x^k{}_{,2} \, du^1 \, du^2.$$

It will be shown next that the left-hand member of (73) takes the form of (76). First write

77) $$\int_{\Gamma} \mathbf{F} \cdot d\mathbf{R} = \int_{\Gamma} (F_i \mathbf{e}^i) \cdot (dx^j \mathbf{e}_j) = \int_{\overline{\Gamma}} F_i x^j{}_{,\alpha} \, du^\alpha \, \delta_j{}^i = \int_{\overline{\Gamma}} F_i x^i{}_{,\alpha} \, du^\alpha.$$

By Green's theorem in the plane (11-4)

78) $$\int_{\overline{\Gamma}} F_i x^i{}_{,\alpha} \, du^\alpha = \int_{\overline{\Gamma}} F_i x^i{}_{,1} \, du^1 + F_i x^i{}_{,2} \, du^2$$

$$= \iint_{\overline{\mathcal{K}}} [(F_i x^i{}_{,2})_{,1} - (F_i x^i{}_{,1})_{,2}] \, du^1 \, du^2.$$

Observe that

$$F_{i,\alpha} = \frac{\partial F_i}{\partial x^l} \frac{\partial x^l}{\partial u^\alpha} = F_{i,l} x^l{}_{,\alpha},$$

and make use of the assumption that $x^i{}_{,12} \equiv x^i{}_{,21}$ to see that the last member of (78) can be written as

79) $$\iint_{\overline{\mathcal{K}}} (F_{k,j} - F_{j,k}) x^j{}_{,1} x^k{}_{,2} \, du^1 \, du^2,$$

which is precisely the form of (76). This concludes the proof.

Although Stokes' theorem was proved for a restricted type of region on a surface, the theorem can be extended to a more general region by inserting suitable curves to obtain a reticulation of the surface for which each subregion satisfies the hypotheses of the theorem. On adding the results obtained by applying the theorem to all of the subregions, the line integrals along the internal boundaries will cancel out, because each such curve is traversed once in each direction. Thus, only the integral around the boundary of the general region remains and the theorem holds for this region. The foregoing argument is sufficient for most situations which occur in practical problems. However, if a region cannot be reticulated by a finite number of subregions a limit process may be necessary in concluding the proof. As in the case of the divergence theorem (Section 11-8), the reader is referred to the book by Kellogg.

A surprising result appears from the following discussion. Consider two surfaces S_1 and S_2 (Fig. 41) which span the closed space curve Γ. The two surfaces S_1 and S_2 make up a single surface S which is separated

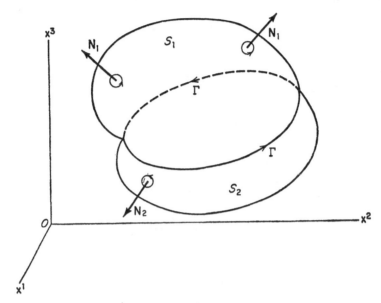

Fig. 41

into S_1 and S_2 by Γ. The outward normal $\mathbf{N}_1$ on S_1 determines the positive direction of advance along the boundary Γ. From Stokes' theorem

80) $$\iint_{S_1} (\operatorname{curl} \mathbf{F}) \cdot \mathbf{N}_1 \, d\sigma = \int_{\Gamma} \mathbf{F} \cdot d\mathbf{R}.$$

The outward normal $\mathbf{N}_2$ to S_2 requires that the positive direction on the bounding curve Γ be opposite to that used for S_1. Hence,

81) $$\iint_{S_2} (\operatorname{curl} \mathbf{F}) \cdot \mathbf{N}_2 \, d\sigma = -\int_{\Gamma} \mathbf{F} \cdot d\mathbf{R}.$$

Addition of (80) and (81) shows that

82) $$\iint_{S_1} (\operatorname{curl} \mathbf{F}) \cdot \mathbf{N}_1 \, d\sigma + \iint_{S_2} (\operatorname{curl} \mathbf{F}) \cdot \mathbf{N}_2 \, d\sigma = \iint_{S} (\operatorname{curl} \mathbf{F}) \cdot \mathbf{N} \, d\sigma = 0.$$

Note that the divergence theorem applied to the surface integral over the total surface S verifies the result in (82). From (82) it follows that

$$83) \quad \iint_S (\text{curl } \mathbf{F}) \cdot \mathbf{N}_1 \, d\sigma = \iint_{S_2} (\text{curl } \mathbf{F}) \cdot \mathbf{N}_2 \, d\sigma$$

provided the direction of the normal to S is taken so that a chosen direction of traverse along Γ is positive (or negative) for both surfaces S_1 and S_2. If Γ is conceived as a closed wire with a soap film spanning it, the film may be blown into a configuration S_1 and then continuously into another configuration S_2 on the same side of Γ. The curve Γ is then a cuspidal edge for S_1 and S_2 together as one surface, and the positive direction along Γ for S_1 is the same as that for S_2. The surface integral in (80) is then the same for all surfaces spanning the curve Γ, which on first consideration seems a surprising conclusion. S_1 is assumed to be deformed into S_2 without passing any singular points of the vector field.

As in the case of the divergence theorem, the theorem of Stokes can be expressed in tensor form. To do this, recall that (Section 10–3) the curl of the vector with components λ_i is $\mu^i = -\epsilon^{ijk}\lambda_{j,k}$. It follows that curl $\mathbf{F}$ in (73) is given by $G^i \equiv -\epsilon^{ijk}F_{j,k}$, where F_j are the covariant components of $\mathbf{F}$. If N_i are components of the unit normal to the surface at any point of the region $\mathcal{K}$ then the normal component of curl $\mathbf{F}$ is expressed by the invariant $G^i N_i$.

If T^i are the components of the unit tangent to Γ at any point, then the invariant $F_i T^i$ is the component of the vector field $\mathbf{F}$ along the tangent to Γ. The tensor form for equation (73) may be written as

$$84) \quad \int_\Gamma F_i T^i \, ds = -\iint_\mathcal{K} \epsilon^{ijk} F_{j,k} N_i \, d\sigma$$

which shows that the theorem of Stokes holds for any coordinate system in three-space, and, indeed, also for a Riemannian space.

The divergence theorem afforded an alternative approach to the meaning of divergence of a vector. In a similar manner it will be shown that the curl of a vector can be given a new interpretation by means of the curl theorem of Stokes. Apply the theorem to a circle C of radius r and center at the point $x_1{}^i$. This circle is spanned by the plane region S. Equation (73) gives

$$85) \quad \int_C \mathbf{F} \cdot d\mathbf{R} = \iint_S (\text{curl } \mathbf{F}) \cdot \mathbf{N} \, d\sigma = \text{curl } \mathbf{F} \cdot \mathbf{N}]_{\bar{x}^i} \iint_S d\sigma$$
$$= (\text{curl } \mathbf{F} \cdot \mathbf{N})_{\bar{x}^i} (\pi r^2),$$

in which the theorem of the mean for integrals is applied and the scalar function curl $\mathbf{F} \cdot \mathbf{N}$ is evaluated at a properly chosen point $\bar{x}^i$ in the circular region S. From (85) it is seen that

86) $$(\text{curl } \mathbf{F}\cdot\mathbf{N})_{x^i} = \frac{1}{\pi r^2}\int_C \mathbf{F}\cdot d\mathbf{R}.$$

As the radius r tends to zero and the circular region therefore shrinks to the point $x_1{}^i$ at the center of C, the point $\bar{x}^i$ must also tend to $x_1{}^i$, so the normal component of the curl of the vector field $\mathbf{F}$ at $x_1{}^i$ is the limit of the right-hand member of (86) as r approaches zero. If $\mathbf{F}$ is the velocity vector for the motion of a fluid, the circulation $\int_C \mathbf{F}\cdot d\mathbf{R}$ around a circle of arbitrarily small radius r is a measure of the rotation of the fluid about the circle in the direction determined by the unit normal $\mathbf{N}$ to the circular region. The limiting form of equation (86) suggests the following interpretation.

> The component, in a given direction $\mathbf{N}$, of the curl of a vector field at a point P, is the limit of the ratio of the circulation of the field (about a circle C surrounding P and normal to $\mathbf{N}$) to the area of the circle C as the radius of C tends to zero at P.

11–11. Applications of the Curl Theorem. As a first application, the converse of Theorem 2 in Section 11–7 will be proved. The statement may be made as follows.

If the functions U_i have continuous derivatives in a *simply* connected region $\mathcal{K}$ of space, and curl $\mathbf{F} \equiv \text{curl } U_i\mathbf{e}^i$ is a zero vector in $\mathcal{K}$, then $U_i\,dx^i$ is the total differential of a scalar function ϕ.

Any simple closed curve Γ in $\mathcal{K}$ can be spanned by a surface S containing a region R enclosed by Γ. By Stokes' theorem,

$$\int_\Gamma \mathbf{F}\cdot d\mathbf{R} = \iint_R (\text{curl }\mathbf{F}\cdot\mathbf{N})\,d\sigma \equiv 0$$

because curl $\mathbf{F} \equiv 0$. Therefore, because $\int_\Gamma \mathbf{F}\cdot d\mathbf{R} = 0$, the line integral $\int_A^B \mathbf{F}\cdot d\mathbf{R}$ is independent of the path, and by Theorem 1, Section 11–7 this implies that $\mathbf{F} = \nabla\phi$.

An application of Stokes' theorem on a two-dimensional Riemannian space, i.e., on a curved surface, will appear in Section 12–7, where the remarkable Gauss-Bonnet theorem is introduced.

Example 1. Verify Stokes' theorem for the field $\mathbf{F} = x\mathbf{i} + (x - 2z)\mathbf{j} + y\mathbf{k}$ over the triangle in the plane $x + 2y + 3z = 6$ where x, y, z are all ≥ 0.

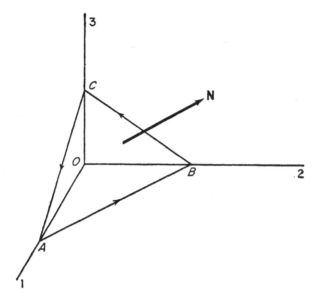

Fig. 42

Solution: Calculate first the left-hand side of equation (73). On the boundary (ABC), (Fig. 42)

$$\int_{ABC} \mathbf{F} \cdot d\mathbf{R} = \int_{ABC} x\, dx + (x - 2z)\, dy + y\, dz$$

$$= \int_{AB} x\, dx + x\left(-\frac{dx}{2}\right) + \int_{BC} -2\left(\frac{6-2y}{3}\right) dy$$

$$+ y\left(-\frac{2}{3}\right) dy + \int_{CA} x\, dx$$

$$= (-9) + (9) + (18) = 18.$$

Now find the value of the right-hand member of equation (78). For this, curl $\mathbf{F} = 3\mathbf{i} + \mathbf{k}$, and the unit normal $\mathbf{N}$ to the plane is given by $(\mathbf{i} + 2\mathbf{j} + 3\mathbf{k})/\sqrt{14}$. The area element $dx\, dy = (3d\sigma)/\sqrt{14}$, so that $d\sigma = \sqrt{14}\, dx\, dy/3$. Hence, (curl $\mathbf{F}$) $\cdot \mathbf{N}\, d\sigma = 2dy\, dx$, and

$$\iint_{\mathcal{K}} (\text{curl } \mathbf{F}) \cdot \mathbf{N}\, d\sigma = \iint_{\mathcal{K}_{xy}} 2dy\, dx = 2\,(\text{area of triangle } OAB) = 18,$$

which agrees with the value obtained for the left-hand side of (73).

The next example illustrates a useful vector technique in applied mathematics.

Example 2. Prove that

$$\int_\Gamma d\mathbf{R} \times \mathbf{G} = \iint_\mathcal{K} (\mathbf{N} \times \nabla) \times \mathbf{G}\, d\sigma,$$

where Γ is a closed space curve spanned by a surface $\mathcal{K}$, $\mathbf{G}$ is a vector field, and $\mathbf{N}$ is the unit normal to $\mathcal{K}$.

Solution: Let $\mathbf{F} = \mathbf{C} \times \mathbf{G}$, where $\mathbf{C}$ is an arbitrary non-zero constant vector. Then

$$\int_\Gamma \mathbf{F} \cdot d\mathbf{R} = \int_\Gamma (\mathbf{C} \times \mathbf{G}) \cdot d\mathbf{R} = \int_\Gamma \mathbf{C} \cdot (\mathbf{G} \times d\mathbf{R}) = -\mathbf{C} \cdot \int_\Gamma d\mathbf{R} \times \mathbf{G},$$

for the evaluation of the left-hand member in Stokes' theorem. For the right-hand member

$$\iint_\mathcal{K} (\text{curl } \mathbf{F}) \cdot \mathbf{N}\, d\sigma = \iint_\mathcal{K} [\nabla \times (\mathbf{C} \times \mathbf{G})] \cdot \mathbf{N}\, d\sigma = \iint_\mathcal{K} \mathbf{N} \cdot [\nabla \times (\mathbf{C} \times \mathbf{G})]\, d\sigma.$$

Now, by use of the identity $\mathbf{A} \times (\mathbf{B} \times \mathbf{C}) = (\mathbf{A} \cdot \mathbf{C})\mathbf{B} - (\mathbf{A} \cdot \mathbf{B})\mathbf{C}$ (see 7-14), it can be shown that

$$\nabla \times (\mathbf{C} \times \mathbf{G}) = \mathbf{C}(\nabla \cdot \mathbf{G}) - (\mathbf{C} \cdot \nabla)\mathbf{G}.$$

(Note that $\nabla \cdot \mathbf{C} \equiv 0$ and that an incorrect result is obtained unless the order $\mathbf{C} \cdot \nabla$ is used. Because the operator ∇ is not actually a vector, care must be exercised in treating ∇ as a vector.) By use of the identity just cited

$$\iint_\mathcal{K} \mathbf{N} \cdot [\nabla \times (\mathbf{C} \times \mathbf{G})]\, d\sigma = \iint_\mathcal{K} \mathbf{N} \cdot [\mathbf{C}(\nabla \cdot \mathbf{G}) - (\mathbf{C} \cdot \nabla)\mathbf{G}]\, d\sigma$$

$$= \mathbf{C} \cdot \iint_\mathcal{K} (\nabla \cdot \mathbf{G})\mathbf{N}\, d\sigma - \mathbf{C} \cdot \iint_\mathcal{K} \nabla(\mathbf{G} \cdot \mathbf{N})\, d\sigma,$$

which may be written in the form

$$-\mathbf{C} \cdot \iint_\mathcal{K} (\mathbf{N} \times \nabla) \times \mathbf{G}\, d\sigma.$$

Hence, equation (73) becomes

$$-\mathbf{C} \cdot \int_\Gamma d\mathbf{R} \times \mathbf{G} = -\mathbf{C} \cdot \iint_\mathcal{K} (\mathbf{N} \times \nabla) \times \mathbf{G}\, d\sigma,$$

or

$$\mathbf{C} \cdot \left[\int_\Gamma d\mathbf{R} \times \mathbf{G} - \iint_\mathcal{K} (\mathbf{N} \times \nabla) \times \mathbf{G}\, d\sigma \right] = 0.$$

Because C is an arbitrary non-zero vector, the second vector in the dot product must be a zero vector, with the consequence that

$$\int_\Gamma d\mathbf{R} \times \mathbf{G} = \iint_\mathcal{K} (\mathbf{N} \times \nabla) \times \mathbf{G}\, d\sigma. \tag{87}$$

As was seen in the foregoing example, there is some degree of uncertainty in manipulating ∇ as a vector. The following example is adduced to show how the tensor form (84) may be employed to obviate a possible phony use of the ∇ operator.

Example 3. Repeat Example 2 using the tensor form (84).
Solution: The components of $\mathbf{F} = \mathbf{C} \times \mathbf{G}$ are $F_i = e_{ilm}C^l G^m$. Replace $\mathbf{F}$ by $\mathbf{C} \times \mathbf{G}$ in (84) to obtain

$$\int_\Gamma e_{ilm}C^l G^m T^i\, ds = -\iint_\mathcal{K} \epsilon^{ijk}(e_{jlm}C^l G^m)_{,k} N_i\, d\sigma. \tag{88}$$

(Note that ϵ^{ijk} reduce to $e^{ijk} \equiv e_{ijk}$ for orthogonal cartesian coordinates.) Remember that $\mathbf{C}$ is a constant vector and carry out the covariant derivative in the last integral to obtain

$$-\iint_\mathcal{K} e^{ijk} e_{jlm} C^l G^m{}_{,k} N_i\, d\sigma = -\iint_\mathcal{K} (\delta_m{}^i \delta_l{}^k - \delta_l{}^i \delta_m{}^k) C^l G^m{}_{,k} N_i\, d\sigma$$

$$= \iint_\mathcal{K} (C^i G^k{}_{,k} - C^k G^i{}_{,k}) N_i\, d\sigma$$

$$= \iint_\mathcal{K} [(C^l N_l) G^k{}_{,k} - C^l N_i G^i{}_{,l}]\, d\sigma.$$

Equation (88) becomes, after a change of dummy indices,

$$C^l \left[\int_\Gamma e_{ilm} G^m T^i\, ds - \iint_\mathcal{K} (N_l G^k{}_{,k} - N_k G^k{}_{,l})\, d\sigma \right] = 0.$$

Because $\mathbf{C}$ is an arbitrary vector it follows that the expression in brackets must vanish. Hence, on noticing that $e_{ilm}G^m T^i$ is $\mathbf{G} \times \mathbf{T}$, one obtains the vector equation

$$\int_\Gamma \mathbf{G} \times \mathbf{T}\, ds \equiv \int_\Gamma \mathbf{G} \times d\mathbf{R} = \iint_\mathcal{K} (N_l \operatorname{div} \mathbf{G} - N_k G^k{}_{,l})\, d\sigma. \tag{89}$$

The right-hand member of (89) can be reduced to the form in (87) as follows. Note that

$$N_k G^k{}_{,l} - N_l \operatorname{div} \mathbf{G} = \nabla(\mathbf{N}\cdot\mathbf{G}) - \mathbf{N}\nabla\cdot\mathbf{G} = (\mathbf{N} \times \nabla) \times \mathbf{G},$$

provided ∇ does *not* operate on $\mathbf{N}$ in $\nabla(\mathbf{N}\cdot\mathbf{G})$. With this restriction (89) reduces to (87).

The next example is provided as a brief indication of the use of Stokes' theorem in electric and magnetic fields.

Example 4. Faraday's law is expressed by

$$\text{90)} \qquad \int_\Gamma \mathbf{E} \cdot d\mathbf{R} = -\frac{\partial \phi}{\partial t},$$

which states that the integral of the tangential component of the electric intensity vector $\mathbf{E}$ around any closed curve Γ is equal to the negative time rate of change of the magnetic flux ϕ through any surface $\mathcal{K}$ which spans Γ. Derive the equation

$$\operatorname{curl} \mathbf{E} = -\frac{\partial \mathbf{B}}{\partial t},$$

where $\mathbf{B}$ is called the magnetic flux density.

Solution: Apply Stokes' theorem to the left-hand member of (90) to obtain

$$\text{91)} \qquad \int_\Gamma \mathbf{E} \cdot d\mathbf{R} = \iint_\mathcal{K} (\operatorname{curl} \mathbf{E}) \cdot \mathbf{N}\, d\sigma = -\frac{\partial \phi}{\partial t}.$$

Now the total flux ϕ through the surface $\mathcal{K}$ is expressed by

$$\phi = \iint_\mathcal{K} \mathbf{B} \cdot \mathbf{N}\, d\sigma.$$

With this expression for ϕ, equation (91) yields

$$\iint_\mathcal{K} (\operatorname{curl} \mathbf{E}) \cdot \mathbf{N}\, d\sigma = -\frac{\partial}{\partial t} \iint_\mathcal{K} \mathbf{B} \cdot \mathbf{N}\, d\sigma = -\iint_\mathcal{K} \frac{\partial \mathbf{B}}{\partial t} \cdot \mathbf{N}\, d\sigma,$$

from which

$$\iint_\mathcal{K} \mathbf{N} \cdot \left(\operatorname{curl} \mathbf{E} + \frac{\partial \mathbf{B}}{\partial t} \right) d\sigma = 0$$

must hold. By the usual argument concerning the continuity of the integrand and the fact that the integral is zero for arbitrary Γ and $\mathcal{K}$, the integrand must vanish. Since $\mathbf{N}$ is non-zero and not necessarily orthogonal to the vector in parentheses it follows that

$$\operatorname{curl} \mathbf{E} = -\frac{\partial \mathbf{B}}{\partial t}.$$

EXERCISES

1) Show that Stokes' theorem in x, y, z orthogonal cartesian coordinates takes the form

$$\int_\Gamma F_1\, dx + F_2\, dy + F_3\, dz$$
$$= \iint_\mathcal{K} \left[\left(\frac{\partial F_3}{\partial y} - \frac{\partial F_2}{\partial z} \right) dy\, dz + \left(\frac{\partial F_1}{\partial z} - \frac{\partial F_3}{\partial x} \right) dz\, dx + \left(\frac{\partial F_2}{\partial x} - \frac{\partial F_1}{\partial y} \right) dx\, dy \right].$$

2) Verify Stokes' theorem for the field $\mathbf{F} = yz\mathbf{i} + zx\mathbf{j} + x^2\mathbf{k}$ over the unit cube with vertices at (0,0,0), (1,0,0), (0,1,0), (0,0,1), etc., where the face of the cube in the xy-plane is removed. *Ans.* The common value is 0.

3) Integrate the normal component of the curl of the field $\mathbf{F} = (x^2 + y)\mathbf{i} + (y^2 + z)\mathbf{j} + (z^2 + x)\mathbf{k}$ over the upper half of the hemisphere $x^2 + y^2 + z^2 = 1$, $z > 0$. *Ans.* $-\pi$.

4) Ampère's law, expressed by $\int_\Gamma \mathbf{H} \cdot d\mathbf{R} = I$ states that the integral of the tangential component of the magnetic intensity vector $\mathbf{H}$ around any closed curve Γ is equal to the current I flowing through any surface $\mathcal{K}$ which spans Γ. Apply Stokes' theorem to Ampère's law to arrive at

$$\text{curl } \mathbf{H} = \mathbf{J},$$

where $\mathbf{J}$ is the current density.

5) Verify Stokes' theorem for the field $\mathbf{F} = (x + 2y)\mathbf{i} + 2yz^2\mathbf{j} + 4y^2z\mathbf{k}$ over the conical surface $h^2(x^2 + y^2) = (h - z)^2$ with $x^2 + y^2 = 1$, $z = 0$ as its boundary. *Ans.* The common value is -2π.

6) (a) The quantities ϵ_{ijk} and ϵ^{ijk} were shown (Section 10-3) to be tensors for a transformation with positive jacobian. Show in a similar manner that

$$\epsilon_{\alpha\beta} \equiv e_{\alpha\beta}\sqrt{g}, \qquad \epsilon^{\alpha\beta} \equiv \frac{1}{\sqrt{g}} e^{\alpha\beta} \qquad (\alpha, \beta = 1, 2),$$

are tensor components, where the quantities $e_{\alpha\beta}$ and $e^{\alpha\beta}$ are defined by

$$e_{11} = e_{22} = e^{11} = e^{22} = 0, \qquad e_{12} = e^{12} = 1, \qquad e_{21} = e^{21} = -1.$$

(b) Show that Green's theorem in the plane (or Stokes' theorem in two-space) can be expressed in the tensor form

$$\int_\Gamma U_\alpha \, du^\alpha = -\iint_\mathcal{K} \epsilon^{\alpha\beta} U_{\alpha,\beta} \, d\sigma.$$

(*Aid:* Start with $\int_\Gamma U_1 \, du^1 + U_2 \, du^2 = \iint_\mathcal{K} \left(\frac{\partial U_2}{\partial u^1} - \frac{\partial U_1}{\partial u^2} \right) du^1 \, du^2$ and then obtain

$$\int_\Gamma U_\alpha \, du^\alpha = -\iint_\mathcal{K} e^{\alpha\beta} U_{\alpha,\beta} \, du^1 \, du^2$$

$$= -\iint_\mathcal{K} \epsilon^{\alpha\beta} U_{\alpha,\beta} \sqrt{g} \, du^1 \, du^2 = -\iint_\mathcal{K} \epsilon^{\alpha\beta} U_{\alpha,\beta} \, d\sigma.)$$

By the tensor form of the theorem it can be seen that it holds for curvilinear coordinates in the plane or for any coordinate system on a curved surface, i.e., a Riemannian space of two dimensions (see Section 12-7).

12

Geodesic and Union Curves

12–1. Two-Dimensional Curved Space. In Chapter 8 a surface was introduced as a vector function of two scalar parameters, that is, by

1) $$\mathbf{R} = x^i(u^1,u^2)\mathbf{e}_i,$$

where the Gauss imbedding equations of the surface in euclidean three-space are

2) $$x^i = x^i(u^1,u^2) \qquad (i = 1, 2, 3).$$

The fundamental metric induced upon the surface as a point set in three-dimensional euclidean space was found to be given by

3) $$ds^2 = g_{\alpha\beta}\, du^\alpha\, du^\beta, \qquad (\alpha, \beta = 1, 2)$$

where the components $g_{\alpha\beta}$ of the fundamental tensor are

4) $$g_{\alpha\beta} = \delta_{ij} \frac{\partial x^i}{\partial u^\alpha} \frac{\partial x^j}{\partial u^\beta}.$$

(*Note:* Latin letters will take the usual range 1, 2, 3, for space coordinates, but Greek letters will be used for surface coordinates with the range 1, 2.)

It has been seen that a surface has an extrinsic geometry relative to the ambient three-dimensional space, and from this point of view the arc length of a curve on a surface is an extrinsic property because the curve lies in the ambient space. However, a fundamental metric may be assigned to a two-dimensional space, and from this point of view the fundamental metric is intrinsic to the geometry of the surface. In general, the metric $g_{\alpha\beta}\, du^\alpha\, du^\beta$ assigned for ds^2 is required to be a positive definite quadratic differential form, which means that the value of the form is positive at all points for all directions du^α. The metric to be used here will be that induced by the ambient space.

12-2. Geodesics as Curves of Shortest Distance.

For much of what follows the intrinsic approach will be used, and for this the vector representation of a surface will not be employed. However, some extrinsic geometry will also appear. It should be realized that the tensor notation to be used for the two-dimensional Riemannian space (surface) extends immediately to an n-dimensional curved space. For this it is only necessary to allow the coordinate range to be $1, \cdots, n$ instead of 1, 2.

The principal aim of the following presentation is to provide some essential preparatory information for the study of the motion of a particle on a surface. All of the results obtained heretofore concerning tensor analysis for general coordinate systems in three-space can be utilized for general coordinates in a curved two-space.

12-2. Geodesics as Curves of Shortest Distance. It will be instructive to see several ways to define geodesic curves on a surface. The first, an intrinsic definition, is incident to the quest for an arc on a surface S which joins two fixed points P_1 and P_2, and which minimizes the distance between the two points. There is a set of infinitely many curves joining P_1 and P_2 and the question arises as to the existence of a curve in this set for which the length from P_1 to P_2 is shorter than that for any other curve of the set. This problem of the calculus of variation was first studied by Johann Bernoulli in 1697. If the surface S is a plane, the unique curve of shortest distance is a straight line. It is to be expected therefore that geodesics on a curved surface play a role similar to that of straight lines in the plane.

Consider a curve G represented by $u^\alpha = u^\alpha(t)$ on the surface S with equations (2). Conditions on the functions $u^\alpha(t)$ are to be found under which G is a minimizing curve for the distance from P_1 where $t = t_1$ to P_2 where $t = t_2$. In order to obtain the necessary conditions, a family of curves joining P_1 and P_2 and in some neighborhood of G will be considered. Let

5) $$U^\alpha(t) = u^\alpha(t) + \epsilon w^\alpha(t)$$

represent the family of curves, under the condition that

6) $$w^\alpha(t_1) = 0, \quad w^\alpha(t_2) = 0 \quad (\alpha = 1, 2)$$

which require a neighboring curve to pass through P_1 and P_2. As the parameter ϵ approaches zero, the neighbor represented by $U^\alpha(t)$ approaches G.

The length l of the curve $G:u^\alpha(t)$ on S is given by (8-53) as

7) $$l = \int_{t_1}^{t_2} (g_{\alpha\beta}\dot{u}^\alpha \dot{u}^\beta)^{1/2}\, dt$$

where a dot is used to indicate differentiation with respect to t. Because $g_{\alpha\beta}$ are functions of position, it is evident that the integrand in (7) is a function of $u^1, u^2, \dot{u}^1, \dot{u}^2$. For convenience, denote the integrand by $\phi(u^1,u^2,\dot{u}^1,\dot{u}^2)$. Thus, the length l is given by

8)
$$l = \int_{t_1}^{t_2} \phi(u^1,u^2,\dot{u}^1,\dot{u}^2)\, dt.$$

The length L of a neighboring curve is then given by

9)
$$L = \int_{t_1}^{t_2} \phi(U^1,U^2,\dot{U}^1\dot{U}^2)\, dt,$$

which is a function of ϵ. The function $L(\epsilon)$ approaches l as ϵ goes to zero, and $L(\epsilon)_{\epsilon=0} = l$. The Taylor series expansion of $L(\epsilon)$ in a neighborhood of $\epsilon = 0$ is

10)
$$L(\epsilon) = L(0) + \epsilon \left(\frac{\partial L}{\partial \epsilon}\right)_{\epsilon=0} + \frac{\epsilon^2}{2!}\left(\frac{\partial^2 L}{\partial \epsilon^2}\right)_{\epsilon=0} + \cdots.$$

By elementary calculus, the value of ϵ for a minimum value of $L(\epsilon)$ must satisfy $\partial L/\partial \epsilon = 0$. Since the minimum value of $L(\epsilon)$ is to be attained for $\epsilon = 0$, it appears that $(\partial L/\partial \epsilon)_{\epsilon=0} = 0$ is a necessary condition on $L(\epsilon)$. In order to impose this condition on $L(\epsilon)$, differentiate equation (9) with respect to ϵ, remembering that the variables U^α and $\dot{U}^\alpha$ in ϕ are given by (5), to find

$$\frac{\partial L(\epsilon)}{\partial \epsilon} = \int_{t_1}^{t_2} \left(\frac{\partial \phi}{\partial U^\alpha} w^\alpha + \frac{\partial \phi}{\partial \dot{U}^\alpha} \dot{w}^\alpha\right) dt = 0.$$

With the condition $\epsilon = 0$, the last equation reduces to the form

11)
$$\int_{t_1}^{t_2} \left(\frac{\partial \phi}{\partial u^\alpha} w^\alpha + \frac{\partial \phi}{\partial \dot{u}^\alpha} \dot{w}^\alpha\right) dt = 0.$$

Next, assuming that the requisite conditions of continuity are satisfied by the functions involved, an integration by parts is effected to obtain

12)
$$\int_{t_1}^{t_2} \frac{\partial \phi}{\partial \dot{u}^\alpha} \dot{w}^\alpha\, dt = \frac{\partial \phi}{\partial \dot{u}^\alpha} w^\alpha \Big|_{t_1}^{t_2} - \int_{t_1}^{t_2} \frac{d}{dt}\left(\frac{\partial \phi}{\partial \dot{u}^\alpha}\right) w^\alpha\, dt,$$

where the first term on the right is zero at both limits by virtue of the stipulation in (6). Use of (12) in (11) gives

13)
$$\int_{t_1}^{t_2} \left[\frac{\partial \phi}{\partial u^\alpha} - \frac{d}{dt}\left(\frac{\partial \phi}{\partial \dot{u}^\alpha}\right)\right] w^\alpha(t)\, dt = 0.$$

Because G is a geodesic, the integral in (13) must vanish for arbitrary choice of $w^\alpha(t)$, which requires that[1]

$$14) \qquad \frac{\partial \phi}{\partial u^\alpha} - \frac{d}{dt}\left(\frac{\partial \phi}{\partial \dot{u}^\alpha}\right) = 0 \qquad (\alpha = 1,2).$$

Equations (14) are the conditions sought for the function ϕ, but it is desirable to express the conditions in terms of quantities in the integrand of (7). With ϕ defined in (7) as

$$15) \qquad \phi = [g_{\alpha\beta}(u^1,u^2)\dot{u}^\alpha \dot{u}^\beta]^{1/2},$$

one calculates

$$16) \qquad \frac{\partial \phi}{\partial \dot{u}^\alpha} = \frac{g_{\alpha\beta}\dot{u}^\beta}{\phi}, \qquad \frac{\partial \phi}{\partial u^\alpha} = \frac{1}{2\phi}\frac{\partial g_{\beta\gamma}}{\partial u^\alpha}\dot{u}^\beta \dot{u}^\gamma,$$

where in the last term a change of umbral indices had to be effected in order to exhibit the derivative of $g_{\beta\gamma}$ with respect to u^α. Use of (16) in (14) yields

$$17) \qquad \frac{d}{dt}\left(\frac{g_{\alpha\beta}\dot{u}^\beta}{\phi}\right) - \frac{1}{2\phi}\frac{\partial g_{\beta\gamma}}{\partial u^\alpha}\dot{u}^\beta \dot{u}^\gamma = 0,$$

and on carrying out the differentiation with respect to t and multiplying by ϕ there results

$$18) \qquad g_{\alpha\beta}\ddot{u}^\beta + \frac{\partial g_{\alpha\beta}}{\partial u^\gamma}\dot{u}^\gamma \dot{u}^\beta - \frac{1}{2}\frac{\partial g_{\beta\gamma}}{\partial u^\alpha}\dot{u}^\beta \dot{u}^\gamma - g_{\alpha\beta}\dot{u}^\beta \frac{\dot{\phi}}{\phi} = 0.$$

Equation (18) can be reduced to the form

$$19) \qquad \frac{d^2 u^\beta}{dt^2} + \Gamma_{\alpha\gamma}{}^\beta \frac{du^\alpha}{dt}\frac{du^\gamma}{dt} - \frac{du^\beta}{dt}\frac{d^2 s/dt^2}{ds/dt} = 0 \qquad (\beta = 1, 2),$$

which are the differential equations of geodesics if s is arc length and t is a general parameter. However, the reduction from (18) is simpler if t is taken as the arc length s. In this case $(\phi)^2 = g_{\alpha\beta}(du^\alpha/ds)(du^\beta/ds)$ which is unity. Hence, $\dot{\phi} = 0$, and equation (18) becomes

$$20) \qquad g_{\alpha\beta}\frac{d^2 u^\beta}{ds^2} + \left(\frac{\partial g_{\alpha\beta}}{\partial u^\gamma} - \frac{1}{2}\frac{\partial g_{\beta\gamma}}{\partial u^\alpha}\right)\frac{du^\beta}{ds}\frac{du^\gamma}{ds} = 0.$$

Now equation (9–62), with Latin letters changed to Greek letter indices, with coordinates u^α, and with the surface metric tensor $g_{\alpha\beta}$ replacing the space metric tensor a_{ij}, reads as

$$21) \qquad \frac{\partial g_{\beta\gamma}}{\partial u^\alpha} = g_{\sigma\beta}\Gamma_{\gamma\alpha}{}^\sigma + g_{\sigma\gamma}\Gamma_{\alpha\beta}{}^\sigma,$$

---

[1] Equations (14) are the Euler-Lagrange equations associated with the integral in (8).

where the Christoffel symbols are now calculated with respect to $g_{\alpha\beta}$ instead of a_{ij}. From (21) it follows, by a change of indices, that

22) $$\frac{\partial g_{\alpha\beta}}{\partial u^\gamma} = g_{\sigma\alpha}\Gamma_{\beta\gamma}{}^\sigma + g_{\sigma\beta}\Gamma_{\gamma\alpha}{}^\sigma.$$

On substituting the expressions in (21) and (22) into (20), one obtains, on using the symmetry in the sum on β and γ,

23) $$g_{\alpha\beta}\left(\frac{d^2 u^\beta}{ds^2} + \Gamma_{\sigma\gamma}{}^\beta \frac{du^\sigma}{ds}\frac{du^\gamma}{ds}\right) = 0.$$

Now multiply (23) by $g^{\alpha\tau}$, sum on α, and make use of the fact that $g^{\alpha\tau} g_{\alpha\beta} = \delta_\beta{}^\tau$ to obtain

24) $$\frac{d^2 u^\tau}{ds^2} + \Gamma_{\sigma\gamma}{}^\tau \frac{du^\sigma}{ds}\frac{du^\gamma}{ds} = 0,$$

or, on changing the indices,

25) $$\frac{d^2 u^\alpha}{ds^2} + \Gamma_{\beta\gamma}{}^\alpha \frac{du^\beta}{ds}\frac{du^\gamma}{ds} = 0 \qquad (\alpha = 1, 2).$$

Equations (25) are the differential equations sought for geodesics on a surface. They were obtained by means of a variational principle. Notice that equations (19) reduce to (25) if the parameter t in (19) is taken as arc length s.

Observe that equations (25) are of the second order. It is therefore true, by the theory of ordinary differential equations, that a unique geodesic passes through a given point with a given direction. Also, if two points on a surface are sufficiently close to each other, there exists a unique geodesic joining the two points.

An example will be adduced to give meaning to equations (25) as applied to a particular type of surface.

Example. Show that the meridian curves on the surface of revolution

$$x^1 = u^1 \cos u^2, \qquad x^2 = u^1 \sin u^2, \qquad x^3 = f(u^1)$$

are geodesics (see Exercise 6 in Section 2-4).

Solution: Obtain $ds^2 = \delta_{ij}\, dx^i\, dx^j = (1 + f'^2)(du^1)^2 + (u^1)^2 (du^2)^2$, where the prime indicates differentiation with respect to u^1. The expressions for $g_{\alpha\beta}$ in $ds^2 = g_{\alpha\beta}\, du^\alpha\, du^\beta$ are therefore $g_{11} = 1 + f'^2$, $g_{12} = g_{21} = 0$, $g_{22} = (u^1)^2$. The Christoffel symbols with superscript 2 are $\Gamma_{11}{}^2 = 0$, $\Gamma_{12}{}^2 = \Gamma_{21}{}^2 = 1/u^1$, $\Gamma_{22}{}^2 = 0$. Therefore, the second of equations (25) is

$$\frac{d^2 u^2}{ds^2} + \frac{2}{u^1}\frac{du^1}{ds}\frac{du^2}{ds} = 0,$$

or

$$\frac{d^2u^2/ds^2}{du^2/ds}\,ds = -\frac{2}{u^1}\frac{du^1}{ds}\,ds = -\frac{2}{u^1}\,du^1,$$

which has the first integral

$$\ln\left(\frac{du^2}{ds}\right) + 2\ln u^1 = \ln c,$$

or, in more compact form,

26) $$\frac{du^2}{ds} = \frac{c}{(u^1)^2}.$$

By dividing the form for $(ds)^2$ by $(ds)^2$, and using (26), it follows that

$$(1 + f'^2)\left(\frac{du^1}{ds}\right)^2 + (u^1)^2\frac{c^2}{(u^1)^4} = 1$$

from which

27) $$\frac{du^1}{ds} = \pm\frac{1}{u^1}\left[\frac{(u^1)^2 - c^2}{1 + f'^2}\right]^{1/2}.$$

On eliminating ds from (26) and (27), one obtains

$$du^2 = \pm\frac{c}{u^1}\left[\frac{1 + f'^2}{(u^1)^2 - c^2}\right]^{1/2}\,du^1,$$

which has the integral

$$u^2 = \pm c\int\left[\frac{1 + f'^2}{(u^1)^4 - c^2(u^1)^2}\right]^{1/2}\,du^1 + d,$$

where c and d are arbitrary constants of integration. With the choice $c = 0$ it is seen that the meridians $u^2 = $ constant are geodesics on a surface of revolution.

EXERCISES

1) (a) Use equations (25) to obtain the equations of geodesics in the plane $x^1 = u^1$, $x^2 = u^2$, $x^3 = 0$. (b) From equations (25) the equations of geodesics in a three-space can be deduced as

$$\frac{d^2x^i}{ds^2} + \Gamma_{jk}{}^i\frac{dx^j}{ds}\frac{dx^k}{ds} = 0$$

in general coordinates. Use these equations to find the equations of geodesics in euclidean space with orthogonal cartesian coordinates (see Exercise 4, Section 9-6).

2) Find the differential equations of the geodesics on the sphere $x^1 = a\sin u^1 \cos u^2$, $x^2 = a\sin u^1 \sin u^2$, $x^3 = a\cos u^1$, and verify that the meridians

given by $u^2 =$ constant are geodesics. (*Aid:* Obtain $ds^2 = \delta_{ij}\,dx^i\,dx^j = a^2[(du^1)^2 + (\sin u^1\,du^2)^2]$, $\Gamma_{12}{}^2 = \cot u^1$, $\Gamma_{22}{}^1 = -\sin u^1 \cos u^1$, and zero for all the other Christoffel symbols.)

3) Show that the transformation of surface coordinates

$$u^1 = \arcsin \frac{U^1}{a}, \qquad u^2 = U^2$$

changes the equations of the sphere in Exercise 2 to the form

$$x^1 = U^1 \cos U^2, \qquad x^2 = U^1 \sin U^2, \qquad x^3 = \sqrt{a^2 - (U^1)^2},$$

which take the form of the equations of a surface of revolution in the illustrative example with $f(U^1) = [a^2 - (U^1)^2]^{1/2}$.

4) Instead of expressing a curve on a surface in parametric form $u^\alpha = u^\alpha(s)$, the parameter s may be eliminated to obtain the equation $u^2 = \phi(u^1)$, giving u^2 as a function of u^1. Eliminate s from the two equations (25) to obtain the differential equation of geodesics in the form

$$\phi'' - \Gamma_{22}{}^1 \phi'^3 + (\Gamma_{22}{}^2 - 2\Gamma_{12}{}^1)\phi'^2 + (2\Gamma_{12}{}^2 - \Gamma_{11}{}^1)\phi' + \Gamma_{11}{}^2 = 0,$$

where the primes indicate differentiation with respect to u^1. (*Aid:* The form of

$$\phi' = \frac{du^2/ds}{du^1/ds}, \qquad \phi'' = \frac{(du^1/ds)(d^2u^2/ds^2) - (du^2/ds)(d^2u^1/ds^2)}{(du^1/ds)^3}$$

suggests the following operations. Multiply the first of (25) by du^2/ds, the second by du^1/ds, subtract, divide by $(du^1/ds)^3$, and use ϕ' and ϕ'' to arrive at the desired equation.)

5) Show that the Gauss equations

$$x^1 = x^1(u^1), \qquad x^2 = x^2(u^1), \qquad x^3 = u^2$$

represent a cylindrical surface and investigate the geodesics on the surface. (*Aid:* The only non-zero Christoffel symbol is

$$\Gamma_{11}{}^1 = \frac{x^{1\prime} x^{1\prime\prime} + x^{2\prime} x^{2\prime\prime}}{(x^{1\prime})^2 + (x^{2\prime})^2},$$

in which $x^{1\prime}$, for instance, means $\dfrac{d}{du^1} x^1(u^1)$. The geodesics (25) yield

$$\frac{d^2 u^1}{ds^2} + \frac{x^{1\prime} x^{1\prime\prime} + x^{2\prime} x^{2\prime\prime}}{(x^{1\prime})^2 + (x^{2\prime})^2} \frac{du^1}{ds}\frac{du^1}{ds} = 0, \qquad \frac{d^2 u^2}{ds^2} = 0$$

which have the integrals

$$\ln\left(\frac{du^1}{ds}\right) + \frac{1}{2}\ln[(x^{1\prime})^2 + (x^{2\prime})^2] = \ln C, \qquad u^2 = as + b.$$

By use of the second of the last two equations the next integral of the first can be written as

$$\alpha \int \sqrt{(x^{1'})^2 + (x^{2'})^2}\, du^1 + \beta u^2 + \gamma = 0.$$

The transformation $U^1 = \int \sqrt{(x^{1'})^2 + (x^{2'})^2}\, du^1$, $U^2 = u^2$, leaves the $u^2 =$ constant curves unaltered and effects only a change of parameter on the $u^1 =$ constant curves. If the surface is rolled onto a plane, the new form $\alpha U^1 + \beta U^2 + \gamma = 0$ of the geodesics shows that these curves meet the lines $U^2 =$ constant at constant angle and are therefore the helices on the given surface.)

6) Show that infinitely many geodesics on the cylinder $x^2 + y^2 = a^2$ connecting the points $A\,(a,0,0)$ and $B\,(a \cos \alpha,\, a \sin \alpha,\, b)$ are given by

$$x = a \cos \theta, \qquad y = a \sin \theta, \qquad z = \frac{b\theta}{\alpha + 2n\pi}.$$

(The geodesic of shortest length is the one for which $n = 0$.)

12-3. The Second Fundamental Form of a Surface. The first fundamental form $ds^2 = g_{\alpha\beta}\, du^\alpha\, du^\beta$ for a surface is involved in questions of intrinsic geometry of the surface. For extrinsic geometry the *second* fundamental form is necessary. It is introduced as follows.

In Section 2-3 an expression was found for the distance from a point $\bar{x}^i$ on a curve to a plane through a neighboring point x^i on the curve. A similar procedure will be employed here to find the distance from a point $\bar{x}^i$ on a surface S to the tangent plane to S at a point P with space coordinates x^i, and surface coordinates u^α.

The tangent plane to S at $P(x^i)$ is given by

28) $$X^i(\xi^i - x^i) = 0,$$

where ξ^i are current point coordinates in the plane and X^i are components of the unit normal vector to the surface. Summation on i from 1 to 3 is understood. Consider a curve C on S through P with equations $x^i = x^i(s)$ in space, and $u^\alpha = u^\alpha(s)$ on the surface. For a neighboring point $\bar{x}^i$ on C (Fig. 43) the Taylor expression gives

29) $$\bar{x}^i = x^i + (x^{i'})_0 s + \tfrac{1}{2}(x^{i''})_0 s^2 + \cdots,$$

where s measures the arc length from P and the subscript on the derivative indicates evaluation at P where $s = 0$. Now from $x^i = x^i(u^1, u^2)$, one has

30) $$\frac{dx^i}{ds} = \frac{\partial x^i}{\partial u^\alpha}\frac{du^\alpha}{ds}$$

and

31) $$\frac{d^2 x^i}{ds^2} = \frac{\partial x^i}{\partial u^\alpha}\frac{d^2 u^\alpha}{ds^2} + \frac{du^\alpha}{ds}\frac{\partial^2 x^i}{\partial u^\beta \partial u^\alpha}\frac{du^\beta}{ds},$$

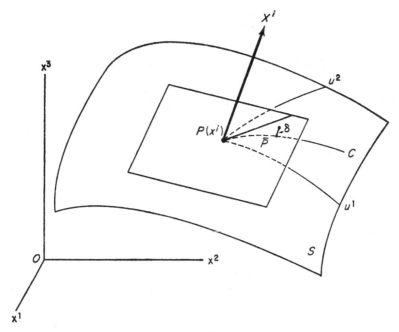

Fig. 43

which can be abbreviated to

32) $$x^{i\prime} = x^i{}_{,\alpha} u^{\alpha\prime}, \qquad x^{i\prime\prime} = x^i{}_{,\alpha} u^{\alpha\prime\prime} + \frac{\partial^2 x^i}{\partial u^\beta \, \partial u^\alpha} u^{\alpha\prime} u^{\beta\prime},$$

because the first covariant derivative of a scalar function is the same as the partial derivative.

The distance δ from $\bar{P}$ to the tangent plane to S at P is obtained merely by substituting the coordinates $\bar{x}^i$ of $\bar{P}$ in (29) for ξ^i in equation (28). This gives

33) $$\delta = X^i[(x^{i\prime})_0 + \tfrac{1}{2}(x^{i\prime\prime})_0 s^2 + \cdots].$$

But $X^i x^{i\prime}$ and $X^i x^i{}_{,\alpha}$ are zero at $s = 0$ because the normal to S at P is orthogonal to all curves through P on S. Hence, by use of the second expression in (32), the formula for δ becomes

34) $$\delta = \frac{1}{2} X^i \left(\frac{\partial^2 x^i}{\partial u^\beta \, \partial u^\alpha} \right)_0 u^{\alpha\prime} u^{\beta\prime} s^2 + \cdots,$$

or

35) $$\delta = \tfrac{1}{2}(d_{\alpha\beta} u^{\alpha\prime} u^{\beta\prime}) s^2 + \cdots,$$

where, for any point on the surface, the quantities $d_{\alpha\beta}$ are defined by the sum

$$36) \qquad d_{\alpha\beta} \equiv X^i \frac{\partial^2 x^i}{\partial u^\alpha \, \partial u^\beta}.$$

It is evident from the definition that $d_{\alpha\beta} = d_{\beta\alpha}$. Note that the covariant derivative of the surface vector $x^i{}_{,\alpha}$ with respect to u^β is given by

$$(x^i{}_{,\alpha})_{,\beta} \equiv x^i{}_{,\alpha\beta} = \frac{\partial}{\partial u^\beta} x^i{}_{,\alpha} - x^i{}_{,\gamma} \Gamma_{\alpha\beta}{}^\gamma,$$

and that, because $X^i x^i{}_{,\gamma} = 0$,

$$37) \qquad X^i x^i{}_{,\alpha\beta} = X^i \frac{\partial^2 x^i}{\partial u^\alpha \, \partial u^\beta},$$

so that the definition of $d_{\alpha\beta}$ in (36) can be written as

$$38) \qquad d_{\alpha\beta} = X^i x^i{}_{,\alpha\beta}.$$

The quantities $d_{\alpha\beta}$ are evidently components of a symmetric covariant tensor of the second order for the surface. This is seen by the fact that both x^i and X^i are scalar functions of surface coordinates u^α, so $x^i{}_{,\alpha\beta}$ is a tensor for each value of i. The sum in (38) where X^i are scalars, is therefore a tensor.

A geometric interpretation of the second fundamental form $d_{\alpha\beta} \, du^\alpha \, du^\beta$ is afforded by equation (35). If terms of order higher than the second in s are neglected, equation (35) states that the second fundamental form $d_{\alpha\beta} \, du^\alpha \, du^\beta$ [where the $d_{\alpha\beta}$ are evaluated at $P(x^i)$] are proportional to the distance from a neighboring point $\bar{P}$ (on a curve with direction $du^1 : du^2$) to the tangent plane to the surface at P. Observe that $d_{\alpha\beta} \, du^\alpha \, du^\beta$ is an invariant. Therefore, the geometric result just found is independent of the surface coordinates employed, which is to be expected for a geometric property.

A query naturally arises here. Suppose the curve C is such that the distance δ in (35) is of higher than the second order in s. This is true if, and only if,

$$39) \qquad d_{\alpha\beta} \, du^\alpha \, du^\beta = 0,$$

which is an invariant differential equation of the first order and second degree in the ratio $du^1 : du^2$. Therefore, through each point of a region of the surface for which $d_{12}{}^2 - d_{11} d_{22} > 0$, there are two curves which satisfy equation (39). Hence, there are two one-parameter families (called a *net*) of curves determined by equation (39). This is the *asymptotic* net on the surface.

By analogy with equation (39) one may wish to investigate the net of curves on the surface given by the vanishing of the first fundamental form, that is, by

$$40) \qquad g_{\alpha\beta} \, du^\alpha \, du^\beta = 0.$$

The curves determined by (40) constitute the net of *minimal* curves, or curves of zero length. They are not real on a real surface, because on such a surface the first fundamental form is positive definite, which requires that $g_{11}g_{22} - g_{12}^2 > 0$.

It can be shown that a surface is completely determined (except for position in space) by its first and second fundamental tensors[2] $g_{\alpha\beta}$ and $d_{\alpha\beta}$.

EQUATIONS OF GAUSS AND WEINGARTEN. To conclude this section, the important differential equations due to Gauss and Weingarten will be introduced. Recall that in the study of space curves the Frenet-Serret formulas were used to express the derivatives of the unit vectors **T**, **N**, **B** with respect to arc length as linear combinations of **T**, **N**, **B**. Here, it is desired to express the derivatives of the three vectors with components X^i, $x^i{}_{,1}$ and $x^i{}_{,2}$ with respect to u^α as linear combinations of these three linearly independent vectors. Note, however, that whereas **T**, **N**, **B** form a mutually orthogonal set of vectors, the surface vectors with components $x^i{}_{,1}$ and $x^i{}_{,2}$ are not orthogonal, in general.

Differentiation of the defining equation for $g_{\alpha\beta}$, namely,

$$41) \qquad g_{\alpha\beta} = \delta_{ij} \frac{\partial x^i}{\partial u^\alpha} \frac{\partial x^j}{\partial u^\beta}$$

with respect to u^γ yields

$$42) \qquad \delta_{ij}\left(\frac{\partial^2 x^i}{\partial u^\gamma \, \partial u^\alpha} \frac{\partial x^j}{\partial u^\beta} + \frac{\partial x^i}{\partial u^\alpha} \frac{\partial^2 x^j}{\partial u^\gamma \, \partial u^\beta}\right) = \frac{\partial g_{\alpha\beta}}{\partial u^\gamma}.$$

Cyclic permutation on α, β, γ gives

$$43) \qquad \delta_{ij}\left(\frac{\partial^2 x^i}{\partial u^\alpha \, \partial u^\beta} \frac{\partial x^j}{\partial u^\gamma} + \frac{\partial x^i}{\partial u^\beta} \frac{\partial^2 x^j}{\partial u^\alpha \, \partial u^\gamma}\right) = \frac{\partial g_{\beta\gamma}}{\partial u^\alpha},$$

and a cyclic permutation on α, β, γ in (43) produces

$$44) \qquad \delta_{ij}\left(\frac{\partial^2 x^i}{\partial u^\beta \, \partial u^\gamma} \frac{\partial x^j}{\partial u^\alpha} + \frac{\partial x^i}{\partial u^\gamma} \frac{\partial^2 x^j}{\partial u^\beta \, \partial u^\alpha}\right) = \frac{\partial g_{\gamma\alpha}}{\partial u^\beta}.$$

Subtraction of equation (42) from the sum of equations (43) and (44)

---

[2] This is a generalization of a theorem which states that a space curve is uniquely determined (up to a rigid motion) by its curvature and torsion, that is, by its two curvatures.

leads to

45) $$\frac{1}{2}\left(\frac{\partial g_{\beta\gamma}}{\partial u^\alpha} + \frac{\partial g_{\gamma\alpha}}{\partial u^\beta} - \frac{\partial g_{\alpha\beta}}{\partial u^\gamma}\right) = \delta_{ij}\frac{\partial^2 x^i}{\partial u^\alpha \partial u^\beta}\frac{\partial x^j}{\partial u^\gamma}.$$

The left-hand member of (45) is the Christoffel symbol of the first kind $\Gamma_{\alpha\beta;\gamma}$ (9–6) for the first fundamental tensor of the surface. Hence, equation (45) can be written as

46) $$\Gamma_{\alpha\beta;\gamma} = \frac{\partial^2 x^i}{\partial u^\alpha \partial u^\beta}\frac{\partial x^i}{\partial u^\gamma}.$$

Write

47) $$\frac{\partial^2 x^i}{\partial u^\alpha \partial u^\beta} = A_{\alpha\beta}{}^\gamma x^i{}_{,\gamma} + B_{\alpha\beta} X^i$$

to express the left-hand member as a linear combination of the basic vectors $x^i{}_{,\gamma}$ and X^i, where the quantities $A_{\alpha\beta}{}^\gamma$ and $B_{\alpha\beta}$ are to be determined. Multiply (47) by X^i, sum on i, and use (37) to see that

48) $$B_{\alpha\beta} = \frac{\partial^2 x^i}{\partial u^\alpha \partial u^\beta} X^i = d_{\alpha\beta}.$$

Next, multiply (47) by $x^i{}_{,\delta}$ and use $x^i{}_{,\gamma} x^i{}_{,\delta} = g_{\gamma\delta}$ to obtain

49) $$A_{\alpha\beta}{}^\gamma g_{\gamma\delta} = \frac{\partial^2 x^i}{\partial u^\alpha \partial u^\beta}\frac{\partial x^i}{\partial u^\delta},$$

or, by use of (46),

$$A_{\alpha\beta;\delta} = \Gamma_{\alpha\beta;\delta}.$$

It follows that equation (47) has the form

50) $$\frac{\partial^2 x^i}{\partial u^\alpha \partial u^\beta} = \frac{\partial x^i}{\partial u^\gamma}\Gamma_{\alpha\beta}{}^\gamma + d_{\alpha\beta} X^i,$$

which exhibits the differential equations of Gauss.

The equations of Weingarten express $X^i{}_{,\alpha}$ as a linear combination of $x^i{}_{,1}$ and $x^i{}_{,2}$. Differentiate the identity $X^i X^i = 1$ with respect to u^α to obtain

$$X^i X^i{}_{,\alpha} = 0 \quad (\alpha = 1, 2),$$

which means that $X^i{}_{,\alpha}$ is a vector in the surface, and therefore is a linear combination of $x^i{}_{,1}$ and $x^i{}_{,2}$. Let

51) $$X^i{}_{,\alpha} = C_\alpha{}^\beta x^i{}_{,\beta},$$

where $C_\alpha{}^\beta$ are to be determined. Multiply (51) by $x^i{}_{,\gamma}$ to obtain

52) $$X^i{}_{,\alpha} x^i{}_{,\gamma} = C_\alpha{}^\beta x^i{}_{,\beta} x^i{}_{,\gamma} = C_\alpha{}^\beta g_{\beta\gamma} = C_{\alpha\gamma}.$$

If the identity $X^i x^i{}_{,\gamma} = 0$ is differentiated with respect to u^α, one has

53) $$X^i \frac{\partial^2 x^i}{\partial u^\alpha \partial u^\gamma} + X^i{}_{,\alpha} x^i{}_{,\gamma} = 0,$$

or, by (37) and (38),

54) $$X^i{}_{,\alpha} x^i{}_{,\gamma} = -d_{\alpha\gamma}.$$

Hence, $C_{\alpha\gamma}$ in (52) are determined as $-d_{\alpha\gamma}$, and

$$C_\alpha{}^\beta = g^{\sigma\beta} C_{\sigma\alpha} = -g^{\sigma\beta} d_{\sigma\alpha},$$

so that equation (51) gives

55) $$X^i{}_{,\alpha} \equiv \frac{\partial X^i}{\partial u^\alpha} = -d_{\alpha\sigma} g^{\sigma\beta} x^i{}_{,\beta},$$

the equations of Weingarten. The equations of Gauss and of Weingarten for surfaces are analogous to the Frenet-Serret formulas for curves.

EXERCISES

1) Find the asymptotic curves on the surface

$$x^1 = u^1 + u^2, \qquad x^2 = u^1 - u^2, \qquad 2x^3 = u^1 u^2.$$

2) Find the differential equation of the asymptotic net on a surface in the Monge form $z = f(x,y)$.

12–4. Normal Curvature of a Surface. Any curve C on a surface S is given by $x^i = x^i(s)$ in space, and by $u^\alpha = u^\alpha(s)$ on the surface, and it has at each point P a trihedron of unit vectors, that is, a tangent **T**, a principal normal **N**, and a binormal **B**. These are independent of the surface. There is also the unit surface normal with components X^i, which is normal to C at P. If the angle between the surface normal (X^i) and the principal normal (β^i) to C at P is ω, then, by use of (2–27),

56) $$\cos \omega = X^i \beta^i = \rho X^i \frac{d^2 x^i}{ds^2},$$

where ρ is the radius of first curvature of C at P. By formulas (31) and (38) in the preceding section

$$X^i \frac{d^2 x^i}{ds^2} = X^i \frac{\partial^2 x^i}{\partial u^\beta \partial u^\alpha} \frac{du^\alpha}{ds} \frac{du^\beta}{ds} = d_{\alpha\beta} \frac{du^\alpha}{ds} \frac{du^\beta}{ds}.$$

Hence, from (56),

57) $$\frac{\cos \omega}{\rho} = d_{\alpha\beta} \frac{du^\alpha}{ds} \frac{du^\beta}{ds} = \frac{d_{\alpha\beta} du^\alpha du^\beta}{ds^2} = \frac{d_{\alpha\beta} du^\alpha du^\beta}{g_{\alpha\beta} du^\alpha du^\beta},$$

which states that the numerical value of the quantity $(\cos \omega)/\rho$ at a point on a curve C is the ratio of the second and first fundamental forms for the surface evaluated at the point. Note that $du^1:du^2$ appearing in the ratio gives the direction of C at P, and that $du^1:du^2$ is the same for all curves on S tangent to C at P. The denominator $g_{\alpha\beta} du^\alpha du^\beta$ is always positive so the sign of $\cos \omega/\rho$ depends upon the sign of the second fundamental form. If C is one of the curves of the net given by $d_{\alpha\beta} du^\alpha du^\beta = 0$, that is, if C is an asymptotic curve on the surface, then either $1/\rho = 0$ or $\cos \omega = 0$. In the latter case the principal normal to C at P is perpendicular to the surface normal, so the osculating plane to C is the tangent plane to the surface. Hence, the asymptotic curves which are not straight lines $(1/\rho = 0)$, may be characterized geometrically as those curves for which the osculating plane at every point coincides with the tangent plane to the surface at the point.

If $\omega = 0$ or π in (57), then the principal normal to C is along the surface normal X^i or in the opposite direction. Let C^* be the plane curve of normal section in the direction $du^1:du^2$ of C, that is, C^* is the intersection of the plane determined by the unit surface normal and the unit tangent $\mathbf{T}$ to C.

For this direction

58) $$\frac{d_{\alpha\beta} du^\alpha du^\beta}{g_{\alpha\beta} du^\alpha du^\beta} = \frac{1}{R},$$

where the numerical value of $1/R$ is the curvature κ_n of the plane curve C^* of normal section. To each curve C there is a unique curve C^* of normal section in the direction of C. The quantity R is called the *radius of normal curvature* of the surface at $P(u^\alpha)$ with respect to the direction $du^1:du^2$. In general, the value of R varies with the direction of the curve C. Only for a plane $(d_{\alpha\beta} = 0)$, and a sphere (for which it can be shown that $d_{\alpha\beta}$ are proportional to $g_{\alpha\beta}$), is the value of $1/R$ independent of the direction of C.

Because the value of $1/R$ in (58) varies with the direction $du^1:du^2$, it is natural to ask for the directions on the surface at point P in which $1/R$ takes on extreme values. To find these directions, write equation (58) in the form

59) $$(g_{\alpha\beta} - Rd_{\alpha\beta}) du^\alpha du^\beta = 0,$$

divide this quadratic form by $(du^2)^2$, and denote the ratio $du^1:du^2$ by λ. The resulting quadratic equation in λ is

60) $$(g_{11} - Rd_{11})\lambda^2 + 2(g_{12} - Rd_{12})\lambda + (g_{22} - Rd_{22}) = 0.$$

Differentiate equation (60) with respect to λ to obtain

61) $$(g_{11} - Rd_{11})\lambda + (g_{12} - Rd_{12}) = 0.$$

If equation (61) is multiplied by λ and subtracted from equation (60), there results

62) $\qquad (g_{12} - Rd_{12})\lambda + (g_{22} - Rd_{22}) = 0.$

Elimination of λ between (61) and (62) yields the quadratic in R

63) $\qquad \begin{vmatrix} g_{11} - Rd_{11} & g_{12} - Rd_{12} \\ g_{12} - Rd_{12} & g_{22} - Rd_{22} \end{vmatrix} = 0,$

or

64) $\quad (d_{11}d_{22} - d_{12}{}^2)R^2 - (d_{11}g_{22} + d_{22}g_{11} - 2d_{12}g_{12})R + (g_{11}g_{22} - g_{12}{}^2) = 0$

the roots of which are the maximum and minimum values of the radii of normal curvature for the surface at P.

Equations (61) and (62), can be rearranged in the form

65) $\qquad \begin{aligned} (g_{11}\lambda + g_{12}) - R(d_{11}\lambda + d_{12}) &= 0, \\ (g_{12}\lambda + g_{22}) - R(d_{12}\lambda + d_{22}) &= 0, \end{aligned}$

from which R can be eliminated to give

$$\begin{vmatrix} g_{11}\lambda + g_{12} & d_{11}\lambda + d_{12} \\ g_{12}\lambda + g_{22} & d_{12}\lambda + d_{22} \end{vmatrix} = 0,$$

or, on replacing λ by $du^1 : du^2$,

66) $\qquad \begin{vmatrix} g_{1\alpha}\, du^\alpha & d_{1\beta}\, du^\beta \\ g_{2\alpha}\, du^\alpha & d_{2\beta}\, du^\beta \end{vmatrix} = 0,$

or, in another form,

67) $\qquad \begin{vmatrix} (du^2)^2 & -du^1\, du^2 & (du^1)^2 \\ g_{11} & g_{12} & g_{22} \\ d_{11} & d_{12} & d_{22} \end{vmatrix} = 0.$

A net of curves on the surface is determined by (67), and the two curves of the net through any point of the surface are in the direction of the curves of normal section having maximum and minimum radii of curvature. The curves of the net (67) are styled *lines of curvature*. It can be shown that they are orthogonal at every point on the surface. There are many interesting properties of the lines of curvature, but for these the reader is referred to more extended treatments of differential geometry.

12–5. Curvature Formulas. Two important measures of the curvature of a surface can be deduced from (64). The *total curvature*, or Gaussian curvature, of a surface at a point is defined as the product of the principal

Sec. 12-5]   GEODESIC AND UNION CURVES   215

normal curvatures at the point. If ρ_1, ρ_2 are the roots of equation (64), the Gaussian or total curvature κ is expressed by

$$68) \quad \kappa \equiv \frac{1}{\rho_1\rho_2} \equiv \frac{d_{11}d_{22} - d_{12}^2}{g_{11}g_{22} - g_{12}^2}.$$

The connection between formula (68) and the Riemann-Christoffel tensor for a surface will be shown next. In Section 9–2 the number of components of the R–C tensor was given as $n^2(n^2 - 1)/12$. For an ordinary surface ($n = 2$), there is just one non-zero component which can be written as R_{1212}. In order to find the value of R_{1212}, advantage will be taken of the fact that it is a tensor, so any coordinate system may be employed. For simplicity then, suppose that the coordinate system on the surface is such that $g_{12} = g_{21} = 0$ which means that the coordinate curves are everywhere orthogonal. The six Christoffel symbols of the first kind (with $g_{12} = 0$), for the fundamental tensor $g_{\alpha\beta}$ of the surface are

$$69) \quad \begin{array}{lll} \Gamma_{11;1} = \tfrac{1}{2}g_{11,1}, & \Gamma_{12;1} = \tfrac{1}{2}g_{11,2}, & \Gamma_{22;1} = -\tfrac{1}{2}g_{22,1}, \\ \Gamma_{11;2} = -\tfrac{1}{2}g_{11,2}, & \Gamma_{12;2} = \tfrac{1}{2}g_{22,1}, & \Gamma_{22;2} = \tfrac{1}{2}g_{22,2}, \end{array}$$

and the symbols of second kind are

$$70) \quad \begin{array}{lll} \Gamma_{11}^1 = \dfrac{g_{11,1}}{2g_{11}}, & \Gamma_{12}^1 = \dfrac{g_{11,2}}{2g_{11}}, & \Gamma_{22}^1 = -\dfrac{g_{22,1}}{2g_{11}}, \\[2mm] \Gamma_{11}^2 = -\dfrac{g_{11,2}}{2g_{22}}, & \Gamma_{12}^2 = \dfrac{g_{22,1}}{2g_{22}}, & \Gamma_{22}^2 = \dfrac{g_{22,2}}{2g_{22}}. \end{array}$$

From formulas (26) and (29) in Section 9–2,

$$71) \quad R_{\alpha\beta\delta\sigma} = g_{\alpha\gamma}R^\gamma{}_{\beta\delta\sigma} = g_{\alpha\gamma}\left(\frac{\partial \Gamma_{\sigma\beta}^\gamma}{\partial u^\delta} - \frac{\partial \Gamma_{\delta\beta}^\gamma}{\partial u^\sigma} - \Gamma_{\delta\beta}^\tau \Gamma_{\sigma\tau}^\gamma + \Gamma_{\sigma\beta}^\tau \Gamma_{\delta\tau}^\gamma\right),$$

where the Greek letters replace the Latin letters, and the Γ's are calculated with respect to the surface tensor $g_{\alpha\beta}$. Remember that $g_{12} = 0$ and obtain

$$72) \quad R_{1212} = g_{11}R^1{}_{212} = g_{11}\left(\frac{\partial \Gamma_{22}^1}{\partial u^1} - \frac{\partial \Gamma_{12}^1}{\partial u^2} - \Gamma_{12}^\tau \Gamma_{2\tau}^1 + \Gamma_{22}^\tau \Gamma_{1\tau}^1\right),$$

which, by use of (70), gives

$$73) \quad R_{1212} = \frac{1}{4g_{11}g_{22}}\left[g_{22}\left(\frac{\partial g_{11}}{\partial u^2}\right)^2 + g_{11}\left(\frac{\partial g_{22}}{\partial u^1}\right)^2 + g_{11}\frac{\partial g_{11}}{\partial u^2}\frac{\partial g_{22}}{\partial u^2}\right.$$
$$\left. + g_{22}\frac{\partial g_{11}}{\partial u^1}\frac{\partial g_{22}}{\partial u^1}\right] - \frac{1}{2}\left[\frac{\partial^2 g_{11}}{\partial u^2 \partial u^2} + \frac{\partial^2 g_{22}}{\partial u^1 \partial u^1}\right].$$

By a rather involved calculation incident to integrability conditions of the Gauss differential equations (50), it can be shown that $d_{11}d_{22} - d_{12}^2$ takes the form of R_{1212} in (73). Observe next that the tensor law of transformation

74) $$\bar{R}_{\rho\tau\mu\sigma} = R_{\alpha\beta\gamma\delta} \frac{\partial u^\alpha}{\partial \bar{u}^\rho} \frac{\partial u^\beta}{\partial \bar{u}^\tau} \frac{\partial u^\gamma}{\partial \bar{u}^\mu} \frac{\partial u^\delta}{\partial \bar{u}^\sigma},$$

gives

$$\bar{R}_{1212} = R_{\alpha\beta\gamma\delta} \frac{\partial u^\alpha}{\partial \bar{u}^1} \frac{\partial u^\beta}{\partial \bar{u}^2} \frac{\partial u^\gamma}{\partial \bar{u}^1} \frac{\partial u^\delta}{\partial \bar{u}^2},$$

which, after summing on the right-hand side, leads to

75) $$\bar{R}_{1212} = R_{1212} \left(\frac{\partial u^1}{\partial \bar{u}^1} \frac{\partial u^2}{\partial \bar{u}^2} - \frac{\partial u^1}{\partial \bar{u}^2} \frac{\partial u^2}{\partial \bar{u}^1} \right)^2 = R_{1212} J^2,$$

where J is the jacobian of the transformation of surface coordinates. On taking the determinant of both sides of the transformation law

$$\bar{g}_{\rho\sigma} = g_{\alpha\beta} \frac{\partial u^\alpha}{\partial \bar{u}^\rho} \frac{\partial u^\beta}{\partial \bar{u}^\sigma}$$

one has

$$|\bar{g}_{\rho\sigma}| = |g_{\alpha\beta}| J^2,$$

so that (75) may be written as

76) $$\frac{\bar{R}_{1212}}{|\bar{g}_{\rho\sigma}|} = \frac{R_{1212}}{|g_{\alpha\beta}|},$$

which means that the function

77) $$\frac{R_{1212}}{g_{11}g_{22} - g_{12}{}^2}$$

is an invariant in surface theory. Because $R_{1212} = d_{11}d_{22} - d_{12}{}^2$, the invariant is identified as the total (or Gaussian) curvature as defined in (68). Because the formula (77) is a function of the components $g_{\alpha\beta}$ and their derivatives, and not of the $d_{\alpha\beta}$, one sees here the proof of the famous

Theorem of Gauss: The total curvature at a point of a surface is determined by the first fundamental tensor $g_{\alpha\beta}$.

It is easy to understand why Gauss called this his Theorema Egregium —"Most Splendid Theorem," because it is a remarkable fact that although the Gaussian curvature was defined in (68) in an *extrinsic* manner, the theorem asserts that the Gaussian curvature is actually an *intrinsic* property of the surface.

The *mean curvature* of a surface at a point is defined as the sum of the principal normal curvatures $1/\rho_1 + 1/\rho_2$ at the point. From equation (64)

the mean curvature κ_m is given by the formula

78) $$\kappa_m \equiv \frac{1}{\rho_1} + \frac{1}{\rho_2} \equiv \frac{g_{11}d_{22} + g_{22}d_{11} - 2g_{12}d_{12}}{g_{11}g_{22} - g_{12}{}^2}.$$

Minimal surfaces are those for which the mean curvature is everywhere zero. A trivial example is the plane for which $d_{\alpha\beta}$ are all zero.

Remark: The Gauss measure of total curvature of a surface at a point is not entirely satisfactory as a measure of curvature because there are curved surfaces (a cylinder, for instance) for which the Gauss measure is zero. To see this, note that for a straight line element of the cylinder the radius of normal curvature is infinite. R. B. Deal, Jr.[3] has proposed a measure of curvature which is zero only if both radii of normal curvature are infinite at the point. If κ_m is the mean curvature and κ is the Gauss total curvature at a point of the surface, then Deal's curvature is given by

79) $$\kappa_m{}^2 - 2\kappa = \frac{1}{\rho_1{}^2} + \frac{1}{\rho_2{}^2}.$$

From formulas (57) and (58) it follows that

80) $$\frac{1}{R} = \frac{\cos \omega}{\rho}$$

which states that $R \cos \omega = \rho$, which may be interpreted as follows. The center of curvature at a point P of a curve C on a surface is the foot of the perpendicular dropped upon the osculating plane to C at P from the center of curvature of the curve of normal section which is tangent to C at P. This is the theorem of Meusnier. A generalization of it will appear in Section 12-8.

EXERCISES

1) Calculate the value of the invariant (77) for the sphere

$$x^1 = a \sin u^1 \cos u^2, \qquad x^2 = a \sin u^1 \sin u^2, \qquad x^3 = a \cos u^1.$$

2) For the helicoid

$$x^1 = u^1 \cos u^2, \qquad x^2 = u^1 \sin u^2, \qquad x^3 = bu^2,$$

find the principal radii of normal curvature at any point, and show that the lines of curvature are given by

$$du^2 = \pm \frac{du^1}{\sqrt{(u^1)^2 + b^2}}.$$

---

[3] R. B. Deal, Jr., "Union Differentiation and Union Correspondence in Metric Differential Geometry," Unpublished Doctoral dissertation, The University of Oklahoma, Norman, Oklahoma, 1953.

12–6. Geodesic Curvature.

Instead of defining a geodesic as the curve of shortest distance joining two points on a surface, one may employ the concept of *geodesic curvature* as in the following

Theorem: A geodesic on a surface is a curve for which the geodesic curvature is zero at every point of the curve.

The proof of the theorem will be immediate, once the concept of geodesic curvature is explained.

The geometrical meaning of geodesic curvature will be stated first. Consider a curve C on a surface S. Project the curve C orthogonally onto the tangent plane to S at a point P of C. The curvature of the resulting plane curve C' in the tangent plane will be denoted by κ_g. The curvature κ_g of C' was styled the *geodesic curvature* of C by Liouville in 1850.

The intrinsic derivative for a vector field was introduced in (9–36). Let λ^α denote the unit tangent vector du^α/ds to a curve C given by $u^\alpha = u^\alpha(s)$ with arc length s as parameter. Because

$$81) \qquad g_{\alpha\beta} \frac{du^\alpha}{ds} \frac{du^\beta}{ds} = 1$$

it follows that

$$82) \qquad g_{\alpha\beta} \lambda^\alpha \lambda^\beta = 1,$$

and intrinsic differentiation of (82) with respect to s gives

$$83) \qquad g_{\alpha\beta} \lambda^\alpha \frac{\delta \lambda^\beta}{\delta s} + g_{\alpha\beta} \lambda^\beta \frac{\delta \lambda^\alpha}{\delta s} = 0,$$

or

$$84) \qquad g_{\alpha\beta} \lambda^\alpha \frac{\delta \lambda^\beta}{\delta s} = 0.$$

Equation (84) states that the derived vector $\delta \lambda^\beta/\delta s$ is perpendicular to λ^α. If μ^α is the unit vector on the surface, and if μ^α is normal to C, the tangent and normal to C may be oriented the same as the u^1 and u^2 coordinate curves. Since $\delta \lambda^\alpha/\delta s$ is in the direction normal to C,

$$85) \qquad \frac{\delta \lambda^\alpha}{\delta s} = \kappa_g \mu^\alpha,$$

and the invariant scalar function κ_g is called the *geodesic curvature*, or surface curvature, of C at P. The reader will observe that this definition is analogous to that of first curvature of a space curve in three-space (Section 8–2). By applying Meusnier's theorem to the curve C as a curve on the cylindrical surface projecting C upon the tangent plane to the surface at P, it is seen that if r is the radius of curvature of C' at P and

ρ is the radius of curvature of C, then

$$\frac{1}{r} = \frac{\cos \psi}{\rho},$$

where ψ is the angle between the osculating plane to C and the tangent plane to S at P. Thus, the numerical value of κ_g is $1/r$.

By (9-36), the intrinsic derivative $\delta\lambda^\alpha/\delta s$ on the surface is

86) $$\frac{\delta \lambda^\alpha}{\delta s} = \lambda^\alpha{}_{,\gamma} \frac{du^\gamma}{ds},$$

where, by (9-35),

87) $$\lambda^\alpha{}_{,\gamma} \equiv \frac{\partial \lambda^\alpha}{\partial u^\gamma} + \lambda^\beta \Gamma_{\beta\gamma}{}^\alpha,$$

and the Γ's here are calculated with respect to $g_{\alpha\beta}$. Now, since $\lambda^\alpha \equiv du^\alpha/ds$, equation (85) may be written in the form

88) $$\frac{d}{ds}\left(\frac{du^\alpha}{ds}\right) + \frac{du^\beta}{ds} \Gamma_{\beta\gamma}{}^\alpha \frac{du^\gamma}{ds} = \kappa_g \mu^\alpha,$$

or

89) $$\frac{d^2 u^\alpha}{ds^2} + \Gamma_{\beta\gamma}{}^\alpha \frac{du^\beta}{ds} \frac{du^\gamma}{ds} = \kappa_g \mu^\alpha.$$

If the geodesic curvature κ_g of C is everywhere zero, (89) gives

90) $$\frac{d^2 u^\alpha}{ds^2} + \Gamma_{\beta\gamma}{}^\alpha \frac{du^\beta}{ds} \frac{du^\gamma}{ds} = 0$$

as the differential equations to be satisfied. These are equations (25) of geodesics on the surface. Therefore, a geodesic curve has the characteristic property that its geodesic curvature is identically zero.

It should be noticed that the geodesic curvature at each point of curve C is zero if and only if the angle ψ is $\pi/2$ which means that the principal normal to C lies along the normal to the surface. An equivalent statement is that the osculating plane to C contains the normal to the surface at all points along a geodesic curve.

The differential equations of geodesics may be found therefore by demanding that the osculating plane of a curve $C: u^\alpha = u^\alpha(s)$ on S contain the normal to S. If $u^\alpha(s)$ are substituted into the Gauss equations $x^i = x^i(u^1, u^2)$ of S, the coordinates of any point on C are $x^i(s)$, and the osculating plane to C at a point P is given by (2-30) as

91) $$e_{ijk}(\xi^i - x^i) \frac{dx^j}{ds} \frac{d^2 x^k}{ds^2} = 0,$$

where ξ^i are current coordinates of a point in the plane. Any point on the normal to S at P has coordinates

92) $$x^i + pX^i,$$

where p is a parameter. Demand that these coordinates satisfy the equation (91) of the osculating plane to obtain the condition

93) $$e_{ijk}X^i \frac{dx^j}{ds}\frac{d^2x^k}{ds^2} = 0,$$

which is true for all values of p. By use of (32), (93) becomes

94) $$e_{ijk}X^i x^j{}_{,a}u^{a'}\left(x^k{}_{,\gamma}u^{\gamma''} + \frac{\partial^2 x^k}{\partial u^\tau \partial u^\sigma}u^{\tau'}u^{\sigma'}\right) = 0,$$

or, on writing (94) as the sum of two determinants,

95) $$(e_{ijk}X^i x^j{}_{,a}x^k{}_{,\gamma}u^{a'}u^{\gamma''} + e_{ijk}X^i x^j{}_{,a}(x^k{}_{,\gamma}\Gamma_{\tau\sigma}{}^\gamma + d_{\tau\sigma}X^k)u^{a'}u^{\tau'}u^{\sigma'} = 0,$$

where the Gauss equations (50) were used to express $\partial^2 x^k/\partial u^\tau \partial u^\sigma$ in terms of the base directions $x^k{}_{,\gamma}$ and X^k. Now because a determinant with two identical rows vanishes, (95) takes the form

96) $$(e_{ijk}X^i x^j{}_{,a}x^k{}_{,\gamma})u^{a'}(u^{\gamma''} + \Gamma_{\tau\sigma}{}^\gamma u^{\tau'}u^{\sigma'}) = 0.$$

The determinant in (96) is zero if $\alpha = \gamma = 1$, or $\alpha = \gamma = 2$, but not zero if $\alpha = 1, \gamma = 2$, or $\alpha = 2, \gamma = 1$. On summing on α and γ in (96) and dividing out the non-zero determinant, there results

97) $$e_{\alpha\beta}u^{a'}(u^{\beta''} + \Gamma_{\tau\sigma}{}^\beta u^{\tau'}u^{\sigma'}) = 0,$$

where $e_{11} = e_{22} = 0, e_{12} = 1, e_{21} = -1$.

Because (97) must hold for arbitrary direction $u^{a'}$, it follows that the curves with the property that the osculating plane contains the normal to the surface at every point satisfy the differential equations

98) $$\frac{d^2u^\alpha}{ds^2} + \Gamma_{\beta\gamma}{}^\alpha \frac{du^\beta}{ds}\frac{du^\gamma}{ds} = 0,$$

but these are the equations (25) of geodesics.

Definition: A doubly infinite set of lines is called a *congruence* of lines.

One instance of a congruence is the set of all lines which intersect both of two fixed lines in three-space. Another instance is the set of lines normal to a surface. Continuous functions of position on a surface can be employed to represent the direction cosines of the lines of a congruence (which is not necessarily normal to the surface). Because a congruence of *curves* may also be defined, one refers to a congruence of straight lines as a *rectilinear congruence*.

Given a congruence of lines not normal to the surface, it is possible to find the differential equations of the curves on the surface which enjoy the property that the osculating plane at each point contains the line of the congruence at the point. Curves with this property are called *union curves* relative to the given congruence. It is evident that geodesics are a particular case of union curves, for the union curves become geodesics if the congruence is normal to the surface. Many of the results concerning geodesics may be generalized to the case of union curves. The differential equations of union curves will be introduced in Section 12–9.

12–7. Geodesics as Auto-Parallel Curves. In Section 9–3 conditions were obtained for euclidean parallelism along a curve for vectors of a vector field in general coordinates in three-space. The components of the vector field, denoted by λ^i, and the coordinates $x^i = x^i(t)$ along the curve were seen to satisfy

$$99) \qquad \frac{d\lambda^j}{dt} + \lambda^r \Gamma_{rp}{}^j \frac{dx^p}{dt} = 0.$$

It was mentioned in Section 9–3 that vectors tangent to a curved surface at points of a curve on the surface are not parallel in the euclidean sense. However, a kind of parallelism can be defined for a curved space which reduces to euclidean parallelism in case the space is flat. Let λ^α be components of a vector field in a surface S and consider vectors of the field at points of the curve $C: u^\alpha = u^\alpha(s)$ on S. The tensor form (99) becomes

$$100) \qquad \frac{d\lambda^\alpha}{ds} + \lambda^\gamma \Gamma_{\beta\gamma}{}^\alpha \frac{du^\beta}{ds} = 0$$

for the surface, where the parameter t is taken as s. Equations (100) can be interpreted as the equations of parallelism on the surface. If λ^α satisfy equations (100) at all points along C, then λ^α are said to be *parallel* in the sense of Levi-Civita. It is understood that the vectors with components λ^α are not necessarily tangent to the curve C.

Another approach to equations (100) is the following: Let λ^α be components of the unit vectors of a field at points of C. (Note that λ^α are not restricted to being du^α/ds as in Section 6.) For the unit vector one has $g_{\alpha\beta}\lambda^\alpha\lambda^\beta = 1$, and the intrinsic derivative of this equation yields

$$101) \qquad g_{\alpha\beta}\lambda^\alpha \frac{\delta\lambda^\beta}{\delta s} = 0,$$

which shows that $\delta\lambda^\beta/\delta s$ is a surface vector which is normal to the vector with components λ^α. If r is the length of the vector $\delta\lambda^\beta/\delta s$, then

$$102) \qquad \frac{\delta\lambda^\beta}{\delta s} = r\nu^\beta,$$

where ν^β is a unit surface vector normal to λ^α. Bianchi called ν^β the vector *associate* to λ^α with respect to the curve C. If it happens that λ^α is tangent to C ($\lambda^\alpha \equiv du^\alpha/ds$), then by (85), $|r| = \kappa_g$, the geodesic curvature of C at the point. If, moreover, C is a geodesic, then the vector $\dfrac{\delta}{\delta s}\dfrac{du^\alpha}{ds}$ is a zero vector at every point of C.

Consider next two unit vector fields, one being that of the unit tangent vectors $\lambda^\alpha \equiv du^\alpha/ds$ to C, and the other an arbitrary field with components μ^α along C. The angle θ between λ^α and μ^α is given by

$$103) \qquad \cos\theta = g_{\alpha\beta}\lambda^\alpha\mu^\beta.$$

Differentiate (103) intrinsically with respect to s to obtain

$$-\sin\theta\,\frac{\delta\theta}{\delta s} = g_{\alpha\beta}\left(\lambda^\alpha\frac{\delta\mu^\beta}{\delta s} + \mu^\beta\frac{\delta\lambda^\alpha}{\delta s}\right).$$

If the angle θ is to remain constant along C, then $\delta\theta/\delta s = 0$, which requires that

$$g_{\alpha\beta}\left(\lambda^\alpha\frac{\delta\mu^\beta}{\delta s} + \mu^\beta\frac{\delta\lambda^\alpha}{\delta s}\right) = 0,$$

or

$$104) \qquad \lambda_\beta\frac{\delta\mu^\beta}{\delta s} + \mu_\alpha\kappa_g\nu^\alpha = 0,$$

by virtue of the fact that λ^α is tangent to C. From (104) it is clear that if C is a geodesic ($\kappa_g = 0$), the condition for θ to be constant requires that

$$105) \qquad \frac{\delta\mu^\beta}{\delta s} = 0.$$

If the surface is a plane, then C is a straight line, and condition (105) means that the vectors μ^β are parallel in the euclidean sense. Because geodesics in a curved space are analogous to straight lines in euclidean space, and because condition (105) ensures that the vectors μ^α make equal angles with a geodesic curve, it is natural to define parallelism of vectors by the condition (105). One may make a further generalization and say with Levi-Civita that if a vector μ^α moves along a curve and satisfies (105) then it moves in a parallel manner.

Consider next λ^α and μ^α as unit vectors of any two vector fields defined along an arbitrary curve C. Let the vectors λ^α and μ^α be moved along C in a parallel manner, that is, such that

$$106) \qquad \frac{\delta\lambda^\alpha}{\delta s} = 0, \qquad \frac{\delta\mu^\alpha}{\delta s} = 0.$$

GEODESIC AND UNION CURVES

Let ϕ be the angle between the two vectors at any point of C. Then
$$\cos \phi = g_{\alpha\beta}\lambda^\alpha \mu^\beta,$$
and
$$-\sin \phi \frac{\delta\phi}{\delta s} = g_{\alpha\beta}\lambda^\alpha \frac{\delta\mu^\beta}{\delta s} + g_{\alpha\beta}\mu^\beta \frac{\delta\lambda^\alpha}{\delta s} = 0,$$

so that ϕ remains constant. Hence, it may be stated that if two vectors are each moved in a parallel manner along a curve, the angle between the vectors remains constant.

Observe that if the *tangent* vector du^α/ds of a curve C is moved along C in a parallel manner, then
$$\frac{\delta}{\delta s} \frac{du^\alpha}{ds} = 0,$$
which, by (100), means that
$$\frac{d^2 u^\alpha}{ds^2} + \Gamma_{\beta\gamma}{}^\alpha \frac{du^\beta}{ds} \frac{du^\gamma}{ds} = 0.$$

Hence, C is a geodesic curve. A straight line in euclidean space is a curve which preserves its direction. A geodesic in a curved space preserves its direction also from the point of view that its tangent moves in a parallel manner. Thus, because the tangent to a geodesic moves parallel to itself, geodesics may be styled *auto-parallel* curves.

It should be pointed out that the condition (100) of parallelism is relative to a curve. If a vector is transported in a parallel manner along a curve C_1 joining two points A and B on a surface, and if the same vector at A is transported parallelly along another curve C_2 to B, the resulting vectors at B are not the same, in general. It can be proved that they are the same if the space is euclidean.

FRENET-SERRET FORMULAS. The next objective is to obtain for a surface curve a set of formulas analogous to the Frenet-Serret formulas in general coordinates for a curve in a two-dimensional linear space (see 10–26 with i and l ranging over 1, 2).

For a vector λ^α of a field at points of a curve C, (101) states that the associate vector ν^α to λ^α is perpendicular to λ^α, that is

107) $$g_{\alpha\beta}\lambda^\alpha \nu^\beta = 0.$$

Differentiation of equation (107) gives

108) $$g_{\alpha\beta}\left(\frac{\delta\lambda^\alpha}{\delta s}\nu^\beta + \lambda^\alpha \frac{\delta\nu^\beta}{\delta s}\right) = 0,$$

or
$$g_{\alpha\beta}(r\nu^\alpha \nu^\beta) + g_{\alpha\beta}\lambda^\alpha \frac{\delta\nu^\beta}{\delta s} = 0,$$

by use of (101). Since ν^α is a unit vector, $g_{\alpha\beta}\nu^\alpha\nu^\beta = 1$, and therefore

$$g_{\alpha\beta}\lambda^\alpha \frac{\delta\nu^\beta}{\delta s} = -r,$$

or

109) $$\lambda_\beta \frac{\delta\nu^\beta}{\delta s} = -r.$$

Now ν^α is a unit vector and $-\lambda^\alpha$ is at right angles to it, so with ν^α playing the role of λ^α in (101),

110) $$\frac{\delta\nu^\beta}{\delta s} = -h\lambda^\beta,$$

where h can be determined by multiplying (110) by λ^β to obtain

111) $$\lambda_\beta \frac{\delta\nu^\beta}{\delta s} = -h\lambda_\beta\lambda^\beta = -h = -r,$$

by virtue of (109). Hence, from (101) and (110),

112) $$\frac{\delta\lambda^\alpha}{\delta s} = r\nu^\alpha, \qquad \frac{\delta\nu^\alpha}{\delta s} = -r\lambda^\alpha,$$

which hold for any vector field along a curve. In particular, if $\lambda^\alpha = du^\alpha/ds$, the vector ν^α is along the normal to the curve in the surface. Denote it by μ^α. One has then

113) $$\frac{\delta\lambda^\alpha}{\delta s} = \kappa_g \mu^\alpha, \qquad \frac{\delta\mu^\alpha}{\delta s} = -\kappa_g \lambda^\alpha,$$

which should be compared with the Frenet-Serret formulas (8–17) and (10–26) for the case of a plane curve (torsion = 0).

GAUSS-BONNET THEOREM. This section will be concluded by exhibiting a result in integral geometry which will be achieved by starting with the local concept of geodesic curvature and then applying Stokes' theorem on a Riemannian space of two dimensions (Exercise 6 in Section 11–11). Let λ^α be a vector which will be transported in a parallel manner around a closed curve on a surface. A measure is desired for the angle between the vector λ^α at a starting point and the vector at this point after transport around the closed curve. Consider first the angle θ which λ^α makes with a u^1-curve (on which u^2 = constant). Assume that λ^α is a unit vector. The unit vector tangent to the u^1-curve has components $(du^1/ds, 0)$. Therefore, the angle θ between λ^α and the tangent to the u^1-curve is given by

$$\cos\theta = g_{11}\lambda^1 \frac{du^1}{ds} = g_{11} \frac{\lambda^1}{\sqrt{g_{11}}} = \sqrt{g_{11}}\,\lambda^1.$$

Without loss of generality, the coordinate curves may be taken as orthogonal ($g_{12} = g_{21} = 0$). It follows that $\sin \theta = \sqrt{g_{22}}\, \lambda^2$, so that

$$\lambda^1 = \frac{\cos \theta}{\sqrt{g_{11}}}, \qquad \lambda^2 = \frac{\sin \theta}{\sqrt{g_{22}}}.$$

Now, because the vector λ^α is transported parallelly,

114) $$\frac{\delta \lambda^\alpha}{\delta s} \equiv \frac{d\lambda^\alpha}{ds} + \lambda^\beta \Gamma_{\beta\gamma}{}^\alpha \frac{du^\gamma}{ds} = 0.$$

It can be shown that a coordinate system can be chosen for which $ds^2 = (du^1)^2 + g_{22}(du^2)^2$. Let this be done in order to reduce the labor of calculation. For $\alpha = 1$, the first equation of parallelism becomes

$$\frac{\delta \lambda^1}{\delta s} \equiv \frac{d\lambda^1}{ds} + \lambda^\beta \Gamma_{\beta\gamma}{}^1 \frac{du^\gamma}{ds} = 0,$$

or

$$-\sin \theta \frac{d\theta}{ds} + \lambda^\beta \Gamma_{\beta\gamma}{}^1 \frac{du^\gamma}{ds} = 0,$$

which, after writing $du^1/ds = 1/\sqrt{g_{11}} = 1$, and expressing λ^1 and λ^2 in terms of $\cos \theta$ and $\sin \theta$, respectively, reduces to

$$\frac{d\theta}{ds} = -\frac{1}{2\sqrt{g_{22}}} \frac{\partial g_{22}}{\partial u^1} \frac{du^2}{ds}.$$

By integrating around the closed contour Γ which is within the neighborhood in which the chosen coordinate system is valid, one finds

$$\int_\Gamma \frac{d\theta}{ds} ds = -\int_\Gamma \frac{\partial \sqrt{g_{22}}}{\partial u^1} \frac{du^2}{ds} ds.$$

If the region within Γ is denoted by $\mathcal{R}$, application of Stokes' theorem to the last integral yields

$$-\int_\Gamma \frac{\partial \sqrt{g_{22}}}{\partial u^1} du^2 = -\iint_{\mathcal{R}} \frac{\partial^2 \sqrt{g_{22}}}{\partial u^1\, \partial u^1} du^1\, du^2$$

$$= -\iint_{\mathcal{R}} \frac{1}{\sqrt{g_{22}}} \frac{\partial^2 \sqrt{g_{22}}}{(\partial u^1)^2} d\sigma.$$

Now, the formula (73) for R_{1212} (with $g_{11} = 1$, $g_{12} = 0$) yields

$$\kappa = \frac{R_{1212}}{\sqrt{g}} = \frac{R_{1212}}{\sqrt{g_{22}}} = -\frac{1}{\sqrt{g_{22}}} \frac{\partial^2 \sqrt{g_{22}}}{\partial u^1\, \partial u^1},$$

so it follows that

$$\int_\Gamma \frac{d\theta}{ds}\, ds = \iint_\mathcal{R} \kappa\, d\sigma.$$

The desired measure of the change in the direction of the vector λ^α on being transported parallelly around a closed circuit is found in the form of the surface integral of the Gaussian curvature over the region enclosed by the circuit.

Next, let the vector λ^α be the unit tangent to the circuit Γ which will be assumed to have no corners. The vector is integrated around the circuit but is not transported parallelly on Γ. In this case

$$\frac{\delta \lambda^1}{\delta s} = \kappa_g \mu^1 = -\kappa_g \sin\theta$$

from the first Frenet formula in (113). [Note that $\mu^1 = \cos(\pi/2 + \theta) = -\sin\theta$.] From the definition of intrinsic derivative (114) one has

$$-\kappa_g \sin\theta = -\sin\theta \frac{d\theta}{ds} + \frac{\partial \sqrt{g_{22}}}{\partial u^1} \sin\theta \frac{du^2}{ds}$$

or, on integrating,

$$\int_\Gamma \kappa_g\, ds = \int_\Gamma d\theta - \int_\Gamma \frac{\partial \sqrt{g_{22}}}{\partial u^1} \frac{du^2}{ds} = \int_\Gamma d\theta - \iint_\mathcal{R} \kappa\, d\sigma.$$

Since there are no corners $\int_\Gamma d\theta = 2\pi$, so the result may be stated as

$$\int_\Gamma \kappa_g\, ds + \iint_\mathcal{R} \kappa\, d\sigma = 2\pi.$$

At a corner with exterior angle θ_1, the tangent vector λ^α makes a turn through the positive angle θ_1. Therefore if there are m corners with exterior angles θ_i the result becomes

$$\int_\Gamma \kappa_g\, ds + \iint_\mathcal{R} \kappa\, d\sigma = 2\pi - \sum_{i=1}^{m} \theta_i.$$

This proves the remarkable

Gauss-Bonnet Theorem: If the Gaussian curvature κ is continuous over a simply connected region $\mathcal{R}$ of a surface, and the geodesic curvature κ_g of the boundary Γ is continuous, then the integral of the geodesic curvature around Γ plus the integral of the Gaussian curvature over $\mathcal{R}$ is equal to 2π less the sum of the exterior angles of Γ, if any.

For a triangular region with interior angles A_1, A_2, A_3 the theorem gives the result

$$\int_\Gamma \kappa_g \, ds + \iint_\mathcal{R} \kappa \, d\sigma = A_1 + A_2 + A_3 - \pi.$$

For the particular case of a geodesic triangle ($\kappa_g = 0$) the result reduces to

$$\iint_\mathcal{R} \kappa \, d\sigma = A_1 + A_2 + A_3 - \pi,$$

which is called the *excess* of the triangle.

12-8. A Generalization of the Theorem of Meusnier. The purpose here is to introduce further elements of the theory of surfaces, some of which will be useful later, and to show how the theorem of Meusnier can be deduced from the generalization obtained.

Consider a congruence of lines with one line of the congruence passing through each point P of the surface S. At each point P the direction cosines λ^i of the line l of the congruence through the point are given as continuous functions of the coordinates u^α of P. The congruence may be said to be *referred* to the surface. Let the line l make an angle θ with the unit normal to the surface at P. The unit vector with components λ^i can be expressed as a linear combination of the surface vectors $x^i_{,\alpha}$ and the unit normal vector X^i in the form

115) $$\lambda^i = p^\alpha x^i_{,\alpha} + q X^i.$$

Because $X^i x^i_{,\alpha} = 0$ ($\alpha = 1, 2$), multiplication of (115) by X^i shows that $q = \lambda^i X^i = \cos \theta$. Because $\lambda^i \lambda^i = 1$, it follows that p^α satisfy $g_{\alpha\beta} p^\alpha p^\beta + q^2 = 1$, and $p^\alpha \sqrt{g_{\alpha\alpha}}$ are the projections of the unit vector with components λ^i upon the tangents to the coordinate curves on the surface at P. If ω is the angle between the line l of the congruence and the principal normal to a curve C through P on S, one finds

116) $$\cos \omega = \lambda^i \beta^i = \lambda^i \left(\rho \frac{d^2 x^i}{ds^2} \right)$$

$$= \rho (p^\sigma x^i_{,\sigma} + q X^i) \left(\frac{\partial^2 x^i}{\partial u^\alpha \partial u^\beta} u^{\alpha\prime} u^{\beta\prime} + x^i_{,\alpha} u^{\alpha\prime\prime} \right),$$

where ρ is the radius of curvature of C at P, and use was made of (31) for $x^{i\prime\prime}$.

Use of the Gauss equations (50) allows (116) to be written as

117) $$\frac{\cos \omega}{\rho} = (p^\sigma x^i{}_{,\sigma} + qX^i)[(\Gamma_{\alpha\beta}{}^\gamma x^i{}_{,\gamma} + d_{\alpha\beta}X^i)u^{\alpha'}u^{\beta'} + x^i{}_{,\alpha}u^{\alpha''}]$$
$$= p^\sigma g_{\sigma\gamma}\Gamma_{\alpha\beta}{}^\gamma u^{\alpha'}u^{\beta'} + q\, d_{\alpha\beta}u^{\alpha'}u^{\beta'} + p^\sigma g_{\sigma\alpha}u^{\alpha''}$$
$$= p_\gamma \Gamma_{\alpha\beta}{}^\gamma u^{\alpha'}u^{\beta'} + p_\alpha u^{\alpha''} + q\, d_{\alpha\beta}u^{\alpha'}u^{\beta'}$$
$$= p_\gamma(u^{\gamma''} + \Gamma_{\alpha\beta}{}^\gamma u^{\alpha'}u^{\beta'}) + q\,\frac{d_{\alpha\beta}\, du^\alpha\, du^\beta}{g_{\alpha\beta}\, du^\alpha\, du^\beta}.$$

By use of equation (58) for the normal curvature for the direction $du^1:du^2$ of the curve C, and with $q = \cos\theta$, equation (117) reads

118) $$\frac{\cos\omega}{\rho} = \frac{\cos\theta}{R} + p_\gamma(u^{\gamma''} + \Gamma_{\alpha\beta}{}^\gamma u^{\alpha'}u^{\beta'})$$
$$= \frac{\cos\theta}{R} + p_\gamma \frac{\delta}{\delta s}\left(\frac{du^\alpha}{ds}\right) = \frac{\cos\theta}{R} + \kappa_g p_\gamma \mu^\gamma,$$

by virtue of (113). By $g_{\alpha\beta}p^\alpha p^\beta = 1 - q^2 = 1 - \cos^2\theta = \sin^2\theta$, the length of the projection p of the vector λ^i upon the tangent plane to S at P is $\sin\theta$. If ψ is the angle between this vector and the normal μ^α to C in the tangent plane, the projection of p upon the normal to C is $\sin\theta\cos\psi$. Hence, (118) becomes

119) $$k \cos\omega = \kappa_n \cos\theta + \kappa_g \cos\psi \sin\theta,$$

where $k \equiv 1/\rho$ is the first curvature, κ_n the normal curvature, and κ_g the geodesic curvature of C at the point P on the surface. Equation (119) may be described by the following

> **Theorem:** If vectors of lengths k, κ_n, κ_g are laid off, respectively, along the principal normal to a curve C at P on a surface S, the normal to S, and the normal to C in the surface, the projection of k on any line l through P is equal to the sum of the projections of κ_n and κ_g on the line.

Another way to state the theorem is that the vector equation **k** = **κ**$_n$ + **κ**$_g$ holds. Observe that the formula (119) gives a means of calculating κ_g (say) from the orientation angles θ, ω, and ψ of the line l.

It should be noted that if $\theta = 0$, in (119), the congruence is normal to S, and Meusnier's theorem (80) results. Other particular cases arise with $\psi = 0$, or $\pi/2$.

The cosine of the angle between l and the tangent to C at P is readily found to be $\cos\phi = \lambda^i(dx^i/ds) = p_\tau(du^\tau/ds)$. Next, the angle δ between l with direction λ^i and the binormal to C at P with direction γ^i will be calculated. One finds

$$\cos \delta = \lambda^i \gamma^i = \lambda^i e_{ijk}\alpha^j \beta^k = \rho e_{ijk}\lambda^i \frac{dx^j}{ds}\frac{d^2x^k}{ds^2}$$

$$= \rho e_{ijk}(p^\tau x^i{}_{,\tau} + qX^i)(x^j{}_{,\sigma}u^{\sigma'})\left(\frac{\partial^2 x^k}{\partial u^\beta \partial u^\alpha}u^{\alpha'}u^{\beta'} + x^k{}_{,\gamma}u^{\gamma''}\right),$$

by means of (32). By use of (50),

$$\frac{\cos \delta}{\rho} = e_{ijk}(p^\tau x^i{}_{,\tau} + qX^i)(x^j{}_{,\sigma}u^{\sigma'})(\Gamma_{\alpha\beta}{}^\gamma x^k{}_{,\gamma}u^{\alpha'}u^{\beta'} + d_{\alpha\beta}u^{\alpha'}u^{\beta'}X^k + x^k{}_{,\gamma}u^{\gamma''})$$

$$= e_{ijk}(x^i{}_{,\tau}x^j{}_{,\sigma}X^k)(p^\tau u^{\sigma'}\kappa_n + qu^{\tau'}\Gamma_{\alpha\beta}{}^\sigma u^{\alpha'}u^{\beta'} + qu^{\tau'}u^{\sigma''}).$$

It can be verified that the determinant $e_{ijk}x^i{}_{,\tau}x^j{}_{,\sigma}X^k = \sqrt{g}\,e_{\tau\sigma} = \epsilon_{\tau\sigma}$, where $\epsilon_{\tau\sigma}$ is a surface tensor (see Exercise 6 in Section 11–11). With this simplification, $\cos \delta/\rho$ is given by

$$\frac{\cos \delta}{\rho} = \epsilon_{\tau\sigma}[\kappa_n p^\tau u^{\sigma'} + qu^{\tau'}(\Gamma_{\alpha\beta}{}^\sigma u^{\alpha'}u^{\beta'} + u^{\sigma''})]$$

$$= \epsilon_{\tau\sigma}\left[\kappa_n p^\tau u^{\sigma'} + qu^{\tau'}\frac{\delta}{\delta s}\left(\frac{du^\sigma}{ds}\right)\right]$$

$$= \epsilon_{\tau\sigma}[-p^\sigma \kappa_n + q\kappa_g \mu^\sigma]u^{\tau'}.$$

Now define new parameters l^σ for the congruence by means of $p^\sigma = l^\sigma q$ (and remember that $q = \cos \theta$) to arrive at

$$\frac{\cos \delta}{\rho} = \cos \theta \, \epsilon_{\tau\sigma} u^{\tau'}(\kappa_g \mu^\sigma - \kappa_n l^\sigma).$$

Let

$$\kappa_u \equiv \frac{1}{\rho_u} \equiv \epsilon_{\tau\sigma} u^{\tau'}(\kappa_g \mu^\sigma - \kappa_n l^\sigma),$$

and call this invariant expression the *union curvature of the curve C relative to the congruence*.[4] One may write

$$\frac{\cos \delta}{\rho} = \frac{\cos \theta}{\rho_u},$$

or

$$\kappa \cos \delta = \kappa_u \cos \theta.$$

The union curvature κ_u of C at P can be calculated by means of the last equation if the curvature k of C at P, and the angles θ and δ are known. Note that if the line l of the congruence makes equal angles (θ and δ)

[4] C. E. Springer, Union Curves and Union Curvature, *Bulletin Amer. Math. Soc.*, vol. 51, pp. 686–691.

with the normal to S at P and the binormal to C at P, then the union curvature is numerically equal to the ordinary curvature of C at P.

Just as geodesic curves on S were defined as curves with geodesic curvature identically zero, curves for which the union curvature is identically zero may be called *union curves*. However, a different property of union curves will be used to define them in the next section.

12–9. Union Curves on a Surface. Consider a congruence of lines referred to a surface S. The direction cosines λ^i of the line l of the congruence through point P on S are given by (115). A *union curve* on S relative to the congruence has the property that its osculating plane at each point of the curve contains the line l of the congruence through the point. If the lines of the congruence are normal to the surface, the union curves become geodesics. This was seen in Section 12–6. The differential equations of union curves on S may be developed as follows.

The osculating plane to C: $u^\alpha = u^\alpha(s)$ on S at $P(x^i)$ has the determinantal equation (2–30)

$$120) \qquad e_{ijk}(\xi^i - x^i)\frac{dx^j}{ds}\frac{d^2x^k}{ds^2} = 0,$$

with ξ^i as current coordinates. On making use of equations (32), and in turn, the Gauss equations (50), (120) can be written in the form

$$121) \qquad e_{ijk}(\xi^i - x^i)(x^j{}_{,\sigma}u^{\sigma\prime})(\rho^\tau x^k{}_{,\tau} + d_{\alpha\beta}u^{\alpha\prime}u^{\beta\prime}X^k) = 0,$$

in which ρ^τ are components of the curvature vector given by

$$122) \qquad \rho^\tau \equiv \kappa_g \mu^\tau = u^{\tau\prime\prime} + \Gamma^\tau_{\alpha\beta}u^{\alpha\prime}u^{\beta\prime}.$$

If the osculating plane (121) to C at P contains the line l, the coordinates $x^i + t\lambda^i \equiv x^i + t(p^\gamma x^i{}_{,\gamma} + qX^i)$ of any point on l must satisfy equation (121) identically in t. Hence, the equation

$$123) \qquad e_{ijk}(p^\gamma x^i{}_{,\gamma} + qX^i)(x^j{}_{,\sigma}u^{\sigma\prime})(\rho^\tau x^k{}_{,\tau} + d_{\alpha\beta}u^{\alpha\prime}u^{\beta\prime}X^k) = 0$$

must obtain. Use of the fact that

$$e_{ijk}x^i{}_{,\gamma}x^j{}_{,\sigma}x^k{}_{,\tau} \equiv 0$$

because γ, σ, τ (each ranging over 1, 2) cannot all be different, and the fact that

$$e_{ijk}X^i x^j{}_{,\sigma}X^k \equiv 0,$$

together with a change of umbral indices, allows equation (123) to take the form

$$124) \qquad (e_{ijk}X^i x^j{}_{,\sigma}x^k{}_{,\tau})(p^\sigma u^{\tau\prime}d_{\alpha\beta}u^{\alpha\prime}u^{\beta\prime} + qu^{\sigma\prime}\rho^\tau) = 0.$$

Recall from (57) that

$$d_{\alpha\beta}u^{\alpha'}u^{\beta'} \equiv \frac{d_{\alpha\beta}\,du^{\alpha}\,du^{\beta}}{g_{\alpha\beta}\,du^{\alpha}\,du^{\beta}} = \kappa_n,$$

the normal curvature of the curve C with direction $u^{\alpha'}$ on S.

Summing on σ and τ in (124), and neglecting the non-zero determinant

$$e_{ijk}X^{i}x^{j}_{,1}x^{k}_{,2},$$

lead to

125) $\quad\quad\quad \epsilon_{\sigma\tau}(p^{\sigma}u^{\tau'}\kappa_n + qu^{\sigma'}\rho^{\tau}) = 0.$

Equation (125) is a differential equation of the second order [by virtue of the definition (122) of ρ^{τ}], and it defines the union curves on S, the surface coordinates being arbitrary. It can be concluded from (125) that if $p^{\sigma} = 0$ ($\sigma = 1, 2$), the resulting equation is

$$\epsilon_{\sigma\tau}u^{\sigma'}\rho^{\tau} = 0,$$

that is, equation (97) which is satisfied by the geodesics. If $q \equiv \cos\theta = 0$ in (125), then

$$\kappa_n \epsilon_{\sigma\tau} p^{\sigma} u^{\tau'} = 0,$$

which means that the line l of the congruence lies in the tangent plane to S at P. In this case, the only union curves are in the direction given by $\epsilon_{\sigma\tau}p^{\sigma}\,du^{\tau} = 0$, and in the directions for which $\kappa_n = 0$, that is, the asymptotic directions given by $d_{\alpha\beta}\,du^{\alpha}\,du^{\beta} = 0$.

Suppose that $q \neq 0$, and write $l^{\sigma} \equiv p^{\sigma}/q$, so l^{σ} may be taken as the two essential parameters of the congruence at any point P on S. With this notation, after an interchange of σ and τ in one of the terms of (125), equation (125) becomes

126) $\quad\quad\quad \epsilon_{\sigma\tau}u^{\sigma'}(\rho^{\tau} - \kappa_n l^{\tau}) = 0.$

Equation (126) is a single differential equation of the second order, but it can be shown to be equivalent to a pair of differential equations each of the second order. To do this, make use of the fact that

127) $\quad\quad\quad g_{\alpha\beta}\rho^{\alpha}u^{\beta'} = 0,$

which obtains because the curvature vector with components ρ^{α} is orthogonal to the direction $u^{\alpha'}$ of C on S. Multiply (127) by $u^{1'}$ and (126) by $g_{2\beta}u^{\beta'}$, subtract and use the fact that $g_{\alpha\beta}u^{\alpha'}u^{\beta'} = 1$ to obtain the first of the following differential equations of union curves on S,

128) $\quad\quad \eta^{1} \equiv \rho^{1} - \kappa_n g_{2\beta}u^{\beta'}e_{\sigma\tau}l^{\sigma}u^{\tau'} = 0,$

$\quad\quad\quad\quad \eta^{2} \equiv \rho^{2} + \kappa_n g_{1\beta}u^{\beta'}e_{\sigma\tau}l^{\sigma}u^{\tau'} = 0.$

The second of the equations (128) is obtained similarly, by interchanging the roles of 1 and 2. The vector with components η^α given by (128) lies in the tangent plane to S at P. It may be called the *union curvature* vector of C at P. A union curve has the property that its union curvature vector is a zero vector at each point of the curve. Equations (128) may be compressed into the single equation

$$\eta^\gamma = \rho^\gamma + \kappa_n e^{\alpha\gamma} g_{\alpha\beta} u^{\beta'} e_{\sigma\tau} l^\sigma u^{\tau'}.$$

It may be verified that $g_{\alpha\beta} u^{\alpha'} \eta^\beta = 0$, which means that the union curvature vector of a curve on a surface is normal to the curve.

Because the geodesic curvature κ_g of a curve C at P is given by

$$\kappa \equiv \epsilon_{\alpha\beta} u^{\alpha'} \rho^\beta,$$

it appears appropriate to define the union curvature κ_u of C at P by

$$\kappa_u \equiv \epsilon_{\alpha\beta} u^{\alpha'} \eta^\beta,$$

which may be written, by use of (127) and (128), in the form

$$\kappa_u \equiv \epsilon_{\sigma\tau} u^{\sigma'} (\rho^\tau - \kappa_n l^\tau) = \kappa_g - \kappa_n \epsilon_{\sigma\tau} u^{\sigma'} l^\tau.$$

Note that along a union curve, ($\kappa_u = 0$), the geodesic curvature is given by

$$\kappa_g = \kappa_n \epsilon_{\sigma\tau} u^{\sigma'} l^\tau.$$

The following geometric interpretation of union curvature can be shown to be valid.

> The union curvature of a curve C at a point P on a surface S relative to a given rectilinear congruence is the curvature of the plane curve obtained by projecting C onto the tangent plane to S at P, the direction of projection being that of the line l of the congruence at P.

It should be remarked that union curves are of extrinsic character while geodesics are intrinsic. This is seen from the fact that the differential equations of union curves contain the components $d_{\alpha\beta}$ of the second fundamental tensor (in the formula for κ_n), whereas these components do not appear in the differential equations of geodesics.

It is important to realize that there is freedom of choice of the congruence. The parameters l^α may be any continuous functions of u^α. Hence, a particular congruence may be specified to satisfy some condition. For instance, one may require the congruence for which the coordinate curves on S are union curves relative to the congruence. In order to determine the l^α which satisfy this condition, write the differential equation (126) with the expressions (122) and (58) for ρ^τ and κ_n inserted. Then for the u^1-curve, set $u^{2'} = 0$, and for the u^2-curve set $u^{1'} = 0$. This procedure yields

129) $$l^1 = \frac{\Gamma_{22}^{\ 1}}{d_{22}}, \qquad l^2 = \frac{\Gamma_{11}^{\ 2}}{d_{11}}.$$

Another instance in which l^α are chosen to satisfy a condition will be seen in the following section.

12–10. Union Curves and Dynamical Trajectories. The motion of a particle on a plane or on a curved surface may be described by a set of differential equations. The path of the particle is determined by the solution for the dependent variables as functions of an independent variable or parameter. Solution paths are called *trajectories*. In the Lagrange equations (Section 10–9), the dependent variables appear as generalized coordinates which are equal in number to the number of degrees of freedom of the dynamical system. A set of k values for the k generalized coordinates determines one possible configuration or state of the system. Although these coordinates may not be actual space coordinates in the usual sense, they may be interpreted as space variables in a k-dimensional space, which may be a Riemannian space. A solution giving the k generalized coordinates as functions of a parameter, say, the time, is called a *trajectory* of the dynamical system.

A paper[5] by Eisenhart on geodesics and dynamical trajectories appeared in 1929 in which he showed that the trajectories of a dynamical system with n degrees of freedom correspond to the geodesics of a Riemannian space of $n + 1$ dimensions in case the potential function is independent of the time. He showed further that if the potential function involves the time, then the trajectories of the system correspond (or are given by) the geodesics of a Riemannian space of $n + 2$ dimensions. It will be shown below that the dimensionality $n + 1$ can be reduced to n for the case in which the potential function does not involve the time, provided that geodesics in the Riemannian space are replaced by union curves relative to a certain congruence. That is, a congruence will be found for which the union curves with respect to it correspond to the trajectories of the dynamical system.

To gain simplicity in the development, a Riemannian space of two dimensions (a surface) will be used. However, the union curves for a Riemannian space of n dimensions can be obtained,[6] and the argument for n dimensions follows closely that for $n = 2$.

First, the union curves for the Riemannian space $(n = 2)$ will be obtained from a mechanical principle instead of the geometric condition

[5] L. P. Eisenhart, Geodesics and Dynamical Trajectories, *Annals of Math.*, Vol. 30, 1929, pp. 591–606.

[6] C. E. Springer, Union Curves of a Hypersurface, *Canadian Jour. of Math.*, Vol. II, 1950, pp. 457–460.

used in Section 9. Consider a particle of mass m constrained to move on a surface S which lies in a field of force. A force of magnitude F acts at each point of the surface. Consider also the reaction R of the surface against the particle. As usual, let S be represented by $x^i = x^i(u^1, u^2)$, the unit normal to S by $X^i(u^1, u^2)$, the unit vector along the line of action of F by $\lambda^i(u^1, u^2)$, and the unit vector in the direction of R by $\mu^i(u^1, u^2)$. Note that F and R are scalar functions of position. The equations of motion of the particle are

130) $$m \frac{d^2 x^i}{dt^2} = F\lambda^i + R\mu^i.$$

Because the particle is on the surface, the direction cosines λ^i and μ^i may be expressed by

131) $$\lambda^i = p^\alpha x^i{}_{,\alpha} + q X^i, \qquad \mu^i = v^\alpha x^i{}_{,\alpha} + w X^i,$$

where $x^i{}_{,\alpha}$ is the partial derivative of x^i with respect to u^α. It is readily seen that

132) $$\frac{d^2 x^i}{dt^2} = \frac{d^2 x^i}{ds^2}\left(\frac{ds}{dt}\right)^2 + \frac{dx^i}{ds}\frac{d^2 s}{dt^2},$$

where s is arc length and t the time. On using in (130) the formulas (131) and (132), with dx^i/ds and $d^2 x^i/ds^2$ replaced by their equivalents in equations (32) and (50), there results

133) $$m\left\{\frac{d^2 s}{dt^2} x^i{}_{,\sigma} u^{\sigma\prime} + \left(\frac{ds}{dt}\right)^2 [\rho^\gamma x^i{}_{,\gamma} + \kappa_n X^i]\right\}$$
$$= F(p^\alpha x^i{}_{,\alpha} + qX^i) + R(v^\alpha x^i{}_{,\alpha} + wX^i).$$

Multiplication of (133) by X^i and use of the facts that $X^i x^i{}_{,\alpha} = 0$, $X^i X^i = 1$, lead to

134) $$m\kappa_n \left(\frac{ds}{dt}\right)^2 = Fq + Rw.$$

It should be noticed here that if the component of the force field vector normal to the surface is known, and if the velocity ds/dt of the particle along its path is known, then the component of the reaction vector normal to the surface can be found from (134).

Next, multiply equation (133) by $x^i{}_{,\tau}$ and use $x^i{}_{,\sigma} x^i{}_{,\tau} \equiv g_{\sigma\tau}$, (where $ds^2 = g_{\alpha\beta}\, du^\alpha\, du^\beta$ gives the linear element on the surface) to obtain

135) $$m\left\{\frac{d^2 s}{dt^2} g_{\sigma\tau} u^{\sigma\prime} + \left(\frac{ds}{dt}\right)^2 \rho^\gamma g_{\gamma\tau}\right\} = Fp^\alpha g_{\alpha\tau} + Rv^\alpha g_{\alpha\tau} = Fp_\tau + Rv_\tau,$$

or

136) $$mg_{\gamma\tau}\left(\frac{ds}{dt}\right)^2 \left[u^{\gamma\prime} \frac{d^2 s/dt^2}{(ds/dt)^2} + \rho^\gamma\right] = Fp_\tau + Rv_\tau.$$

Use (134) to write (136) in the form

$$137) \qquad g_{\gamma\tau}\left[\rho^\gamma + u^{\gamma\prime}\frac{d^2s/dt^2}{(ds/dt)^2}\right] = \frac{Fp_\tau + Rv_\tau}{Fq + Rw}\kappa_n,$$

or, by multiplying by $g^{\tau\sigma}$, in the form

$$138) \qquad \rho^\sigma + u^{\sigma\prime}\frac{d^2s/dt^2}{(ds/dt)^2} = \frac{Fp^\sigma + Rv^\sigma}{Fq + Rw}\kappa_n.$$

Eliminate the second derivative of s from the two equations (138) to arrive at the single equation

$$139) \qquad \rho^1\frac{du^2}{ds} - \rho^2\frac{du^1}{ds} = \frac{\kappa_n}{Fq + Rw}\left[(Fp^1 + Rv^1)\frac{du^2}{ds} - (Fp^2 + Rv^2)\frac{du^1}{ds}\right],$$

or, in another form,

$$140) \qquad \frac{du^1}{ds}(\rho^2 - \kappa_n l^2) - \frac{du^2}{ds}(\rho^1 - \kappa_n l^1) = 0,$$

where l^α are defined by

$$141) \qquad l^\alpha = \frac{Fp^\alpha + Rv^\alpha}{Fq + Rw}.$$

Equation (140) is the differential equation of motion of the particle on the surface under the specified forces. A congruence of lines relative to the surface is determined by the parameters l^α in (141), and (140) is satisfied by the union curves relative to this congruence (see 126). Note that if the resultant of the field vector and the reaction vector along the normal to S at every point is zero, then $l^\alpha = 0$ and the union curves become geodesics.

The equations (140) are equivalent (see Section 9), to the two equations

$$142) \qquad \rho^\alpha - \kappa_n(l^\alpha - g_{\sigma\tau}l^\sigma u^{\tau\prime}u^{\alpha\prime}) = 0 \qquad (\alpha = 1, 2).$$

Now to see how union curves correspond to dynamical trajectories, consider the case in which a dynamical system with two degrees of freedom, expressed by the generalized coordinates u^α, is such that the potential function U and the constraints do not involve the time t. The kinetic energy T of a unit mass is given by

$$143) \qquad 2T = g_{\alpha\beta}\frac{du^\alpha}{dt}\frac{du^\beta}{dt} \equiv g_{\alpha\beta}\dot{u}^\alpha\dot{u}^\beta,$$

where a dot indicates a time derivative. On substituting $L \equiv T - U$ into Lagrange's equations

$$144) \qquad \frac{d}{dt}\left(\frac{\partial L}{\partial \dot{u}^\alpha}\right) - \frac{\partial L}{\partial u^\alpha} = 0,$$

the equations

145) $$g_{\alpha\beta}\ddot{u}^\beta + \Gamma_{\beta\gamma;\alpha}\dot{u}^\beta\dot{u}^\gamma + \frac{\partial U}{\partial u^\alpha} = 0$$

are obtained. Multiplication of (145) by $g^{\alpha\sigma}$ gives

$$\ddot{u}^\sigma + \Gamma_{\beta\gamma}{}^\sigma\dot{u}^\beta\dot{u}^\gamma + g^{\alpha\sigma}\frac{\partial U}{\partial u^\alpha} = 0,$$

which, by use of the intrinsic derivative, may be written as

146) $$\frac{\delta \dot{u}^\alpha}{\delta t} + g^{\alpha\beta}\frac{\partial U}{\partial u^\beta} = 0,$$

which are the equations of the trajectories of the system. By use of

$$\rho^\alpha \equiv u^{\alpha\prime\prime} + \Gamma_{\beta\gamma}{}^\alpha u^{\beta\prime}u^{\gamma\prime}$$

and a procedure similar to that used in arriving at equations (128) from (126), it can be shown that the equation (140) of the union curves can be written in the form

147) $$\frac{\delta}{\delta s}\left(\frac{du^\alpha}{ds}\right) + \kappa_n g_{\sigma\tau}\frac{du^\sigma}{ds}\left(l^\tau\frac{du^\alpha}{ds} - l^\alpha\frac{du^\tau}{ds}\right) = 0.$$

By the substitution $s = s(t)$, equations (147) become

148) $$\ddot{x}^\alpha + \dot{x}^\alpha(\dot{s})^2\frac{d^2 t}{ds^2} + \Gamma_{\beta\gamma}{}^\alpha \dot{u}^\beta \dot{u}^\gamma + \kappa_n g_{\sigma\tau}\dot{u}^\sigma(l^\tau \dot{u}^\alpha - l^\alpha \dot{u}^\tau)\left(\frac{dt}{ds}\right)^2 = 0.$$

Now equations (146) of the trajectories of the dynamical system and equations (148) of the union curves of the Riemannian space are identical in case

149) $$\frac{d^2 t}{ds^2} = 0$$

and

150) $$g^{\alpha\beta}\frac{\partial U}{\partial u^\beta} = \kappa_n g_{\sigma\tau}\dot{u}^\sigma(l^\tau \dot{u}^\alpha - l^\alpha \dot{u}^\tau)\left(\frac{dt}{ds}\right)^2.$$

From (149) it follows that $t = as + b$, in which b may be taken as zero without loss of generality, and a is a constant. On changing the parameter t to s by means of $t = as$, equations (150) become

151) $$a^2 g^{\alpha\beta}\frac{\partial U}{\partial u^\beta} = \kappa_n g_{\sigma\tau}\frac{du^\sigma}{ds}\left(l^\tau\frac{du^\alpha}{ds} - l^\alpha\frac{du^\tau}{ds}\right),$$

or

152) $$l_\gamma\left(\frac{du^\gamma}{ds}\frac{du^\sigma}{ds} - g^{\sigma\gamma}\right) = \frac{a^2}{\kappa_n}g^{\sigma\gamma}\frac{\partial U}{\partial u^\gamma}.$$

Multiply (152) by $g_{\sigma\tau}(du^\tau/ds)$ to have

$$153) \qquad l_\sigma \frac{du^\sigma}{ds} - l_\tau \frac{du^\tau}{ds} = \frac{a^2}{\kappa_n} \frac{\partial U}{\partial u^\tau} \frac{du^\tau}{ds}.$$

Because of the vanishing of the left-hand member of (153) it follows that

$$154) \qquad \frac{\partial U}{\partial u^\tau} \frac{du^\tau}{ds} = \frac{dU}{ds} = 0,$$

which is true because of the hypothesis that the potential U is independent of the time and therefore of s. Equations (152) may be multiplied by $g_{\sigma\tau}$ to obtain the form

$$155) \qquad g_{\alpha\tau} \frac{du^\alpha}{ds} l_\sigma \frac{du^\sigma}{ds} - l_\tau = \frac{a^2}{\kappa_n} \frac{\partial U}{\partial u^\tau}.$$

It can be verified that

$$156) \qquad l_\tau = -\frac{a^2}{\kappa_n} \frac{\partial U}{\partial u^\tau}$$

are solutions of equations (155). This means that if l_τ take the values given in (156), the equations (147) of the union curves are identical with the equations (146) of the trajectories. Thus, one has the

> **Theorem:** The union curves with respect to the congruence with parameters l_τ given by (156) correspond to the trajectories of a dynamical system with two degrees of freedom in case the potential function does not involve the time.

EXERCISES

1) Show that for the congruence specified by (129) the geodesic curvature of a union curve through P with direction $u^{\alpha'}$ is given by

$$\kappa_g = \sqrt{g}\left(\frac{\Gamma_{11}^2}{d_{11}} u^{1\prime} - \frac{\Gamma_{22}^1}{d_{22}} u^{2\prime}\right) d_{\alpha\beta} u^{\alpha\prime} u^{\beta\prime}.$$

2) Find the differential equations of the union curves on the sphere

$$x^1 = a \sin u^1 \cos u^2, \qquad x^2 = a \sin u^1 \sin u^2, \qquad x^3 = a \cos u^1$$

relative to the congruence of lines parallel to the x^3-axis.

Ans. $\quad \dfrac{d^2 u^1}{ds^2} - \tan u^1 \left(\dfrac{du^2}{ds}\right)^2 = 0, \qquad \dfrac{d^2 u^2}{ds^2} + 2 \csc 2u^1 \dfrac{du^1}{ds} \dfrac{du^2}{ds} = 0.$

3) Show that if a particle moves on a sphere under forces, the resultant of which at each point of the sphere is in the direction of the congruence in Exercise 2, then the particle must follow a union curve relative to this congruence.

Index

Acceleration vector, 89, 93, 107
 transformation of, 134
Affine mapping, 4, 6
Ampère's law, 199
Angle
 between curves on a surface, 103
 between vectors, 59
Applicable surfaces, 110
Arc, length of
 on a curve, 15
 on a surface, 46
Associate
 tensors, 56
 vector, 125, 222
Asymptotic curves, 213
Asymptotic net on a surface, 209
Auto-parallel curves, 223

Base vectors, 70, 73
Bianchi, L. (1856–1928), 222
Bianchi associate vector, 222
Binormal to a curve, 17

Cauchy-Schwarz inequality, 68
Characteristic line of a plane, 91
Christoffel, E. B. (1829–1900), 109
Christoffel, symbols of, 111, 112, 115, 138, 140, 215
Circulation of a vector, 156
Cogredience, 36
Congruence of lines, 220
Conservative field, 136, 154
Contact, order of, 17
Continuity equation in hydrodynamics, 187
Contraction of indices, 41
Contragredience, 36
Contravariant variables, 38
Coordinate curves on a surface, 23

Coordinates
 cylindrical, 53, 131, 188
 dual, 36
 general transformation of, 45
 paraboloidal, 146
 spherical, 28, 46, 64, 68, 139, 143, 188
Covariant differentiation, 116
 of a covariant tensor, 119
 of a general tensor, 120
Covariant variables, 38
Cramer's rule, 32
Curl of a vector, 129
Curl theorem, 189
 application of, 194, 199, 224
 tensor form, 193
Curvature of a curve, 90
 first curvature, 90
  radius of, 91
 mean curvature, 216
 second curvature (torsion), 91
  radius of, 91
 total curvature of a surface, 215
 union curvature, 229
Curved space, 104, 200
 class of, 104

Deal, R. B., Jr. (1919–   ), 217
Degrees of freedom, 13, 17, 235
Derived vector, 125
Descartes, R. (1596–1650), 106
Determinants
 derivative of, 30
 product of, 33
Differentiation
 covariant, 116
 of tensors, 118
 of vectors, 88
Directional derivative, 61, 125
Discriminant of a form, 35

239

INDEX

Divergence of a vector, 125
  in cylindrical coordinates, 131
  in general coordinates, 126
  meaning of, 186
Divergence theorem, 166, 181
  application of, 186
  invariance, 185
Double pendulum, 140
Dyadic, 52, 74
Dynamical trajectories, 233

e^{ijk}, 19, 76
e_{ijk}, 75, 83, 230
Einstein, A. (1878–1955), 115
Eisenhart, L. P. (1876–   ), 233
Eliminant, 18
Equations of motion, 93, 135
Equivalence of forms, 109
Excess of a triangle, 227
Extent
  one-dimensional, 12
  two-dimensional, 13
Extrinsic geometry, 103

Faraday's law, 198
Fields
  conservative, 154
  scalar, 40, 124
  tensor, 40, 125
  vector, 40, 124
First curvature of a curve, 90
  radius of, 90
First fundamental form for a surface, 101, 200, 207, 210
Flat space, 115
Fluid flow, 127
Flux of a vector field, 157, 158
Form
  bilinear, 38
  fundamental quadratic, 47
  linear, 36
  positive definite, 60
  quadratic, 36, 51
Frenet-Serret formulas, 90, 144
  for a curve on a surface, 223
  in general coordinates, 134
Functional dependence, 21

Gauss, C. F. (1777–1855), 21
  theorem of, 216
Gauss-Bonnet theorem, 226
Gauss equations, 21, 159, 200, 210
Gaussian curvature, 215
Geodesic curvature, 218
Geodesic curves, 201, 205, 206
  as auto-parallel curves, 223
Global geometry, 105
Gradient vector field, 49, 61, 69, 124

Green, G. (1793–1841), 162
Green's theorem
  in the plane, 162, 180, 199
  in space, 181
Group, 4
  commutative (abelian), 5
  of mappings, 5
  rotation, 7

Harmonic function, 189
Helicoidal surface, 145
Helix, 15, 20
Homologous curves, 170

Independence of path, 152, 174, 179
  test for, 176
Index
  dummy, 29
  free, 29
Inner product of two vectors, 53, 58
Instantaneous axis of rotation, 91
Integral
  line, 147
  surface, 158
  volume, 161
Integral geometry, 105
Intrinsic
  derivative, 117, 119
  geometry, 103
Invariant, 39, 53
Irrotational vector, 130

Jacobian, 22, 27, 34, 45

Kellogg, O. D. (1878–1932), 184, 191
Kepler, J. (1571–1630), 97
  laws of, 98
Kronecker delta, 10, 30, 112

Lagrange, J. L. (1736–1813), 137, 203
Lagrange equations, 138, 235
Lagrange identity, 19
Laplacian, 128, 144, 146
Leibniz, G. W. (1646–1716), 106
Levi-Civita, T. (1873–1941), 221
  parallelism of, 221, 222
Line integral, 147
  vector form, 153
Linear dependence
  of planes, 20
  of vectors, 79, 87
Lines of curvature on a surface, 214
Liouville, J. (1809–1882), 218
Local geometry, 105

Mapping, 4
  affine, 4, 6
  identity, 5

INDEX

inverse, 5
product of mappings, 5
Matrix, 8, 46
Mean curvature, 216
Metric tensor, 48
Meusnier, J. B. (1754–1793), 217
  theorem of, 217
  generalization of, 227
Minimal
  curves, 210
  surfaces, 217
Mixed tensor, 39
Moment of a force, 82
Monge form of equation, 21
Multiple point on a curve, 153

N-tuple, 38
Newton, I. (1642–1727), 93, 97, 106
Newton's laws of motion, 93, 135
Normal center of curvature of a surface, 212
Normal line to a surface, 24
Normal plane to a curve, 18; *see also* Principal normal to a curve, Binormal to a curve

One-dimensional extent, 12
Order of contact, 17
Orthogonal trajectories of curves on a surface, 108
Orthogonality
  conditions, 8
  of surfaces, 60
  of vectors, 59
Osculating plane, 17, 90, 219, 230

Paraboloidal coordinates, 146
Parallelism of vectors, 118
Parameters of a congruence, 231, 235
Physical components of a vector, 130
Plane
  tangent to a curve, 16
  tangent to a surface, 105, 207
Positive definite form, 60
Potential function, 136
Principal normal to a curve, 17
Product of vectors, 74
  cross, 75
  dot, 74
  triple scalar (box), 77
  triple vector, 78

Quadratic form, 36, 45, 51
Quotient law, 42, 64

Radius
  of first curvature, 91
  of normal curvature, 213
  of torsion, 91
Reciprocal sets of vectors, 70
Rectifying plane to a curve, 18
Region, 153
  closed, 154
  connected, 154
  multiply connected, 154
  simply connected, 154, 168
  on a surface, 154
Ricci tensor, 115
Riemann, G. F. B. (1826–1866), 112
Riemann, symbols of, 114, 115
Riemann-Christoffel tensor, 112, 122, 215
Riemannian geometry, 101
Riemannian metric, 101
Riemannian space, 101, 233
Rotation of a rigid body, 83

Scalar
  field, 40, 49
  product of two vectors, 53, 58
Second curvature, 91
Second fundamental form for a surface, 207
  geometrical interpretation of, 209
Simple closed curve, 153
Singular
  curve on a surface, 22
  point of a surface, 22
Skew-symmetry of tensors, 43
Solenoidal vector field, 128
Space of immersion, 104
Sphere, equations of, 25
Spherical coordinates, 28, 46, 64, 68, 139, 143
Springer, C. E. (1903–   ), 229, 233
Stokes, G. G. (1819–1903), 165
Stokes' theorem, 165, 189, 198, 225
  tensor form, 193
Summation convention, 9
Surface
  first fundamental form for, 101, 200, 207, 210
  Gauss equations, 21
  helicoidal, 145
  integral, 156
  Monge form of equation, 21
  normal, 105
  of revolution, 25, 107, 204
  second fundamental form for, 207, 209
  total curvature, 215
Symmetry of tensors, 43

Tangent
  line to a curve, 15
  plane to a curve, 16
  plane to a surface, 23, 105, 207

INDEX

Taylor series, 15, 127, 202, 207
Tensors
  absolute, 52
  addition, 40
  algebra, 40
  associate, 56
  components, 39
  contraction, 41
  contravariant, 48
  covariant, 48
  field, 40, 125
  first view, 39
  general definition, 52
  mixed, 39
  product, 41
  quotient law, 42, 64
  relative, 52
  weight, 52
Torsion of a space curve, 91
Total curvature of a surface, 215
Trajectories, 125
  dynamical, 233
  orthogonal on a surface, 108
  union curves as, 233
Transformations
  of coordinates, 45
  to curvilinear coordinates, 26
  linear, 26
  rotation in the plane, 6
  rotation in space, 7
  translation, 3
Transitivity of tensors, 42
Two-dimensional extent, 13

Union curvature, 229
Union curvature vector, 232
Union curves, 221, 230, 237

Unit tangent vectors on a surface, 101

Vector
  equation of a hyperplane, 82
  equation of a line, 80
  equation of a plane, 82
  equation of a surface, 100
  field, 40, 124
  function of one variable, 88
  methods in geometry, 84
  projection, 61
  unit, 58, 101
Vector identities, 78, 132
Vectors
  contravariant components, 50
  covariant components, 48
  fixed, 73
  free, 73
  inner (scalar) product, 53
  length, 57
  linear dependence, 79, 87
  meaning of contravariant components, 66, 69
  meaning of covariant components, 67, 70
  in mechanics, 82, 93
  orthogonality, 59
  parallelism, 116, 118
  products, 74–78
  reciprocal sets, 70
  sliding, 73
  as tensors of first order, 39
Velocity vector, 89
  transformation of, 98
Volume integral, 161

Weingarten, J. (1836–1910), 210
Weingarten equations, 210, 212